North-South Grain Markets
and Trade Policies

North-South Grain Markets and Trade Policies

EDITED BY

David Blandford
Colin A. Carter
Roley Piggott

Routledge
Taylor & Francis Group

LONDON AND NEW YORK

First published 1993 by Westview Press, Inc.

Published 2018 by Routledge
52 Vanderbilt Avenue, New York, NY 10017
2 Park Square, Milton Park, Abingdon, Oxon OX14 4RN

Routledge is an imprint of the Taylor & Francis Group, an informa business

Library of Congress Cataloging-in-Publication Data
North-South grain markets and trade policies / edited by David Blandford,
   Colin A. Carter, and Roley Piggott.
      p. cm.
   Includes bibliographical references.
   ISBN 0-8133-8642-X
   1. Grain trade—Government policy—Case studies. 2. Grain trade—
Case studies. I. Blandford, David. II. Carter, Colin Andre.
III. Piggott, Roley.
HD9030.6.N67 1993
382'.4131—dc20   93-9268
                                                                               CIP

ISBN 13: 978-0-367-01106-2 (hbk)
ISBN 13: 978-0-367-16093-7 (pbk)

This book is dedicated to
Pascal Fotzo and Davendra Tyagi,
whose untimely deaths created intense sadness

Davendra Tyagi was Chairman for the Commission for Agricultural Costs and Prices, Government of India. He was shot dead May 10, 1992, at the doorstep of his official residence by two suspected militants only hours before his daughter's marriage.

Pascal Fotzo was a Senior Lecturer in Rural Economy at the Dschang University Centre, Dschang, Cameroon. He was killed in a plane crash, January 1987, en route to a KIFP/FS seminar in Brazil.

We will each remember Davendra and Pascal in our own special way. God rest their souls.

Contents

Preface

Assuring the adequacy of grain supplies has always been high on the list of priorities for many governments. With rapid growth in the world's population, the challenge of meeting the need for grain has become a complex balancing act requiring interactions among technologies, institutions, and policies. The purpose of this book is to show how a set of countries, some highly economically developed (the North) and others at lower levels of economic development (the South), attempts to deal with the organization and management of their grain sectors.

The book is composed of a series of case studies. The countries included reflect the interest and experience of the authors who collaborated in preparing the volume. No attempt was made to provide representative coverage based upon a comprehensive classification of countries, which is why there are no chapters dealing with such exporters as Argentina or Thailand or importers such as Egypt or Japan. Despite the somewhat eclectic geographical mix, many of the fundamental issues that face the North and the South, both individually and collectively, are illustrated by the case countries. We would argue that there is much to be learned about the effective implementation of policy choices and the constraints that policymakers face by looking at individual country experiences, rather than by attempting to generalize on the basis of an abstract theoretical framework. There is a dearth of information on what countries actually do in managing domestic grain markets.

We emphasize that the focus of this book is on domestic policies relating to the most basic of agricultural commodities—grains. The focus is not on "international grain trade" or "world grain markets"—again, this would have required a different, and probably a larger, selection of case studies, as well as a unifying conceptual framework. Too, although it is acknowledged in the book that the position of individual countries needs to be viewed in a global market context, there is no primary focus on the implications for international markets of the policies pursued by the countries analyzed. This emphasis was deliberate, because most policymakers are primarily concerned about achieving domestic objectives first and meeting international commitments second. We implicitly accept this ranking but are careful to stress the implications of interdependency through trade.

The case studies describe grain policies and their effects, highlighting successes and failures. Each case study has a similar order and presentation, permitting, for example, market structures and principal instruments of policy to be identified and compared. The synthesis chapter draws out the lessons to be learned, using information from the case studies mixed with some conventional wisdom. It shows that there is often substantial similarity among policy objectives and the mix of policy instruments used in different countries. In a nontechnical way it evaluates the choices facing governments in using alternative policy instruments in attempting to implement their objectives. A conscious attempt is made to be nondogmatic and to provide a balanced assessment of the pros and cons of government intervention. The chapter recognizes that success or failure is difficult to evaluate objectively, because it depends on the criteria by which success is judged, but makes a number of clear and practical recommendations on what conditions must be met to ensure a healthy and dynamic domestic grain market.

The operation of grain markets and policies is frequently complicated, making economic analysis correspondingly complex. The authors have attempted to keep the analysis understandable to the intelligent lay person, rather than requiring the reader to have a doctorate in economics. It is nevertheless hoped that both specialists and nonspecialists will find this work of interest. The book has two target audiences: 1) policymakers and those engaged in advising policymakers, particularly in developing countries, and 2) students of agricultural policy. For the former group it provides a series of benchmarks against which to evaluate needs, objectives, and potential strategies for their own countries. It provides a set of laboratory examples of what has worked and what has failed and why. For the second target group it provides a set of readings, which can supplement more theoretical courses on agricultural policy dealing with either developed or developing countries.

David Blandford
Colin A. Carter
Roley Piggott

Acknowledgments

In 1986, the Kellogg Foundation of Battle Creek, Michigan, established a three-year, nondegree program to advance professional leadership involved in bringing about improvements in food systems in developing countries. The International Fellowship Program in Food Systems (KIFP/FS) was administered by Michigan State University. A total of 32 mid-career professionals from 23 countries were chosen to participate in a series of activities, including individual and collaborative research projects, seminars, and study tours designed for professional enrichment. The group held annual meetings in Brazil, Zimbabwe, and Thailand during the life of the program. Additional subgroup activities were held in Australia, China, Ghana, India, Indonesia, Jordan, Kenya, Mexico, Tanzania, Sudan, and the United States.

At an initial organizational meeting in East Lansing, Michigan, in July 1986, several of the Kellogg Fellows met to discuss their common interests in grain marketing and policies. Subsequently, this core group expanded and embarked on collaborative research. A number of meetings were held to share knowledge and exchange ideas, including a seminar at Cornell University in Ithaca, New York, in September 1987. This book grew from that seminar.

The work would not have been possible without the Kellogg Foundation's generous financial support and the extensive assistance of the KIFP/FS administration at Michigan State University. Particular thanks are due to Harold Riley, Darrell Fienup, and Ardell Ward for their constant encouragement and their willing help with finances and logistics. Mike Weber and Jim Shaffer, also of Michigan State, provided useful input into group discussions and helped to shape ideas on the substance and purpose of the project. Alex McCalla of the University of California at Davis attended meetings of the Kellogg Fellows in Recife and at Cornell, and his input to this book is gratefully acknowledged. At Cornell University, Beth Rose was tireless and painstaking in her tasks as editorial assistant. Beth's outstanding efforts were complemented by the skilled test processing of Carol Peters, who edited successive drafts. Thanks go to Nancy Ottum for typesetting the text and graphics of this manuscript in camera-ready form. Whilst acknowledging the help of these individuals, the editors and authors remain responsible for the contents of the book.

In the view of the participants, the Kellogg Fellowship Program was successful in broadening our understanding of the complexities of the global food system. It is our hope that this book will contribute to increasing public understanding of the problems in meeting that most basic of human requirements—the need for food.

D.B.
C.A.C.
R.P.

1

An Overview of the World Grain Economy

David Blandford

Grains have long been mankind's primary food source. Through their domestication and cultivation, grains helped to lay the foundation for the development of early civilizations around the world. Strains of barley and wheat dating to 6,500 B.C. were found in the archaeological remains of early societies in the Middle East. Maize or Indian corn was first cultivated about 5,000 B.C. in Mexico. The so-called "drought staples," including millet and sorghum, were domesticated in East Africa (Sudan and Ethiopia) around 2,500 B.C. Likewise, rice and sorghum appear in the archaeological records of China and Southeast Asia by 5,000 B.C. (Chicago Board of Trade, 1977).

With the exception of buckwheat, grains belong to the grass family (*Gramineae*).[1] The major cereal grains (barley, maize, millet, oats, rice, rye, sorghum, and wheat) provide the bulk of the food supply in terms of calories (and oftentimes protein) in most countries of the world. The term *foodgrains* is sometimes used to describe wheat, rice and rye, indicating that these grains are primarily consumed directly by humans. The term *feedgrains* is typically applied to maize, barley, oats, and sorghum because much of the world's supplies of these grains are fed to livestock.[2] However, these distinctions are often blurred. Maize and sorghum are food staples in many African and Latin American countries; barley and other feedgrains are widely used in the brewing and distilling industry; and a substantial amount of wheat is fed to animals in Europe, Canada, and the USSR. For these reasons, the generic term *coarse grains* is used to denote the aggregate of barley, maize, millet, oats, rye, and sorghum.[3] A summary of major characteristics of the cereal grains in Table 1.1 provides additional information.

World Grain Production

The production of grain is a major part of the world's agriculture. Grains occupied roughly 50 percent of the world's harvested cropland in 1987-89, or

1

TABLE 1.1 Major Characteristics of the Cereal Grains

Grain	Characteristics
Barley *(Hordeum)* *vulgare)*	Probably one of the first grains to be domesticated, it is primarily a temperate zone crop, growing best where average summer temperatures are less than 20°C and annual rainfall is below 900 mm. Both winter- and spring-sown varieties exist. In Asia, barley is consumed directly by humans in flours and soups. Ground barley is fed to livestock, particularly hogs. Barley is also widely used in the distilling and brewing industry. Brewing byproducts (e.g., brewers grains) are fed to livestock.
Maize *(Zea mays)*	Now grown worldwide, maize is the only major grain to originate in the Americas, where it fed pre-Columbian civilizations. Maize is distinguished by color (e.g., yellow or white) and by type (e.g., flint, dent, sweet, or popcorn). It requires warm (20°-25°C), moist conditions (650-1,300 mm of rainfall) for highest yields. Maize is a major foodstuff in Latin America and parts of Africa, where it is eaten roasted or boiled, or is ground and made into pancakes (e.g., tortillas) or gruel. In North America, maize is a major livestock feed. Wet milling is used to produce starch and derivatives (such as sweeteners), oil, and livestock feed from maize. Dry milling produces meal, flour, and other products used in the food industry and a variety of industrial processes. Maize is also used to produce distilled spirits and beer.
Millet *(Numerous* *species)*	Produces higher yields than most other grains under poor soil conditions, high temperatures (over 25°C), and low rainfall (less than 500 mm). Has a short growing season. Often used as an insurance crop where rainfall is uncertain. Widely considered an inferior grain, but provides needed calories in Africa and Asia, where it is eaten as meal or gruel. Also used to make beer.
Oats *(Avena sativa)*	Common oats grow best in cool, moist regions, such as Northern Europe; red oats grow in warmer regions, such as the Mediterranean countries. Oats require more moisture than any other major grain crop and tolerate less heat. Both winter and spring varieties exist. Winter oats are sometimes used as a cover crop. Oats are principally used as livestock feed. Rolled oats and oat flour are used in the production of breakfast cereals and baked goods.
Rice *(Oryza sativa)*	Principal foodcrop of Monsoon Asia; generally restricted to areas with warm temperatures (21°C and above during the growing season) and plentiful water (over 1,000 mm). Paddy rice is grown on flooded lowland areas; upland rice, which is relatively unimportant, is grown under rainfed conditions. Paddy yields are usually higher than upland yields. There are more than 8,000 varieties of rice. The three types of cultivated rice are the long-grain indica, the short-grain japonica (also known as Sinica), and Javanica

(continues)

Table 1.1 (*continued*)

Grain	Characteristics
	(grown in areas of Indonesia). Rice is usually milled and boiled prior to consumption, and is also used to make various food products and alcoholic beverages. Except for the bran removed during milling, rice is not generally fed to livestock.
Rye (*Secale cereale*)	A cool climate grain grown in areas where rainfall ranges from 500-800 mm. Winter varieties are the hardiest of all cereals and are grown in high latitudes (e.g., northern Russia). Tolerates poor soils and is sometimes used as a cover crop. In Europe, rye is primarily used as a food, particularly in breads. In North America, it is grown for livestock feed and used to make whiskey.
Sorghum (*Sorghum spp.*)	Grows well where it is hot and rainfall is insufficient for other grains, such as maize or rice (less than 500 mm). There are four types: (a) grain sorghum is used as an animal feed but consumed by humans in the form of pancakes or gruel in parts of India, Africa and China; it is also used to make beer. The two most common types are kafir from southern Africa and milo from east-central Africa. Grain sorghum is used as an industrial raw material in the production of such materials as starch, dextrose, and alcohol; (b) sweet sorghum or sorgo is used to produce syrup and for animal feed; (c) grassy sorghum is used for silage, grazing, and hay; and (d) broomcorn is used to make brooms.
Wheat (*Triticum spp.*)	Yields best in temperate areas with annual rainfall of 600-900 mm, but can be grown with as little as 375 mm of precipitation. Winter wheat is planted in the fall and harvested in the spring. It requires cold winter temperatures for proper development. Spring wheat is planted in the spring and harvested in the summer. Common wheat is used to make bread; durum wheat is used to make pasta. In the Middle East, wheat is parboiled, dried, and cracked and used to make bulgar, which is cooked like rice. Wheat is also fed to livestock, as are its milling byproducts (e.g., bran). Wheat is also used to produce industrial products such as starch, gluten, and distilled spirits.

over 700 million hectares (FAO, 1989). The three most important grains are wheat, maize, and rice, which together occupy approximately 70 percent of the world's grain area and account for almost 75 percent of world production (Table 1.2). Wheat is the leading grain, claiming over 32 percent of the world's harvested grain area and 29 percent of total production. Rice is second most important in terms of cultivated area, but average yields are below those of maize, resulting in lower total production.[4] Maize is, on average, the most productive grain, with a global yield of about 3.4 metric tons per hectare. Millet,

TABLE 1.2 World Production of Major Cereal Grains, 1987-89

	Area		Yield	Production	
	(million ha.)	(percent)	(tons/ha.)	(million tons)	(percent)
Wheat	222	32	2.3	519	29
Maize	128	18	3.4	441	24
Rice	145	21	3.4	488	27
Barley	76	11	2.3	172	10
Sorghum	45	6	1.4	62	3
Oats	23	3	1.8	41	2
Rye	16	2	2.1	34	2
Millet	37	5	0.8	29	2
Total cereals[a]	700	100	2.6	1,800	100

Source: FAO, Production Yearbook, 1989.
Rice figures are in unmilled form.
[a]Includes minor cereals not listed.

typically grown on less fertile soils under poorer growing conditions, is least productive, with an average yield of less than 1 ton per hectare.

Asia, with roughly 60 percent of the world's population, accounts for over 40 percent of the world's grain area and production. North and Central America, with 14 percent of the world's grain area and 8 percent of its population, account for approximately 20 percent of production (Table 1.3). This reflects the high productivity of the region's two dominant grain producing and exporting countries, Canada and the United States. In contrast, Africa, with 10 percent of the world's harvested grain area and 12 percent of its population, accounts for less than 6 percent of global production. Productivity in many of the grain producing areas in the African continent is quite low. Europe and the USSR are the second largest grain producers after Asia. Oceania, with only 2 percent of the world's grain area, generates only 1 percent of the world's grain output, but its leading producer, Australia, plays an important role in world wheat trade.

Wheat is the major grain in Europe and the USSR (42 percent of regional area) and in Oceania (67 percent) (Figure 1.1). Other temperate zone grains such as barley, rye, and oats account for most of the rest of production in the European region, which also dominates world output of these grains (Figure 1.1). Wheat accounts for an important portion of North America's grain production (39 percent), but maize claims an almost equivalent share. This reflects the importance of maize historically in the region and the climatic suitability of certain areas, such as the Corn Belt of the midwestern United States, for cultivation of the crop.

Maize also dominates grain output in the southern part of the Americas, its original heartland. Despite its relatively recent (post-Columbian) introduction to the rest of the world, maize is now the most widely distributed grain globally.

TABLE 1.3 Regional Cereal Production, 1985-87

	Area		Production		Population[a]	
	million ha.	percent	mill. m. tons	percent	million	percent
Geographical						
Africa	73	10	87	5	610	12
N. & C. America	94	13	329	18	417	8
S. America	39	6	81	4	285	6
Asia	303	43	799	44	2,995	59
Europe and USSR	176	25	482	27	783	15
Oceania	14	2	22	1	26	1
Economical						
Developed market economies	148	21	545	30	833	16
Developing market economies	317	45	574	32	2,683	52
Centrally planned economies	234	34	681	38	1,599	31
World	699	100	1,800	100	5,116	100

Source: FAO, Production Yearbook, 1989.
[a]1987

Many different strains allow the crop to be cultivated under a wide variety of conditions (Leonard and Martin, 1963). Its range extends between latitudes 58° north in Canada and the Soviet Union and 40° south in Argentina. Maize can also be grown on many kinds of soils. With the exception of Oceania, all the regions in Figure 1.1 have shares of 10-30 percent of the world maize area.

Rice is the principal grain crop in Asia, accounting for almost 90 percent of the world's harvested area of the crop. The sub-tropical, monsoon climate of many parts of Asia is ideally suited to rice, which requires warmer growing conditions and more moisture than other grain crops. The Asian continent is also an important producer of most other grains because of its size, climatic diversity, and large population. Asia accounts for a substantial share of the world's wheat (37 percent) and maize areas (29 percent), in addition to millet and sorghum (47 percent). The latter two grains make up over 40 percent of Africa's grain area, while over 25 percent of the remainder is planted to maize.

There has been an enormous increase in grain production over the last 30 years. World production almost doubled between the early 1960s and the late 1980s, rising from an average of 860 million metric tons in 1960-64 to 1,659 million metric tons in 1985-90 (Table 1.4).[5] During this period, global production grew at an average annual rate of about 2.6 percent (Table 1.5), about .6 percent faster than the 2 percent annual increase in world population (UN,

6

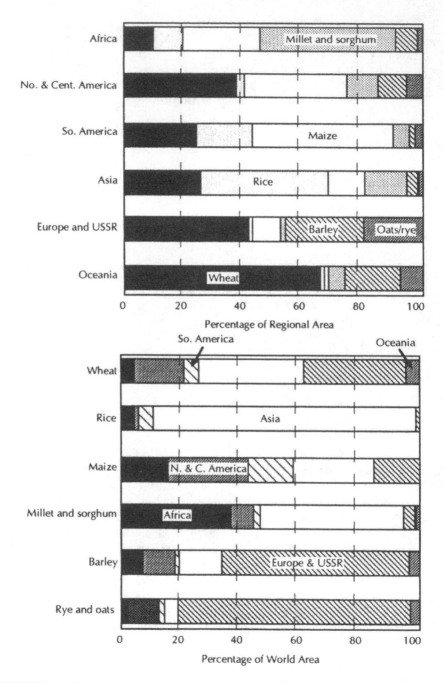

FIGURE 1.1 Harvested Cereal Area by Region (1985-87). *Source:* FAO Production Yearbook, 1987.

TABLE 1.4 Global Grain Balances

	Area Harvested (mill. ha.)	Yield (m. tons/ha.)	Production (mill. m. tons)	Beginning Stocks (mill. m. tons)	Imports (mill. m. tons)	Exports (mill. m. tons)	Feed Use (mill. m. tons)	Apparent Consumption (mill. m. tons)	Ending Stocks (mill. m. tons)
Total grain									
1960-64	648.0	1.3	859.6	193.5	85.7	90.4	292.6	856.0	192.4
1965-69	666.5	1.5	1,019.5	199.9	103.5	108.3	375.6	1,007.9	206.7
1970-74	677.9	1.8	1,187.0	201.9	134.8	139.9	459.8	1,187.4	196.4
1975-79	714.9	1.9	1,368.4	261.2	180.2	185.5	509.9	1,340.1	284.1
1980-84	718.4	2.1	1,524.4	313.0	224.6	228.4	576.2	1,511.1	322.6
1985-90	699.5	2.4	1,659.3	377.4	217.5	220.7	638.4	1,667.4	369.3
Coarse grains									
1960-64	320.5	1.4	455.1	104.9	31.6	33.3	266.1	454.4	103.9
1965-69	321.1	1.6	530.0	90.9	42.5	44.9	325.8	527.7	90.8
1970-74	329.5	1.9	619.9	88.7	63.2	66.6	387.7	618.0	87.2
1975-79	343.6	2.1	710.9	113.0	93.7	95.6	435.0	698.8	123.2
1980-84	340.3	2.2	757.5	142.2	108.5	108.6	486.8	756.9	142.6
1985-90	322.6	2.6	823.9	177.6	101.0	101.1	546.2	827.7	116.1
Rice									
1960-64	120.5	1.3	160.6	12.8	6.8	7.1		160.4	13.0
1965-69	127.4	1.5	187.8	19.9	6.9	7.6		186.1	21.7
1970-74	134.9	1.6	218.8	27.1	8.1	8.2		218.4	27.5
1975-79	142.7	1.8	250.9	40.5	9.9	10.4		246.0	45.4
1980-84	143.6	2.0	292.8	46.6	12.0	12.2		292.3	47.1
1985-90	144.9	2.3	327.9	51.1	12.2	13.1		327.8	51.1

(continues)

Table 1.4 (continued)

	Area Harvested (mill. ha.)	Yield (m. tons/ha.)	Production (mill. m. tons)	Beginning Stocks (mill. m. tons)	Imports (mill. m. tons)	Exports (mill. m. tons)	Feed Use (mill. m. tons)	Apparent Consumption (mill. m. tons)	Ending Stocks (mill. m. tons)
Wheat									
1960-64	207.0	1.2	243.9	75.8	47.3	50.0	26.5	244.2	75.5
1965-69	218.0	1.4	301.7	89.2	54.0	55.8	49.8	296.7	94.2
1970-74	213.6	1.6	348.3	86.2	63.5	65.1	72.1	352.7	81.8
1975-79	228.6	1.8	406.7	107.7	76.6	79.5	74.8	398.8	115.6
1980-84	234.5	2.0	474.2	124.3	104.2	107.5	89.3	465.5	132.9
1985-90	225.5	2.3	527.4	148.7	104.2	106.5	102.7	530.9	145.2

Source: U.S. Department of Agriculture, Economic Research Service, PS&D View Database, 1990.
Note : Figures are averages for the years shown. Rice is milled rice. World exports and imports may not be equal because of discrepancies in recorded trade. Export and import figures for the world and developed countries include trade between the member states of the European Community.

TABLE 1.5 Growth in World Grain Production, 1960-89 (Average annual percentage change)

	Area	Yield	Production
By type			
Coarse grains	0.2	2.1	2.3
Rice[a]	0.7	2.1	2.8
Wheat	0.4	2.6	3.0
Total grains	0.4	2.2	2.6
By economic region			
Developed market economies	0.5	1.8	2.3
Developing market economies	0.8	1.9	2.7
Centrally planned economies	-0.1	3.0	2.9
World	0.4	2.2	2.6

Source: Computed from U.S. Department of Agriculture, Economic Research Service, PS&D View Database, 1989.
[a] Yield and production data for rice are in milled form.

1989). Most of the growth has been due to improved yields. The world's harvested grain area increased at less than one-half percent annually since 1960. In most regions of the world there is little scope for expanding cropland, and there are limits on diverting land from other crops to grain. The fastest rate of growth in area has been in developing economies, particularly for rice, stimulated by increases in population.[6] The most rapid growth in yields has been for wheat, primarily because of technological innovations. Regionally, grain yields have increased fastest in the centrally planned economies, largely because of enormous technological advances in China. Productivity gains in Russia and Eastern Europe have been less dramatic than in China, whose share of total grain production in the centrally planned economic grouping rose from roughly 35 percent in 1960-64 to almost 50 percent in 1985-89. Irrigation has played an important role in increasing grain yields in China and other Asian countries. Between 1971 and 1986 the world's irrigated area rose from 173 to 228 million hectares, with 60 percent of this increase in Asia.

The Importance of Technological Change

Technological change, in particular the development of higher-yielding varieties, has had a major impact on world grain production. Through selection, plant breeders developed short-stemmed varieties of rice and wheat that bear heavier crops of grain without toppling over (lodging). Breeders also focussed on the development of varieties with improved resistance to pests and diseases. International research centers in the Philippines for rice (IRRI) and Mexico for wheat (CIMMYT) were instrumental in the development of these new strains during the 1960s. However, fertilizers and sufficient water are major factors in realizing the yield potential of these varieties. Their introduction, coupled with the increased use of purchased inputs and improved water management, led to a "Green Revolution" and vastly higher yields in a number of developing countries, particularly in Asia (for example, chapters 10 and 11). The new varieties or their derivatives were also adopted widely by farmers in economically developed countries.

The development of hybrids has not been a major factor in increasing rice and wheat yields but has been extremely important for maize. Maize hybrids were first developed in the United States in the 1920s and have since contributed to a more than tripling of average yields. Hybridization, the crossing of different plant varieties, can yield a plant superior to both parents in terms of size, yield, or general vigor. Breeders exploit this phenomenon to produce higher-yielding strains. Unfortunately, farmers cannot simply use seed saved from their fields to produce subsequent crops because new hybrid seed must be purchased to maintain vigor. Such seed is generally obtained from specialized seed producers, presenting difficulties in situations where agricultural input markets are not well developed. Often hybrids are more demanding in terms of growing conditions. Well-aerated soils, adequate water and nutrients, typically provided by chemical fertilizers, are necessary to realize their full yield potential. This has limited the development and use of hybrids in less favorable areas, where the supply of inputs is often a problem, such as Africa.

Maize production is almost totally dominated by hybrids in the Northern countries, such as the United States, where the necessary infrastructure exists to produce and deliver the inputs required to exploit the productive potential of hybrids. In many developing countries, however, open-pollinated varieties of maize are still commonly planted. Plant breeders have made some efforts to improve the productivity of open-pollinated varieties, but without the spectacular results obtained with hybrids.

Technological progress in other grains important to developing countries, particularly millet, has also been modest. Partly this is because millet is regarded as an inferior cereal, grown under soil and climatic conditions that are not conducive to high yields. Because of limited uses and low yields, only a small acreage is planted in the developed industrial countries. Varieties grown in the developing world (finger and pearl millet) are produced in minor quantities in industrial countries. Consequently, millet has not received much attention from commercial plant breeders in these countries. More effort has been directed to sorghum, which is an important crop in drier areas of the United States and Australia. In the United States, hybrids introduced in 1956 rapidly replaced most other sorghum varieties (Leonard and Martin, 1963). In some developing countries where sorghum is used for feed and a hybrid infrastructure exists, the use of higher-yielding sorghum hybrids is important (see Chapter 12).

Technological change has played a major role in increasing global grain production. It is difficult to estimate the contribution of genetic improvements to the increase in yields as distinct from other factors such as the expanded use of chemical fertilizer. A recent review of studies for the United States indicates a median estimate of the gain from genetic improvement since 1930 of 50 percent of the yield increase in wheat and 60 percent for maize (Duvick, 1987). Since the world's land area is limited, increasing yields is the major way in which production can keep pace with the growth in the demand for grain created by an expanding world population and higher consumer incomes.

Grain Consumption

Grains are the major staple for the majority of the world's population. They are an important source of carbohydrates and also contain 10-16 percent protein on a dry matter basis (Kent, 1975). Grains are also an important source of vitamins, particularly those in the B complex, and vitamin E. Grains and their byproducts are widely used as feed for livestock, and are consequently an important element in the production of animal proteins for human consumption. Both directly and indirectly, grains are a vital part of the human food chain.

In per capita terms, the global availability of grain increased at 0.8 percent per year since the early 1970s (Figure 1.2). The rate of growth in the industrial and centrally planned countries was over 1 percent, but in the developing countries it was only 0.6 percent. The much lower figure for developing countries reflects their rapid increase in population (roughly 2.4 percent per year) compared to centrally planned economies (1.4 percent) and the developed economies (0.8 percent). In Asia, where Green Revolution technologies were most effective, the rate of growth in per capita cereal production was a healthy 1.4 percent, despite an annual increase in population of 1.9 percent, and this growth followed a regular upward trend without the peaks and valleys found in South America and Africa (Figure 1.2). In contrast, per capita cereal production in Africa fell by over 1 percent per year between the early 1970s and the late 1980s. A population growth rate of almost 3 percent, the lack of a suitable package of technology to increase grain yields, wars, droughts, inappropriate government policies, and the degradation of natural resources through population pressure all contributed to the dismal performance of grain production in the African subcontinent.

Global usage of grain for livestock feed accounts for close to 40 percent of total grain use (Figure 1.3).[7] The amount of grain fed to livestock has roughly doubled since 1960 to 600 million metric tons, but is approximately the same percentage of total use as in 1960. However, while the centrally planned and developing country share of fed grain has risen, that of the developed world actually declined (Figure 1.3). In 1960, developed countries accounted for

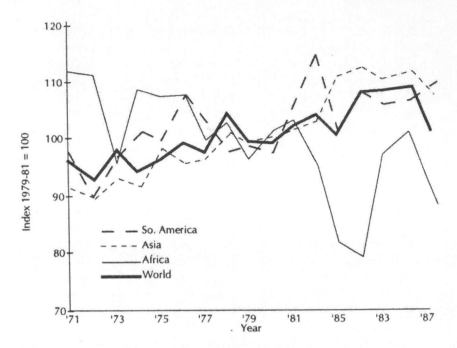

FIGURE 1.2 Per Capita Cereal Production in Major Developing Country Regions. *Source:* FAO Production Yearbook, various years.

about 75 percent of the grain fed to livestock worldwide, but by the late 1980s, the proportion had fallen to roughly 50 percent. The increased importance of centrally planned and developing countries in the global use of grain for feed reflects the high population growth rates in these countries, and the effects of rising consumer incomes on the demand for animal protein.

The percentage of grain fed to livestock is closely correlated with a country's per capita income. As income grows, total food consumption increases (Figure 1.4). The per capita use of grain also rises, as does the proportion of grain fed to livestock. Overall, demand for grain responds positively to growth in incomes, resulting in expanded demand for livestock products and boosting grain consumption. In most countries, a larger and larger proportion of available grain is consumed indirectly through livestock products (Blandford, 1984). In the economically developed countries as a whole, roughly 60 percent of total grain used is fed to livestock, while in developing countries, the proportion is less than 15 percent.

In richer countries, grains are processed into many food products, such as breakfast cereals and cookies. Grains are also an important source of raw

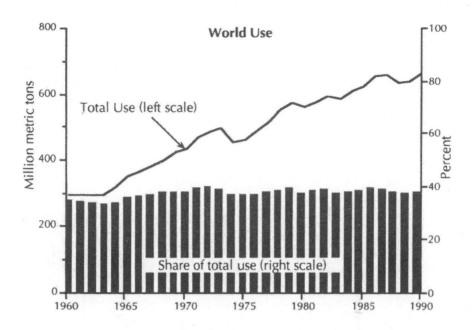

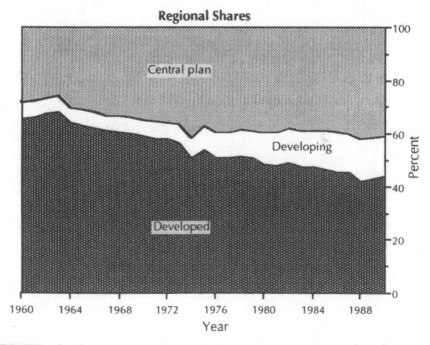

FIGURE 1.3 Trends in the Use of Grain for Livestock Feed. *Source:* U.S. Depart. of Agriculture, Economic Research Service, PS&D View Database, 1989.

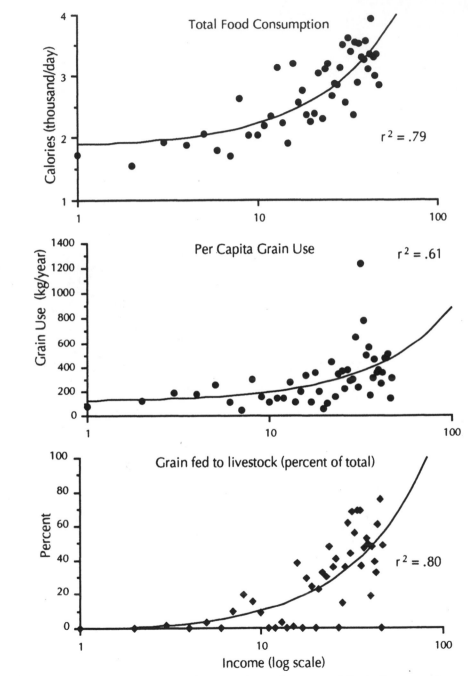

FIGURE 1.4 Relationship Between per Capita Income and Grain Consumption in 47 Countries. *Source:* USDA, ERS. World Agricultural Trends and Indicators, 1970-88, 1989. PS&D View database.

materials for industry and are used in the manufacture of a variety of industrial products, including starches, oils, and sugars. Grains are fermented and distilled to produce alcohol for fuel and beverages for human consumption. In general, as income rises, demand for food and industrial products expands, as does the industrial use of grains. The proportion of grains used in industrial processes is difficult to estimate, but in the early 1980s, the ratio was roughly 5 percent in industrial countries (OECD, 1985).

Trade in Grains

The unprecedented growth in world grain production and consumption resulted in an even more rapid growth in the volume of international grain trade, which increased by over 150 percent between the early 1960s and the late 1980s. This compares to an expansion of roughly 90 percent in world grain consumption. The swift increase in trade is partly attributable to growth in world income, which stimulated consumer demand for grain, partly to pressures placed on domestic supplies in some countries because of rapid population growth, and partly to the fact that major exporters have supplied grain at increasingly competitive prices (see the later discussion of prices in this chapter). However, the relationship between income and trade is particularly important; growth in imports of coarse grains (used primarily for animal feed in most countries) has been especially large. World imports of coarse grains more than tripled between 1960-64 and 1988-89, and the volume of trade is now equivalent to about 12 percent of world consumption. Trade in wheat is also significant. In the late 1980s, about 20 percent of world wheat production was traded. This compares to the thin rice market, where less than 5 percent of world production is traded. Much of the world's export capacity for grains is in developed industrial countries in the temperate zone, which primarily produce wheat and coarse grains rather than rice.

Trade in most grain byproducts is limited. However, in recent years a substantial volume of trade has developed in maize byproducts, primarily between the United States and the European Community. In 1988, for example, the United States exported over 6 million metric tons of maize byproducts, 95 percent of which were destined for the EC (USDA, FAS, 1989). Maize byproducts are dominated by maize gluten meal, which is a byproduct from the manufacture of maize sweeteners. Trade in this commodity is aided by policies in both countries. The protection of sugar manufacturers in the United States stimulates the production of sugar substitutes, such as maize sweetener, boosting the output of byproducts. Further, high cereal prices also caused by policy in the European Community spur tariff-free imports of cereal substitutes for livestock feed (see the EC chapter). This is an example of the important effects that government policies have on domestic and international cereal markets, a major focus of this book.

World grain exports are overwhelmingly dominated by the developed industrial countries, accounting for over 80 percent of the total volume (Figure 1.5). Global trends in supply and demand make it likely that this situation will continue for the foreseeable future. In the early 1960s, the developed countries were the largest grain importers, but rapid growth in developing and centrally planned economies led to a substantial change in import shares. In 1960-64, developed countries accounted for roughly 55 percent of global grain imports. By 1985-89, their share had fallen to 40 percent.[8] Developing countries claimed 34 percent of global imports in the late 1980s, compared to 25 percent in the early 1960s. The centrally planned economies now import 26 percent of traded grain supplies, compared to less than 20 percent in the early 1960s. The grain production systems in the Soviet Union and eastern European countries have had difficulty in keeping pace with the demand for grain created by growing domestic consumption of meat and livestock products; consequently, grain imports have expanded. The demand side of world grain trade is increasingly dominated by events in the developing and centrally planned countries.

The United States is the world's leading grain exporting nation. In the late 1980s, it accounted for over 45 percent of world grain exports. The United States dominates trade in coarse grains, claiming over 60 percent of world exports (mostly maize and sorghum) (Table 1.6). Next largest exporters are Argentina and Canada, each with 6-7 percent shares. In the late 1980s, the United States accounted for over 35 percent of world wheat exports, and in earlier periods its share topped 47 percent. Four countries -Australia, Canada, the European Community, and the United States - collectively accounted for over 85 percent of world wheat exports in the late 1980s (Table 1.6). The United States is also a major player in the world rice market. Its share of world exports has declined slightly in recent years, but was still over 20 percent in the late 1980s (Table 1.6). Changes in U.S. grain trade are discussed in more detail in the U.S. chapter. Thailand is the largest rice exporter, with over 40 percent of world exports. Thailand is highly competitive in the rice export market, capturing market share from the traditional exporters Burma, China, and the United States. Because a small number of countries dominate world grain exports, world grain markets may not be competitive. Exporters may be able to exploit their relative size in setting world prices. A concentrated market may also make importers vulnerable to export restrictions imposed for political or other reasons. At the very least, the small number of exporters makes world markets more susceptible to the effects of weather-induced instability, particularly in North America, although this is limited to some extent by grain stocks (see the discussion of prices below).

The degree of concentration on the import side of the trade equation is less marked than on the export side. However, as a group, the developing countries and, to a lesser extent, centrally planned economies, increasingly dominate

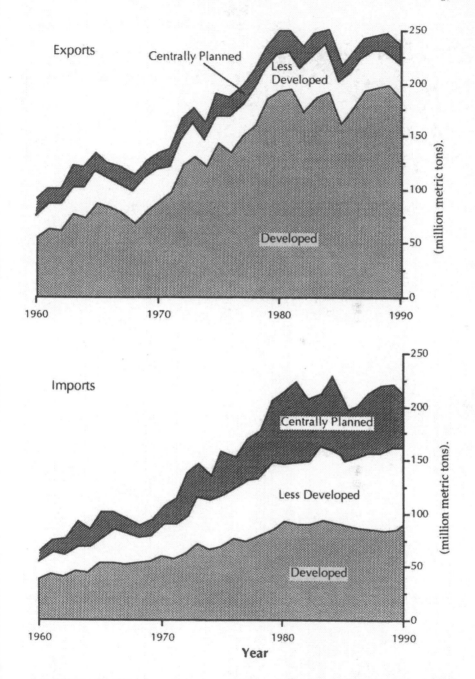

FIGURE 1.5 World Grain Trade by Region. *Source:* USDA, ERS. PS&D View database, 1989.

TABLE 1.6 World Grain Exports

Coarse grains

Years	United States	Canada	Argentina	South Africa	Thailand	European Community	Rest of world (percent)	Total exports (mill. m. tons)
1960-64	52.5	3.5	12.8	6.1	2.5	0.0	22.6	29.7
1965-69	51.1	4.3	15.9	4.9	3.6	0.0	20.2	39.2
1970-74	58.8	6.8	11.9	4.3	3.5	0.0	14.7	58.8
1975-79	68.4	5.2	11.0	3.6	2.5	0.0	9.3	86.7
1980-84	61.5	6.2	12.4	2.3	3.5	0.3	13.9	90.3
1985-89	61.6	6.0	6.9	2.8	2.7	4.6	15.4	83.0

Rice

Years	United States	China	Thailand	Pakistan	Burma		Rest of world (percent)	Total exports (mill. m. tons)
1960-64	17.6	10.2	24.7	1.9	23.9		21.7	6.5
1965-69	24.5	20.0	19.1	2.1	10.1		24.1	6.7
1970-74	23.3	21.9	18.6	6.4	6.1		23.7	7.5
1975-79	24.2	10.3	25.1	9.7	6.2		24.4	9.8
1980-84	21.2	7.2	35.0	9.7	6.0		21.0	11.1
1985-89	21.6	6.9	43.1	10.5	5.1		12.8	10.9

Wheat

Years	United States	Canada	Argentina	Australia	European Community	Rest of world (percent)	Total exports (mill. m. tons)
1960-64	42.6	24.2	6.7	13.9	0.0	12.4	45.9
1965-69	40.0	23.6	5.4	14.6	0.0	16.3	48.4
1970-74	47.2	22.4	3.6	13.4	1.2	12.1	56.4
1975-79	47.0	20.5	5.5	14.3	5.4	7.3	70.7
1980-84	41.5	20.2	6.9	12.8	12.8	5.8	93.5
1985-89	36.2	20.5	4.6	13.9	16.9	7.9	91.8

Source: U.S. Department of Agriculture, Economic Research Service, PS&D View Database, 1989.

Note: Figures are averages for the years shown and exclude trade between member states of the European Community.

markets for the principal foodgrains —rice and wheat (Table 1.7). Developed countries still import substantial amounts of coarse grains for livestock feed, but the share of world imports going to developing and centrally planned economies is rising rapidly. This is partly explained by growth in consumer income, accompanied by rising demand for animal products, and partly through the inability of domestic production to keep pace with demand. The movement of import demand away from developed countries has significant implications for exporters and world trade patterns. Shifts in demand produced by fluctuations in economic growth in the developing and centrally planned economies, or changes in imports caused by their agricultural and trade policies, increasingly determine the state of the world grain market. Such policies are the major focus of this book.

The bulk of world grain trade takes place on commercial terms. Exporting countries often provide credit and financing arrangements for developing or centrally planned importers as an incentive for purchase. The United States, for example, operates this type of program (see the U.S. chapter in this volume). However, such sales are generally for cash. Richer countries also provide grains to poorer countries on a concessional or food aid basis. Food aid is provided to help alleviate the effects of natural disasters, such as floods or drought, or on a longer term basis to meet development or foreign policy objectives. It is also used to dispose of public stocks built up as the result of price support programs in rich countries. Shipments are either made directly by donor countries or through multilateral programs, such as those coordinated through the Food and Agriculture Organization of the United Nations. In recent years, 6-7 percent of world trade in grains was on concessional terms, and 75 percent was bilateral in nature (FAO, 1989). The major donor of food aid cereals is the United States, accounting for over 60 percent of the total. The two largest recipients are Bangladesh and Egypt, which collectively receive over 25 percent of the total. Many developing countries are wary of the long-term effects of dependence on food aid. Imports of grain under food aid programs may depress domestic grain prices and production in a recipient country. The availability of bilateral food aid is also often influenced by political considerations. The grain policies of some countries (see, for example, India in this volume) have been strongly affected by a desire to reduce dependence on food aid shipments.

A small number of private and largely American-based companies play a critical role in world grain trade. Estimates of the proportion of U.S. grain exports handled by the five or six largest companies range from 60 to 90 percent (Butler, 1986; Cramer and Heid, 1983; and Gilmore, 1982). The companies also handle substantial amounts of exports from other major exporting countries. They operate on a global basis and frequently own related businesses, such as grain processing facilities. The firms are secretive about their activities because

TABLE 1.7 World Grain Imports

Years	Developed (percent)	Centrally planned (percent)	Less developed (percent)	Total imports (mill. m. tons)
Coarse grains				
1960-64	79.7	13.1	7.3	28.1
1965-69	82.7	8.0	9.3	35.6
1970-74	70.8	19.3	10.0	52.1
1975-79	60.9	27.9	11.2	81.3
1980-84	51.3	28.3	20.4	92.9
1985-89	47.8	25.7	26.5	83.6
Rice				
1960-64	27.4	8.4	64.2	6.9
1965-69	32.0	19.1	48.9	7.3
1970-74	21.3	17.0	61.7	7.7
1975-79	22.0	12.1	65.9	9.7
1980-84	21.9	10.5	67.5	11.3
1985-89	22.3	14.1	63.6	11.7
Wheat				
1960-64	14.8	13.2	15.2	43.2
1965-69	12.6	13.3	20.3	46.2
1970-74	13.6	16.5	23.6	53.7
1975-79	13.9	20.3	30.6	64.8
1980-84	15.3	38.3	39.1	92.7
1985-89	14.2	32.9	43.3	90.4

Source: U.S. Department of Agriculture, Economic Research Service, PS&D View Database, 1989.
Note: Figures are averages for the years shown and exclude trade between member states of the European Community.

access to market information is an important part of their commercial success. Consequently, they are reluctant to release proprietary data. Only a few studies of the operations of the grain firms have been conducted. Some were highly critical (Gilmore, 1982; and Morgan, 1980), while others concluded that the firms are broadly competitive (Caves, 1977-78; 1982).

Prices and Price Formation

One of the most striking features of the world grain situation is the sustained decline in real grain prices. After adjusting for inflation, world prices of the major grains declined by roughly 1 percent per year since the early 1960s (Figure 1.6). Despite increased demand, large supplies kept grain prices from rising in real terms. Lower prices have made grain more affordable for consumers worldwide, contributing to growth in grain consumption and the expansion of trade. The impact of technological change on supply has been an important factor in the price decline, but import restrictions and export subsidies in some countries have also been a factor (see the chapters on the European Community and the United States).

Despite the favorable trend in prices from consumers' perspective, world grain prices have become more unstable, particularly since the early 1970s. Between 1972 and 1974, for example, international grain prices rose rapidly. During this three-year period, the price of rice rose by over 250 percent in nominal terms, and wheat prices climbed by over 150 percent. Several factors contributed to these price increases (Blandford, 1983), including crop failures engendered by adverse weather and strong consumer demand in some importing countries. The particular combination of factors that underlay the jump in grain prices in the early 1970s is unlikely to reoccur, but world grain markets are still susceptible to shocks, particularly the effects of natural disasters on supply. Experience since the early 1970s demonstrates the important relationship between fluctuations in world grain prices and world stocks. When the ratio of stocks to use is low, prices tend to rise; conversely, when stocks are plentiful, prices decline (Figure 1.7). This relationship is particularly marked for wheat. Periods of low stocks, such as 1973-75, 1979-81, and 1987-88, were associated with increasing world wheat prices. Periods of high stocks, such as 1970-71, 1976-78, and 1983-86, coincided with declining world prices. A number of studies demonstrate how importing and exporting country policies that insulate domestic markets from international grain markets contribute to instability in world grain prices (Blandford, 1983; Blandford and Schwartz, 1983). This issue is discussed in more detail in the concluding chapter.

Typically, 30-40 percent of the world's grain stocks are held in North America, mainly in the United States. Because the United States is the world's leading grain exporting nation, its stocks are more accessible to importers than those in other countries. The level of U.S. stocks is influenced strongly by the country's price and income support policies for farmers. These policies (discussed later in this volume) have significant implications for the world grain market, as well as the U.S. domestic market.

The U.S. grain market plays a central role in international price formation. This is because of its sheer size and the country's important commodity exchanges where grains are traded—for example, the Chicago Board of Trade, the Kansas City Board of Trade, and the Minneapolis Grain exchange. Trading on both a cash and futures basis takes place on the exchanges and is the principal way in which the open market price of grain is determined. Cash

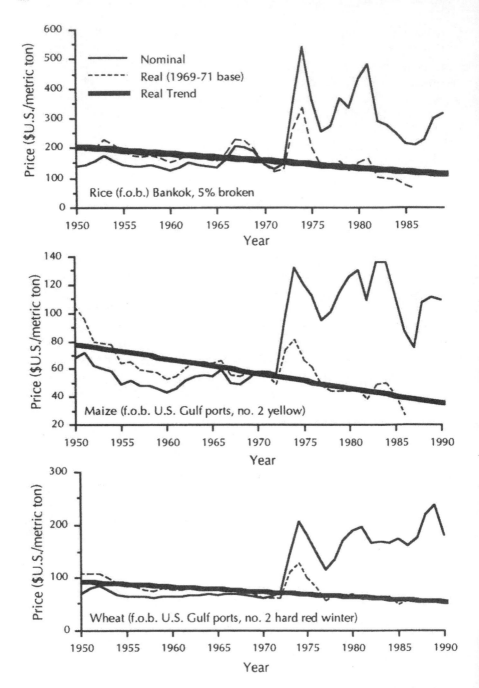

FIGURE 1.6 Trends in World Grain Prices. *Source:* Data from the World Bank (1988), updated by the author.

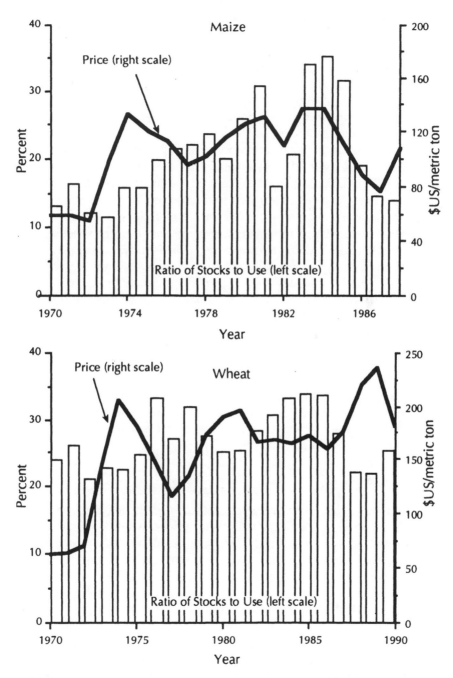

FIGURE 1.7 Relationship Between World Grain Stocks and Prices. *Source:* Data in Figure 1.6 and the PS&D View database.

transactions provide a reference point for current prices, and futures prices are an important indicator of expectations about price trends, not only for farmers, handlers, and consumers in the United States, but also worldwide. Transactions in futures contracts provide a means for traders to hedge their risks in marketing grain or to make (or lose!) money by speculating on price movements. Despite the effects of U.S. government policies on grain marketing in the United States, price determination on the exchanges is extremely important for the formation of global prices.

International sales are made primarily through private negotiation between purchasers and sellers or though tenders. Exporters base their offer prices on futures prices in U.S. commodity exchanges. Frequently, they limit their risk of losing money on forward transactions by hedging on the exchanges. The firm's profits are derived from the price at which they actually obtain grain, minimization of transportation costs, and sometimes through speculation.

As indicated above, there has been some debate on whether international grain markets are competitive. In addition to the small number of private trading companies that dominate world trade in grains, there are several public agencies, such as the Australian and Canadian wheat boards, handling substantial quantities of exports. A number of authors have argued that this leads to oligopolistic behavior on the part of exporters (e.g., McCalla, 1966; and Alaouze, Watson, and Sturgess, 1978). Others have attested that the policies of importers and their state trading organizations allow them to capitalize on market power (e.g., Carter and Schmitz, 1979). Undoubtedly, grain traders sometimes exploit their positions in grain markets. Sinner, Blandford, and Robinson (1987) documented such behavior by the Canadian wheat board, particularly in Japan. However, a study by Caves (1982) suggests that world grain markets as a whole exhibit competitive behavior.

Domestic Policies and World Grain Markets

There are few, if any, countries where the government does not intervene in some fashion in the domestic grain market. In richer countries, governments often protect grain farmers from foreign competition. Governments use various measures to restrict imports, or they subsidize exports to enable farmers to compete internationally. In poorer countries, where sources of government revenue are limited, producer prices are reduced through taxes. Prices may also be kept low to reduce food costs, either to reduce pressure on wages to foster industrial growth or simply to satisfy vocal urban consumers. Grain markets in rich and poor countries are also influenced strongly by concerns about "food security." Interpretation may differ among countries but usually implies insuring an adequate supply of staple foods at affordable prices. Since grains typically make up a major portion of basic foods, measures to control the flow and prices of grain are often an important part of food security policies.

Governments use a wide variety of tools to achieve their grain policy objectives, and the objectives and the tools are not always consistent with one another. Some policy measures may not be effective, nor reflect economic rationality. In addition to economic considerations, historical and political factors are important determinants of grain policies. However, policy measures frequently affect how domestic grain markets function. The policies of the large players can have important implications for other countries because of their effects on international trade. Policies of several of the countries in this volume – the European Community and the United States in the North, and India, Indonesia and Mexico in the South – influence the volume of world trade and prices. A primary objective of this book is to examine how different governments manage their grain sectors and to determine what can be learned from their experiences.

Grain Markets and the Countries in this Book

The countries included in this study differ widely from one another in size, population, wealth, and in the importance of grain in their economies (Table 1.8). They range from the small Caribbean country of the Dominican Republic with a population of 6 million, to the world's most populated country, China, with 1.1 billion people. Per capita incomes stretch from less than $US 350 in India, to over $US 17,000 in Canada and the United States. In some of the countries, such as India, agriculture accounts for a substantial share of national output (GDP). In others, such as the United States and Mexico, agriculture accounts for only a small portion of national output. The grain economies of the countries also differ significantly from one another. The Dominican Republic and Cameroon produce less than 1 million metric tons of grain each year, while the United States and China produce more than 300 million metric tons. Per capita utilization of grain extends from less than 150 kg per year in Cameroon, Colombia, and the Dominican Republic, to over 850 kg in the United States and more than 1,000 kg in Canada, where substantial amounts of grain are fed to livestock.[9]

Despite their diversity, there are important similarities among the two major groups of countries, which we broadly divide into "North" and "South." Per capita incomes in the richer countries of the North all average at least five times as great as the poorer countries of the South. In many cases, the degree of difference between North and South is much greater. The richer Northern countries are also generally the leading grain exporters, with self-sufficiency ratios (ratio of production to consumption) greater than 100 percent and sometimes substantially higher. The poorer countries of the South are primarily importers, with self-sufficiency ratios below 100 percent, although India is an important exception. Through international trade, the grain markets and

TABLE 1.8 Comparative Statistics for the Study Countries, 1986

	Population		Income		Food consumption[a] (calories/ day/cap.)	Production (mil. m. tons)	Per capita availability (kg/yr.)	Cereals		
	Total (mil.)	Rate of growth (%/yr.)	Per capita GNP ($US)	GDP in agriculture (percent)				Self-sufficiency ratio (percent)	Exports (mil. $US)	Imports (mil. $US)
The North										
Australia	16	1.3	11,890	7[b]	3,336	25.2	437	374	2,674.8	11.7
Canada	26	1.0	14,090	4	3,458	57.0	1,017	224	2,619.6	113.9
European Comm.	333	0.3	10,740	4	3,416	153.8	415	114	3,048.2	1,187.0
United States	242	1.0	17,500	3	3,632	316.1	886	145	7,104.9	159.8
The South										
Cameroon	10	2.7	910	21	2,061	0.9	123	80	0.5	37.4
China	1,059	1.3	300	—	2,611	349.2	299	95	882.3	982.3
Colombia	30	2.1	1,230	19	2,480	3.1	133	77	1,025.0	121.2
Dominican Rep.	7	2.5	710	17	2,327	0.6	141	39	0.0	72.6
India	784	2.1	270	28	2,138	164.3	190	101	103.6	7.5
Indonesia	177	2.1	500	23	2,470	45.6	215	95	29.9	292.9
Mexico	80	2.1	1,850	9	3,157	22.5	360	81	0.0	600.9

Sources: U.S. Depart. of Agriculture, World Agricultural Trends and Indicators, 1970-88, 1989; OECD, 1988; and European Communities, 1989.
[a]Ratio of domestic production to consumption.
[b]1985.

policies of these countries come into contact with one another and are interdependent. A further objective of this book is to determine what this interdependence implies for the countries as a whole.

In the case studies in this volume, the grain marketing systems and policies in the countries listed in Table 1.8 are described. Experiences of the countries in attempting to pursue their policies and the impact of these policies on domestic markets and international trade are discussed. In the final chapter, the wealth of information contained in the case studies is synthesized to identify important lessons to be drawn from country experiences.

Notes

1. Buckwheat is a hardy plant related to rhubarb and dock weed with few pests and diseases. It can be grown on poor soils or land that is too wet for most other cereal grains. Buckwheat was first domesticated in Asia and is still an important crop in some parts of Asia, for example, China. In the United States, it is a minor crop, produced largely in the eastern states. The seeds are ground into flour or are fed to livestock. Buckwheat is not included in the grain statistics in this chapter.

2. According to the USDA, FAS (1988), feedgrains also include wheat and rye of feed quality.

3. Data in this chapter are drawn primarily from the FAO, *Production Yearbook*, and USDA, Economic Research Service, *PS&D View Database*. For details on data definitions and conventions, see FAO (1989) and Webb and Gudmunds (1989).

4. In many parts of the world where rice is produced, two and sometimes three crops can be grown within a single year. Because of multiple cropping, rice often yields the most grain per unit of land.

5. Rice is included in milled form.

6. The economic regions used in this chapter correspond to those used by the United Nations (see FAO, 1987).

7. The U.S. Department of Agriculture estimates annual feed usage of grains for most countries. Grains are produced and fed to livestock on the same farm in many countries, and the quantity used cannot be accurately assessed. Livestock sometimes consume unharvested grain, and some cereal crops are grown specifically for forage. Many grain byproducts, such as rice hulls and wheat bran, are fed to livestock. These uses are not captured fully in the estimates, and it is likely that the contribution of grains to livestock production is understated.

8. Trade figures include intra-trade between members of the European Community. If this were removed, the fall in the developed country share would be even more pronounced.

9. These utilization or consumption estimates are based on an estimated balance sheet which includes production, trade, and changes in stocks. Thus, consumption estimates represent the "apparent use" or "disappearance" of grains including direct human consumption, grain fed to livestock, and industrial uses. No allowance is made for waste or seed.

28   David Blandford

References

Alaouze, C.M., A.S. Watson, and N.H. Sturgess. "Oligopoly Pricing in the World Wheat Market." *American Journal of Agricultural Economics* 60(1978): 173-85.

Blandford, D. "Instability in World Grain Markets." *American Journal of Agricultural Economics* 34(1983): 379-94.

———. "Changes in Food Consumption Patterns in the OECD Area." *European Review of Agricultural Economics* 11(1984): 44-64.

Blandford, D. and N.E. Schwartz. "Is the Variability of World Wheat Prices Increasing?" *Food Policy* 9(1983): 305-12.

Butler, N. *The International Grain Trade: Problems and Prospects.* New York: St. Martin's Press, 1986.

Carter, C. and A. Schmitz. "Import Tariffs and Price Formation in the World Wheat Market." *American Journal of Agricultural Economics* 61(1979): 517-22.

Caves, R. "Organization, Scale and Performance of the Grain Trade." *Food Research Institute Studies* 16(1977-78): 107-24.

———. "New Evidence on Competition in the Grain Trade." *Food Research Institute Studies* 18(1982): 261-74.

Chicago Board of Trade. *Grains: Production, Processing, Marketing.* Chicago, 1977.

Cramer, G.L and W.G. Heid, Jr. (eds.). *Grain Marketing Economics.* New York: Wiley, 1983.

Duvick, D.N. "Increasing Yield and Quality of Cereals through Breeding." In J. Dupont and E.M. Osman (eds.), *Cereals and Legumes in the Food Supply.* Ames: Iowa State University Press, 1987.

European Communities. *The Agricultural Situation in the Community, 1988 Report.* Brussels, 1989.

Food and Agriculture Organization (FAO). *Production Yearbook.* Rome, various issues.

———. *Food Aid in Figures.* Rome, 1989.

Gilmore, R. *A Poor Harvest: The Clash of Policies and Interests in the Grain Trade.* New York: Longman, 1982.

Kent, N.L. *Technology of Cereals.* Second Edition. Oxford: Pergamon Press, 1975.

Leonard, W.H. and J.H. Martin. *Cereal Crops.* New York: MacMillan, 1963.

McCalla, A. "A Duopoly Model of World Wheat Pricing." *Journal of Farm Economics* 48(1966): 711-27.

Morgan, D. *Merchants of Grain.* New York: Penguin Books, 1980.

Organization for Economic Cooperation and Development (OECD). *Food Consumption Statistics, 1973-82.* Paris, 1985.

———. *Historical Statistics 1960-86.* Paris, 1988.

Sinner, J.A., D. Blandford, and K.L. Robinson. *Canadian Wheat Export Behavior and Its Implications for the United States.* Department of Agricultural Economics, Cornell University, A.E. Research 87-23, Ithaca NY, August 1987.

United Nations. *World Economic Survey.* 1989.

U.S. Department of Agriculture, Economic Research Service. *World Agricultural Trends and Indicators, 1970-88.* Washington, D.C., 1989.

———, Foreign Agricultural Service. *Dictionary of International Agricultural Trade.* Agriculture Handbook No. 411. Washington, D.C., 1988.

———. *Foreign Agricultural Trade of the United States: Calendar Year Supplement.*

Washington, D.C., 1989.
Webb, A. and K. Gudmunds. *PS&D View Users Manual and Database*. Washington D.C.: United States Department of Agriculture, Economic Research Service, 1989.
World Bank. *Commodity Trade and Price Trends*. Baltimore: Johns Hopkins University Press, 1988.

2

Australia

Roley Piggott and Brian Fisher

Introduction

Australia is similar in size to the continental United States but has a population of only 17 million. The populace is highly concentrated in the major cities, making it perhaps the most urbanized country in the world. Gross domestic product (GDP) (market prices) in 1988/89 was about $A 338 billion in total, or about $A 20 thousand per capita. During the 1980s, the average annual growth in real GDP was 3.3 percent (as compared to a 2.8 percent average for all industrialized countries), and the average annual inflation rate was 8.7 percent (as compared to a 6.0 percent average for all OECD countries). On average, GDP was divided among farming at 4.3 percent, mining at 6 percent, manufacturing at 18 percent, and other sectors at 71.7 percent. The current account was overdrawn throughout the 1980s, with an average deficit of 4.8 percent of GDP (Australian Bureau of Agricultural and Resource Economics, Commodity Statistical Bulletin, 1990). Unemployment was generally less than 10 percent.

Australia's topography, climate, and soil types vary greatly across the country. Approximately 500 million hectares, or less than two-thirds of Australia's land area, is suitable for grazing or cropping. Much of the remaining area is too arid or rugged to support viable agriculture. In the case of the land that can be farmed, soil moisture patterns and their impact on the length of growing seasons are a crucial determinant of land use. About 22 million ha are used for extensive and intensive crop production, while about 26 million ha are planted to improved pastures for animal grazing. On the remaining 90 percent of agricultural land, which does not receive enough rainfall to permit cropping, animals are grazed on native unimproved pastures. Soils are generally phosphorous and nitrogen deficient, and about 25 million ha of land that are used for cropping or animal grazing regularly receive applications of chemical fertilizer (National Farmers' Federation, 1986).

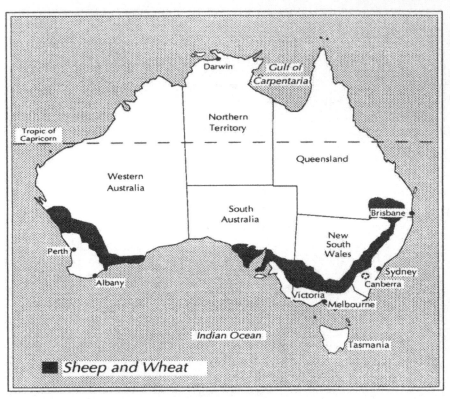

FIGURE 2.1 Australia's Principal Grain-Producing Areas

The principal grain-producing areas, shown in Figure 2.1, are in the so called "wheat-sheep zone." Climatic conditions in these areas (periodic droughts and undependable rainfall) help make grain production a high-risk activity for most farmers. Depending on location, rainfall ranges from 200 to 800 mm per year, and the length of the growing season varies from 5 to 9 months.

Australian farms are specialized in comparison to European farms because the bulk of their revenues comes from just two or three commodities (Davidson, 1982). In the case of grain farms, wool and/or beef production are often complementary enterprises. Australian farms tend to be considerably larger than farms in most other countries, but yields are low, with wheat yields averaging 1.4 tons/ha during the 1980s (Table 2.1). However, such figures mask considerable variability across farms. The average wheat farm is about 1,200 ha in size, but only about 200 ha are actually sown to wheat each year (National Farmers' Federation, 1987). Other notable features of grain farms in Australia include a high proportion of family-

TABLE 2.1 Output and Disappearance of Australian Wheat

Crop year[a]	Area ('000' ha)	Yield (tons/ha)	Production ('000' tons)	Domestic sales (Kt)	Exports (Kt)	Closing stocks (Kt)
1970/71	6,478	1.22	7,890	2,654	9,049	3,404
1971/72	7,138	1.19	8,510	2,703	7,760	1,451
1972/73	7,604	0.87	6,590	3,426	4,137	478
1973/74	8,948	1.34	11,987	3,165	7,418	1,882
1974/75	8,308	1.37	11,357	3,031	8,550	1,658
1975/76	8,555	1.40	11,982	2,742	8,233	2,665
1976/77	8,956	1.30	11,667	2,432	9,763	2,137
1977/78	9,955	0.94	9,370	2,629	8,098	780
1978/79	10,249	1.77	18,090	2,531	11,693	4,646
1979/80	11,153	1.45	16,188	3,369	13,197	4,268
1980/81	11,283	0.96	10,856	3,466	9,614	2,044
1981/82	11,885	1.38	16,360	2,560	11,068	4,776
1982/83	11,520	0.77	8,876	4,087	7,280	2,285
1983/84	12,931	1.70	22,016	2,631	14,152	7,518
1984/85	12,078	1.55	18,666	3,081	14,647	8,456
1985/86	11,736	1.38	16,167	2,791	16,026	5,806
1986/87	11,261	1.49	16,778	3,154	15,658	3,772
1987/88	9,063	1.36	12,369	3,430	9,962	2,750
1988/89	8,827	1.59	14,060	2,835	11,375	2,600
1989/90	9,004	1.58	14,214	3,154	10,760	2,900

Source: Australian Bureau of Agricultural and Resource Economics, Commodity Statistical Bulletin, 1990: Table 82.

[a] Years up to and including 1980/81: December 1 to November 30; 1981/82: December 1 to September 30; 1982/83 and subsequent years: October 1 to September 30.

owned farms and a low labor/land ratio (Davidson, 1982).

Agriculture's contribution to Australia's gross domestic product has steadily dwindled over the post-WWII period, falling from about 24 percent in 1949-50 to about 4 percent in 1989/90. Of particular significance is the decline in the relative importance of agricultural products as export earners. The decline (from about 80 percent of export earnings in 1951-52 to about 30 percent in 1989-90) came about mainly because of rapid growth in Australia's mineral exports (particularly coal and iron ore, but including several others).

Despite the drop in its relative contribution to export earnings, the Australian agricultural economy is still heavily dependent on export markets. The ratio of gross value of agricultural exports to gross value of agricultural production averaged 66 percent during the 1980s (Table 2.2). In the case of grains, export dependency was even higher, averaging about 85 percent during the 1980s. The reliance on export markets and the fluctuation of world grain prices is the second major source of risk (after the climate) confronting Australian farmers in general and grain producers in particular. Income instability generated by variability in yields and export prices

TABLE 2.2 Relative Gross Values of Agricultural Production and Exports for Major Commodity Groups, Australia

Year	Total Agriculture ($A mill.)	Wheat (percent)	Other Grains & Oilseeds (percent)	Total Grains & Oilseeds (percent)	Wool (percent)	Meat[a] (percent)	Dairy (percent)
Production							
1980-81	11,550	14.6	7.3	21.9	14.5	30.2	7.3
1981-82	12,644	20.6	7.2	27.8	14.2	26.1	7.8
1982-83	11,627	13.5	5.6	19.1	15.1	29.7	9.6
1983-84	15,435	23.4	8.6	32.0	13.1	22.8	7.5
1984-85	15,533	20.6	8.3	28.9	15.7	24.5	7.1
1985-86	15,535	17.3	8.8	26.1	17.3	25.1	7.6
1986-87	17,291	14.2	7.5	21.7	19.3	26.5	7.9
1987-88	20,219	10.0	7.7	17.7	27.3	24.8	7.2
1988-89	23,005	12.9	7.4	20.3	25.7	22.9	7.3
1989-90	23,513	11.8	6.8	18.6	24.3	24.3	7.6
Average	15,635	15.9	7.5	23.4	18.7	25.7	7.7
Exports							
1980-81	8,204	21.8	6.6	28.4	24.1	19.5	3.5
1981-82	7,914	21.9	9.0	30.9	24.2	18.7	3.9
1982-83	7,408	18.8	5.6	24.4	25.4	22.5	4.6
1983-84	8,380	21.8	9.9	31.7	24.5	17.8	4.5
1984-85	10,517	27.3	10.3	37.6	24.8	14.3	4.0
1985-86	11,663	25.6	9.7	35.3	26.2	14.8	3.7
1986-87	12,159	17.9	6.6	24.5	31.9	18.6	4.0
1987-88	14,247	12.6	4.9	17.5	40.8	17.7	3.8
1988-89	15,122	14.2	5.6	19.8	39.5	14.8	3.9
1989-90	14,864	18.1	7.2	25.3	26.0	19.1	5.0
Average	11,052	20.0	7.5	27.5	28.7	17.8	4.1

Source: Australian Bureau of Agricultural and Resource Economics, Quarterly Review of the Rural Economy, 3/3 (1991) and previous volumes.
[a]The meat category includes exports of live animals for meat.

has always been a key issue in Australian agriculture.

Data in Table 2.2 show that grains accounted for about 23 percent of the gross value of agricultural production and 28 percent of the gross value of agricultural exports during the 1980s. Grain, wool, and meat production together constitute the "big three" of Australian agriculture. Wheat dominates the grain sector.

Grain Production

General Characteristics

While all the Australian states except the Northern Territory produce some wheat, New South Wales and Western Australia account for the bulk of production (Figure 2.1). Only spring wheat is grown in Australia's temperate climatic conditions. It is sown in the autumn months (usually starting about April) and harvested in early summer after a five- to six-month growing period. Unlike the North American and European winter wheats, there is no dormant growth stage. High-protein, hard wheat (at least 11 percent protein at 11 percent moisture) is planted in the northern part of New South Wales and Queensland, while softer wheats are produced on the southern part of the continent. Australian wheat is classified by protein percentage and chemical and physical characteristics, with about 70 percent falling into the Australian Standard White category.

Barley is a winter crop grown in the same general areas as wheat, although production is concentrated in the state of South Australia. The majority (about 90 percent) of the crop is two-row barley. Oats for grain is a winter crop planted in southern wheat-producing areas. Yellow field maize is not an important crop in Australia, although it is occasionally planted during the summer season along the eastern coastal area and in some irrigated areas. Sorghum is now the most important summer cereal and is grown in northern New South Wales and southern Queensland.

Most of Australia's rice crop, which (unlike other grains) is generally irrigated, is produced in the southern plains area of New South Wales and in Queensland. Climatic conditions allow for only one (summer) crop each year in New South Wales, but two crops per year are possible in tropical northern Queensland. A mixture of medium- and long-grain varieties are grown.

The area sown to wheat grew rapidly in the late 1970s and early 1980s because of favourable relative prices (Table 2.1). The increase between 1982/83 and 1983/84 was particularly large because of the ending of a drought period. Australian Wheat Board (see later discussion of marketing institutions) quotas on wheat deliveries and depressed wool returns helped to elevate the importance of coarse grains in the late 1960s and early 1970s. Data in Table 2.3 indicate that coarse grain area increased rather markedly in the late 1970s and early 1980s. As the 1980s progressed, shorter cropping rotations, coupled with year-round production in those areas where the climate permits, helped to maintain the level of coarse grain production (National Farmer's Federation, 1986). However, wheat production also escalated in the late 1970s (Table 2.1), primarily because extra area was sown to wheat in response to depressed beef prices. As a result, wheat continues to dominate the grains sector.

Sown rice area rose substantially between the late 1970s and 1984/85 (Table 2.4) because farmers diverted irrigated land away from livestock enterprises to rice in response to relatively high international rice prices. The reverse occurred in 1985/86 when rice prices fell, and rice area has remained relatively constant since then. Rice area grew steadily from 1970/71 through 1981/82, while yields increased sporadically, resulting in a tripling of production since 1971/72.

TABLE 2.3 Coarse Grain Output and Disappearance

Crop year	Area ('000' ha)	Yield (m. tons/ha)	Production (Kt)	Domestic Use (Kt)	Exports[a] (Kt)
1970/71	4,191	1.31	5,473	3,277	2,196
1971/72	4,492	1.29	5,782	2,644	3,138
1972/73	3,891	0.93	3,620	1,966	1,654
1973/74	3,664	1.27	4,671	2,932	1,739
1974/75	3,285	1.35	4,421	1,538	2,883
1975/76	3,868	1.44	5,575	2,438	3,137
1976/77	3,901	1.29	5,019	1,725	3,294
1977/78	4,318	0.98	4,217	2,290	1,927
1978/79	4,663	1.51	7,063	4,229	2,834
1979/80	4,200	1.48	6,210	1,751	4,459
1980/81	4,284	1.22	5,211	2,477	2,734
1981/82	4,847	1.38	6,687	3,242	3,445
1982/83	4,527	0.88	3,966	2,176	1,790
1983/84	5,795	1.64	9,497	4,968	4,529
1984/85	5,525	1.59	8,771	2,445	6,326
1985/86	5,307	1.53	8,144	2,205	5,909
1986/87	4,502	1.58	7,093	3,401	3,692
1987/88	4,626	1.58	7,320	4,795	2,525
1988/89	4,758	1.57	6,881	4,312	2,569
1989/90	3,946	1.77	6,969	3,796	3,173

Source: Australian Bureau of Agricultural and Resource Economics, Commodity Statistical Bulletin, 1990: Table 95.
Note: Includes barley, oats and sorghum to 1959/60, maize added in 1960/61, followed by triticale in 1979/80.
[a]Exports are on a July-June basis, including grain and the grain equivalent of malt exported.

Supply Response

Several analysts who have studied the supply response of grains to price changes have emphasised the multiproduct nature of farms which produce grain and the tendency for farmers to switch resources among products (especially among different grains) in response to relative price changes. Estimates of supply elasticities vary across different production areas reflecting differences in the scope for product substitution. In the way of example of supply elasticities, the one–year response in planned output of wheat to a one percent change in expected wheat price has been estimated to be about 0.5 per cent for the wheat–sheep zone. A similar figure has been estimated for barley and the cross–price elasticity of the supply of barley with respect to the price of wheat has been estimated to be about –0.3 (Johnson, Powell and Dixon, 1990).

TABLE 2.4 Rice Production and Exports

Crop year	Area '000' ha	Yield (paddy) m. tons/ha	Production (paddy) Kt	Exports (milled) Kt
1970/71	41	7	301	96
1971/72	40	6	248	169
1972/73	45	7	309	152
1973/74	67	6	409	125
1974/75	75	5	387	160
1975/76	74	6	416	215
1976/77	92	6	528	254
1977/78	91	5	490	273
1978/79	110	6	692	239
1979/80	116	5	613	450
1980/81	106	7	759	276
1981/82	127	7	857	558
1982/83	77	7	519	384
1983/84	119	5	632	241
1984/85	122	7	866	327
1985/86	106	7	716	519
1986/87	97	6	608	359
1987/88	107	7	740	314
1988/89	97	8	748	338
1989/90	105	8	846	424

Source: Australian Bureau of Agricultural and Resource Economics, Commodity Statistical Bulletin, December 1990: Table 105.
Note: Production data are based on rice industry figures calculated on a crop year basis. For example, the 1985/86 data are for the NSW crop harvested in April 1986, the Queensland summer crop harvested in December 1985, and the Queensland winter crop harvested in June 1986. Exports are based on Australian Bureau of Statistics data and are generally for exports of the crop harvested in the previous year.

Domestic Markets

Wheat marketed domestically, usually less than 20 percent of each year's crop, is used mainly for human consumption, but also for animal feed and industrial purposes (e.g., starch, gluten and glucose). The relative importance of different domestic uses varies considerably from year to year. This is due in large part to changes in the demand for stockfeed wheat caused by variations in seasonal conditions and relative prices of alternative feedgrains. The coarse grain industry is less dependent on export sales than is the wheat industry (Table 2.3). Domestic livestock consume a considerable portion of output, and the brewing industries are also important market outlets (Table 2.3). Rice producers, like wheat farmers, are heavily dependent on export sales, with about 60 percent, on average, of the crop exported during the 1980s (Table 2.4).

Because of Australia's dependence on export markets, domestic consumption trends are relatively unimportant insofar as wheat and rice are concerned. However, in the case of coarse grains, the domestic market is growing because of an expanding intensive livestock sector (mainly hogs and poultry) and its requirements for feedgrains. Most Australian beef is produced under extensive conditions, but there are periodic surges in demand for coarse grains when drought destroys pastures.

A set of domestic farm–level demand elasticity and price flexibility estimates for wheat and other feedgrains is reported in Table 2.5. The demand for wheat for human use tends to be highly inelastic but substitution possibilities make for a relatively elastic demand for wheat used as stockfeed.

Marketing Systems and Institutions

A prominent feature of agricultural marketing in Australia, including the marketing of grains, is the importance of marketing boards (or corporations) and other forms of statutory authority. These institutions were first created in the 1920s following the failure of voluntary cooperatives to increase and stabilize producer returns. Campbell and Fisher (1982: 103) explain that the government readily provided the necessary legislative backing to establish what were essentially compulsory cooperatives. These eliminated the necessity of highly visible and expensive price support programs, while the cost to the government for their creation was minimal.

Some marketing boards operate at the state level, while others function at the federal level. Many of the state boards were first established under omnibus legislation, such as the 1927 New South Wales Marketing of Primary Products Act. The basic nature of the legislation has not changed very much. Federal boards each have their own legislation. For example,

TABLE 2.5 Domestic Demand Elasticities and Price Flexibilities for Wheat

Elasticity or flexibility	Estimate
Own-price elasticity of domestic demand for human use	-0.17
Cross-price elasticity of domestic demand for human use with respect to the retail price of potatoes	0.05
Own-price elasticity of domestic demand for stockfeed use	-2.37
Cross-price elasticity of domestic demand for stockfeed use with respect to the price of other feedgrains	1.58
Own-price flexibility of the demand for other feedgrains	-0.46 (-2.17)[a]
Cross-price flexibility of the demand for other feedgrains with respect to the quantity of stockfeed wheat	-0.16

Source: Myers, Piggott and MacAulay, 1985: Table A1.
[a]Figure in parenthesis is the inverse of the corresponding price flexibility.

the Commonwealth Wheat Marketing Act of 1989 provides the statutory basis for the Australian Wheat Board, which markets most of the wheat crop. Farmers contribute to the costs of operating state and federal marketing boards and corporations through a check–off system.

The activities of national and state boards are distinctly different because of the division of powers between the states and the Commonwealth set down in the Australian constitution. When the constitution was framed, the Commonwealth took control over export trade, but the states retained the power to set prices. Section 92 of the constitution provided for free trade among the states. As a result, national–level marketing boards tend to be involved in facilitating and regulating international trade (e.g., overseas promotion and licensing of exporters) rather than in trading commodities, while state–level boards and authorities control marketing within state boundaries.

The Australian Wheat Board is an exception to this general pattern in that it has powers to trade in wheat (and other grains) on the domestic and international markets. Until 1989, the Wheat Board had a monopoly on both the domestic and international sale of Australian wheat. Its monopoly on domestic sales was removed in 1989 but it remains the sole exporter of Australian wheat.

One of the powers generally available to state marketing boards is compulsory acquisition from farmers of the products for which they have responsibility. This power was also available to the Australian Wheat Board until 1989, being enforced through a system of complementary state and federal legislation. Section 92 of the Australian constitution has, over the years, been a real obstacle to the exercise of this power by marketing boards. Farmers can seek to avoid the strictures of compulsory acquisition by selling their output across state borders, seeking immunity under the free interstate provision of Section 92. State marketing boards involved with selling grain still face the possibility of challenges to their compulsory acquisition powers.

In addition to the involvement of marketing boards in the sale of grain, state–level statutory authorities or grower–cooperatives with statutory backing are involved in the handling, transport and storage of grain. Until 1989, grain producers had little choice in how their grain was handled after it left the farm gate. Generally speaking, it had to be delivered to a state–owned storage facility and it was transported by state–owned railways. If the grain was exported it was shipped through ports controlled by statutory authorities. Various pathways of grain movement are shown in Figure 2.2. Private enterprise had little role in the grain handling system.

During the early 1980s there were rapid increases in charges levied on growers for the handling of their grain. For example, wheat storage, handling and transport charges rose by more than 50 percent between

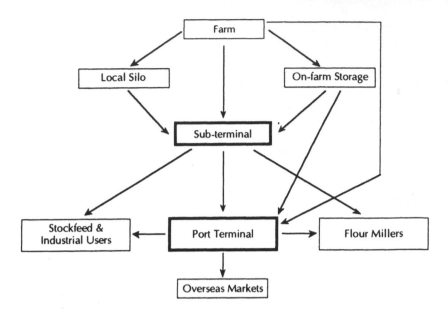

FIGURE 2.2 Alternative Grain Marketing Channels. *Source:* Adapted from Royal Commission into Grain Handling, Storage and Transport (1988a:p. 71).

1979/80 and 1984/85. Concern over the escalating costs led to the establishment of a Royal Commission (an enquiry in which a Royal Commissioner is given wide investigatory powers) to investigate the matter. The Royal Commissioner's report, presented in February 1988, recommended wide-scale deregulation of the grain storage, handling and transport system (Royal Commission into Grain Storage, Handling and Transport, 1988a).

The grain handling, storage and transport system now in place is still dominated by state–owned instrumentalities but, compared with the system which was in place until 1989, it is much less restrictive in terms of choices available to growers and the Australian Wheat Board. Deregulation has progressed most in New South Wales. In the way of examples of what has occurred, the Australian Wheat Board is now required to use the most cost–effective methods of storage, handling and transport and it can be exempted from state laws which would have the effect of increasing the costs it incurs. Whereas previously some state rail authorities had virtual monopolies on grain transportation, private road haulage is gradually increasing in importance. Too, in New South Wales, the former grain bulk handling authority has been "corporatised" (i.e., given much more commercial flexibility in how it operates) and is in the process of being privatised.

State level marketing boards involved with grain selling have also been under review over the past couple of years and some restructuring has occurred. For example, in New South Wales individual grain boards (except the NSW Rice Marketing Board) are being amalgamated. The powers given to state boards to compulsorily acquire crops has also been under examination but it remains a power still generally available to boards, albeit subject to closer scrutiny by state government officials and industry representatives in some cases.

International Trade

Trading Patterns

When the United Kingdom entered the European Community in 1973, Australia was forced to seek new customers to replace its major wheat market. Australia now depends on sales to Russia, China, the Middle East, Japan (one of Australia's traditional buyers), and a number of developing countries. In the decade ending 1983/84, Australia's share of world wheat exports averaged about 10 percent, but it increased significantly between 1984/85 and 1986/87 in response to increases in Australian production (Table 2.6). It has since declined again and Australia's future market share will be dependent on EC and US policies with respect to domestic price supports and subsidized exports. During the 1980s, the Middle East was the principal export market for barley, and Japan was the main export market for sorghum. Australia's share of the world markets for barley and sorghum have fluctuated in a similar pattern as for wheat (Table 2.6).

Developing countries close to Australia, together with Hong Kong,

TABLE 2.6 Estimated Shares of Australian Exports in World Grain Trade (percent)

	Wheat	Barley	Sorghum	Rice
1982/83	8.6	9.7	3.4	4.5
1983/84	10.6	21.1	6.2	2.3
1984/85	15.3	23.9	12.2	2.9
1985/86	19.8	24.5	12.9	4.0
1986/87	16.8	14.0	10.3	3.2
1987/88	11.7	11.8	4.8	2.6
1988/89	11.0	11.8	2.8	3.5
1989/90[a]	12.7	15.6	3.4	3.0
Average	13.3	16.6	7.0	3.3

Source: Compiled from data in Australian Bureau of Agricultural and Resource Economics, Commodity Statistical Bulletin, December 1990: Tables 91 (wheat); 109 (barley); 110 (sorghum); 121 (rice).
Note: Data for rice are on calendar year basis (e.g., 1982/83 = 1982 calendar year).
[a]Preliminary.

constitute the principal export markets for Australian rice, along with a variety of "short-term" customers. Australia has traditionally had only a small share of the world rice market (Table 2.6).

The Response of Export Demand to Price

Despite the relative importance of export markets as outlets for Australian grain, relatively few estimates of export demand elasticities have been calculated. Researchers have usually viewed Australia as a price-taker in world agricultural markets. (The exception is the apparel wool market, which Australia dominates.) This view is implicit in Scobie and Johnson (1979: 65): "... We would argue that, in building economic models of the Australian export sector, analysts will distort reality little by treating export prices as exogenous." Scobie and Johnson report export elasticities of demand for wheat of -4.7, -28.4 for barley, -12.8 for oats, and -254.2 for rice.

Public Policy: Description

The discussion in this section is focused on wheat because it is by far the most important grain crop in Australia, and aspects of its price policy are generally reflected in the policies for other grains. In particular, the exclusion of free market forces from the determination of prices and the pooling of both costs and returns have been common features of all grain marketing schemes in Australia.

Major policy changes with respect to wheat marketing came into effect in 1989 and were deregulatory in nature. However, it is worth tracing some of the history of wheat policy in Australia in order to appreciate the forces which led to the 1989 policy changes and because it may convey useful information to policy makers in other countries.

The 1948–1989 Period

The nature of government intervention in the grains sector changed throughout the 1948–1989 period but, in the case of wheat, there was continuous intervention. The types of interventions which were in place at the end of this period are summarised in Table 2.7. The table is an oversimplification in the sense that the grain industry is affected by many policy measures in place elsewhere in the economy, including in other agricultural industries which complete with grains in resource use. One important example is the implicit tax on grain exports resulting from protection afforded to manufacturing industries. Another is the government's contribution to agricultural research.

The prevalence of statutory marketing in the grain industry has already been discussed. Powers exercised by marketing boards during this period,

TABLE 2.7 Major Grain Policy Instruments in Australia

Instrument	Commodity Barley and Sorghum	Wheat[a]	Rice[b]
PRODUCTION/CONSUMPTION			
Producer guaranteed price		X^c	
Deficiency payments			
Input subsidies			
- credit			
- fert/pest			
- irrigation			
- machinery/fuel			X
- seed			
Crop insurance			
Controlled consumer price		X^d	X^d
TRADE			
Imports			
- tariff			
- quota			
- subsidies			
- licensing			
- state trading			
Exports[e]			
- taxes			
- restrictions			
- subsidies	X^f	X^f	X^f
- licensing			
- state trading	X	X	X
OTHER			
Marketing subsidies			
- storage			
- transport			
- processing			
State marketing	X^g	X^g	X^g

[a]The domestic wheat market was largely deregulated in July 1989 so that domestic wheat prices are now determined by supply and demand forces. The deficiency payments mechanism for wheat was also abolished, being replaced with a government guarantee on a portion of Australian Wheat Board borrowings to finance the first advance payment.
[b]Statutory marketing arrangements allow a grower cooperative to dominate domestic and export sales.
[c]With pooled returns.
[d]Implicit consumer tax through price discrimination.
[e]Strict quarantine requirements for imported grains effectively eliminate imports.
[f]Export sales insurance.
[g]With pooled marketing costs.

together with prohibitions or other controls on imports (e.g. strict quarantine laws), usually resulted in domestic prices exceeding export prices.

Production controls have not been an important instrument in securing higher prices for grains except in the case of rice where rationing of irrigation water has been used as a supply restriction mechanism. Input subsidies (e.g. fertilizer subsidies) were in place for several years but there has also been some taxing of inputs through tariffs on material and capital inputs. Importantly, the Royal Commission report referred to earlier indicated that the regulated system of grain storage, handling and transport was resulting in costs to growers which were about $10 ton higher than the costs under a deregulated system.

As will be explained in some detail later, state authorities practised pooling of costs and pooling of revenues. This caused cross-subsidization among producers. That is, for some producers, pooling resulted in a net subsidy, whereas others paid a net tax above and beyond other intervention measures.

During the 1948–1989 period public policy for wheat was implemented through "wheat marketing plans". There were eight such plans, with the last expiring on 30 June, 1989. Each plan covered five years with the exception of the fifth, which covered the crop years 1968/69 through 1973/74. Because the implementation of a marketing plan required the passage of complementary state and federal legislation, considerable effort went into planning and drafting it. Towards the end of the period Australia's Industries Assistance Commission (now Industry Commission; a Federal government agency that, among other things, makes recommendations to the federal government on levels of assistance to industries), was required to conduct a public assessment of current wheat industry policy before a new plan was formulated. The Commission received submissions from interested parties at public hearings, including the Australian Bureau of Agricultural and Resource Economics, the Australian Wheat Board, bulk handling authorities, state departments of agriculture, grower representatives, and stockfeed manufacturers. Based on this information and its own economic analysis, the Commission recommended changes to the existing plan, although the federal government was not required to follow these suggestions. Some aspects of wheat marketing plans specifically required state sanction, necessitating debate by the Australian Agricultural Council, which is made up of federal and state ministers of agriculture and fisheries.

Two policy measures formed the cornerstone of the first six wheat marketing plans. The first, a "a two-price" or "home-consumption" pricing arrangement, a form of price discrimination, kept domestic prices above export prices, although domestic prices were occasionally lower in years when export prices were particularly high (Table 2.8). Costs of production figured prominently in the determination of domestic prices. The second instrument was a buffer fund used to stabilize returns from the export market. In the seventh plan (1979/80 to 1983/84), the price of wheat for domestic milling was fixed using a formula that tracked trends in export prices, but maintained the domestic price at an average of 20 percent above pool export prices. Domestic stockfeed and industrial wheat, however, were priced at the Australian Wheat Board's discretion. Growers also received a government-guaranteed minimum price, with any deficiency between the grower and guaranteed minimum prices being met by government contributions. This was fundamentally a deficiency payment scheme and replaced the earlier buffer–fund arrangement. The last plan

TABLE 2.8 Australian Wheat Prices ($A/ton)

Year[a]	GMP[b]	Export[c]	Human Consumption[d]	Industrial Use[e]	Stock Feed[f]
1975/76	76.55	106.39	99.32	—	—
1976/77	76.29	96.54	105.40	—	—
1977/78	80.94	116.48	111.16	—	—
1978/79	91.96	137.63	116.61	—	—
1979/80	114.71	153.18	130.78	133.08	140.50
1980/81	131.92	152.05	156.12	151.67	151.37
1981/82	141.55	152.50	187.20	151.15	149.78
1982/83	141.32	179.20	203.46	174.16	184.11
1983/84	150.00	167.70	219.41	170.34	175.24
1984/85	145.35	197.00	210.73	199.09	204.36
1985/86	148.87	184.40	213.89	188.39	193.82
1986/87	139.83	163.20	188.92	164.96	170.12
1987/88	144.29	183.40	193.46	185.42	189.54
1988/89	153.37	217.95	221.07	219.30	221.84
1989/90	170.80	200.32	—	—	—

Source: Australian Bureau of Agricultural and Resource Economics, Commodity Statistical Bulletin, December 1990: Table 86.
[a]Crop years up to and including 1980-81: 1 December to 30 November, 1981-82: 1 December to 30 September, 1982-83 and subsequent years: 1 October to 30 September.
[b]The guaranteed minimum price replaced the stabilization price in 1979/80. For the season 1984-85 to 1988-89, a preliminary guaranteed minimum price was set around September, and a final guaranteed minimum price set in late February. From the 1989-90 season, the preliminary and final guaranteed minimum prices have been replaced by the AWB's harvest and post-harvest payments respectively.
[c]Average of daily asking prices of Australian standard whitewheat FOB eastern states for the relevant crop yea
[d]Prior to 1979-80, domestic prices were not differentiated. Administered pricing of domestic wheat ceased after the 1988-89 season.
[e]Preliminary.
[f]Australian standard white, 10 percent protein, Sydney cash market (all purposes), January-July 1990 average.

was quite similar to the seventh. Domestic prices for human consumption were based on export parity plus a margin of $A 15 per ton. The guaranteed minimum price was 95 percent of the average of the gross pool returns in the subject year and the two lowest pool returns in the previous three seasons, less pool costs for the subject year.

Federal goals for the wheat marketing plans were never (legislatively) spelled out in much detail. Equity among growers and general industry financial stability (including stable prices) were mentioned as goals from time to time.

Two pricing practices were common during this period. Mention was already made of the two-price or home–consumption pricing policy. One justification advanced for this policy was that domestic prices could be manipulated counter–cyclically to export prices, making the "equalized" or weighted average producer price more stable than the export price. Another was that a home–consumption price protects domestic consumers from extreme fluctuations in export prices. In most years domestic prices were above export prices (Table 2.8) and it seems reasonable to conclude that this pricing arrangement transferred income from domestic consumers to producers.

The second and perhaps most distinctive feature of grain pricing arrangements was pooling, both of prices received and handling and transport costs. This was only possible because statutory authorities dominated storage, handling, transport and selling of grain.

Within broad grades, the Australian Wheat Board pooled the revenue from all sales of wheat from a particular crop year regardless of where or when it was sold. As already mentioned, export and domestic prices were not the same. Returns from both sources were pooled for particular grades, and all farmers contributing to the pool received an equalized price. Other forms of pooling also occurred at the national level. Port and shipping costs were pooled by the Wheat Board, with the result that export prices were the same for a particular grade regardless of source, except for a small adjustment made to compensate Western Australian growers for the freight advantage they enjoy to Middle East markets. All marketing costs were pooled by the Wheat Board.

Grain handling and storage systems were characterized by extensive cost pooling, both between handling facilities and among the users of a particular facility within a state. It should be noted that, for many years, the costs associated with the handling and storage of grain were pooled between, as well as within, states. Thus, all growers were charged the same amount per ton, regardless of the actual cost of handling their grain. There was also evidence of cost–pooling and cross-subsidization in the provision of transport services. Almost the entire Australian wheat crop is moved to port by rail. The freight rate that a particular grower was charged may have contained elements of subsidy from other growers, other rail users, or taxpayers. Under-pricing of transport services on low volume branch lines which carried nothing but grain was another major source of cross-subsidization.

The rationale for pooling is unclear. It is administratively convenient for institutions that essentially replace competitive markets as the determinants of prices and costs. Pooling also redistributes income among producers, spreading risk. Perhaps, therefore, some notion of equity provides part of the rationale. It is probably fair to say that grower members of statutory marketing and handling authorities are generally pressured to promote policies that are viewed as "equitable," even if they are not in the interests of all growers.

The Current Situation

The 1989 Wheat Marketing Act reflects many of the changes recommended by the Royal Commission and the Industries Assistance Commission (1988) in its five-yearly report on the wheat industry. The most important changes were: (a) removal of the Australian Wheat Board's monopoly of the domestic market; (b) provision for the Board to trade in grains other that wheat; (c) provision for the Board to be exempted from restrictive state regulations which have the effect of increasing handling, storage and transport costs; (d) replacement of the guaranteed minimum price with a government guarantee on a portion of Board borrowings to finance the first-advance (harvest) payment; and (e) the granting of much greater flexibility to the Board in relation to the number of pools it operates in any year and the payment options it can offer growers (e.g. a grower can opt for a cash payment rather than having wheat, or other grains, placed into a pool). Unlike previous wheat marketing acts, the 1989 Act does not have a 'sunset' clause requiring the passage of new legislation after five years. In short, the current wheat marketing arrangements are much less regulatory than those in place over the 1948–1989 period. However, the Australian Wheat Board still has a monopoly on the export of Australian wheat. Another feature of the new arrangements is that the Board is required to disaggregate its costs more fully.

In the period since the introduction of the new legislation private trading in wheat on the domestic market has been increasing with the establishment of small grain trading companies. Too, more wheat is being transported by road and there has been an expansion in private storage facilities.

Public Policy: Evaluation

In this section attention is focused on the evaluation of wheat policy which was in place until mid–1989. Various criticisms of wheat policy of this period contributed to the deregulatory changes embodied in the 1989

Wheat Marketing Act. Because policy in relation to other grains was similar to that for wheat (particularly pooling of costs and returns), much of the discussion in this section is applicable to them.

For the sake of convenience, the discussion is divided into two parts — pooling and stabilization. In reality, this is somewhat artificial because pooling of returns from different markets could be viewed as part of the overall stabilization package.

Pooling

In an early paper, Longworth (1967) described both inter-temporal transfers and those likely between growers at a given point in time as a result of pooling. Similar observations were made in an Industries Assistance Commission (1983) report. A detailed analysis of the transfer effects of pooling was conducted by the Royal Commission into Grain Storage, Handling and Transport (1988b). In a sample of sites in Western Australia, it was found that some growers received income transfers in the order of $A 3 per ton, while others paid an additional $A 2 per ton for grain handling.

As previously mentioned, the Australian Wheat Board pooled across wheat types, which they classified as broad grades, such as Australian Standard White. This masked price signals to growers with respect to grain quality, having a distinct and obvious effect on farm-level investment decisions. However, it also camouflaged signals to the plant breeding industry, distorting research investments. Little attention has been given to the trade-offs involved in choosing the optimum number of grain classifications, given the demand characteristics for Australian wheat and the costs of handling, transporting, and storing additional grades.

The effects on efficiency can be considered, first, in terms of cost pooling among sites and, second, in terms of pooling among users at a given site. In the case of the first question, researchers usually postulate that transport cost differentials outweigh those for handling. Thus, it is always cheapest to deliver grain to the nearest receiving point. For example, Spriggs, Geldard, Gerardi, and Treadwell (1987) make this assumption, which would apply, in particular, to pooling among widely separated regions when crop acquisition is compulsory. Assuming that the elasticity of grain supply is 0.8 and that handling charges represent approximately 15 percent of the grain price, Quiggin and Fisher (1988) showed that the welfare loss associated with a 10 percent distortion in grain handling charges is approximately 0.02 percent of the value of the crop. If the farm-gate price is $A 100 per ton, the welfare loss would be $A 0.02 per ton. Given such small efficiency losses for widely separated sites, it is noteworthy that cost pooling among states was discontinued well before serious consideration was given to suspension of pooling within states. This was likely because of the associated large income transfers among states.

In the case of cost pooling among nearby receiving points, welfare losses arise when deliveries are made to points with high marginal costs rather than to those with low marginal costs. If, for example, the transport costs to two sites are equal, then the social loss is equal to the difference in marginal costs. If the high-cost site is closer, then the loss from pooling is equal to the difference in marginal costs less the savings in transport costs. Effects of this kind are only relevant within regions where variations in marginal costs are greater than the additional transport costs. In the case of typical cost differences among sites in New South Wales, Quiggin and Fisher (1988) showed that welfare losses in the order of $A 1.50 per ton may not be uncommon when handling costs are pooled among nearby sites.

Pooling among the users of a particular elevator may also give rise to efficiency losses. Some deliveries will always be more costly to process than others because of differences in delivery time, amount unloaded (both in total and per truck), moisture content, and qualitative factors, such as the number of segregations required.

Other factors also affect system costs. For example, there are likely some benefits, such as reduced administrative charges, associated with large deliveries, which may be reflected in bulk discounts. It is also probable that variations in the size of delivery trucks will affect the cost at the receiving point. Since some overhead is associated with unloading, costs per ton will decline as truck size increases up to some limit set by the design of the receiving facilities. The absence of bulk discounts, together with cost pooling among users at a particular site, discourages the use of larger than average size trucks. It is also worth noting that the resource costs per ton per kilometre associated with a typical double-axle farm truck used for delivery from farm to elevator is some three times greater than that of six-axle articulated units (Bureau of Transport Economics, 1987). If full resource costs are not recovered from truck operators, pooling across users at a particular receiving point causes losses, not only to the handling facility, but also to society at large.

There are also indirect costs associated with pooling. These may arise, for example, because the performance of the grain handling system is not adequately evaluated as a result of a lack of publicly available information in a system dominated by statutory authorities.

Stabilization

Two elements of wheat marketing plans that remained intact over the years were home-consumption pricing and the payment of an equalized return to producers from domestic and export sales. We illustrate the general effects of these components in Figure 2.3. DD is the domestic demand function for wheat, while the export demand function is assumed to be perfectly elastic at the world price level, P_w. The domestic supply

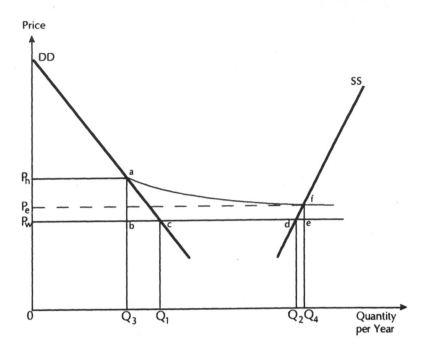

FIGURE 2.3 Price Discrimination with Equalization

schedule is SS, and under free-market pricing, production would be Q_2, with Q_1 being sold on the domestic market and Q_2-Q_1 on the export market. Under home-consumption pricing, domestic prices were set at some premium above world prices, say P_h, which results in a reduction in domestic sales to Q_3. This strategy increased revenue because of the lower elasticity of demand on the domestic as compared to the export market. Revenues from domestic and export sales were pooled, and growers were paid an equalized or weighted average return. Thus, the effective average revenue curve was the locus Daf. The equalized price received by producers, given the supply function SS, was P_e. This results in production increasing from Q_2 to Q_4, with exports represented by Q_4-Q_3.

The net social costs associated with these pricing arrangements are given by area abc plus area def. The first area can be viewed as that part of the reduction in (domestic) consumer utility (Q_3acQ_1) not offset by increased export earnings (Q_3bcQ_1). The second area arises because, under equalization, additional domestic resources valued at Q_2dfQ_4 were used to produce added output which earned only Q_2deQ_4 in export revenue.

In those years in which domestic prices were kept below export prices, the results described above were more or less reversed. Income was transferred from domestic producers to domestic consumers, and, given equalization, less resources were employed to produce wheat than if producers had been paid the true marginal value of output.

Because relatively little of the wheat crop is sold on the domestic market, the welfare effects described above would be relatively small. Moreover, the notion implicit in the early plans, that domestic prices could be manipulated counter–cyclically to export prices to stabilize producer returns, lacks credibility given the small proportion of output sold on the domestic market.

Longworth and Knopke (1982) calculated that the cumulative effects of wheat stabilization price policy over the period 1948/49 to 1978/79 constituted a net social welfare loss of $A 677 million in constant (1979) dollar terms. They estimated this to be about equivalent to the returns from one average-size crop. They found that producers were net losers over the period, while domestic consumers were net gainers. This occurred because the authors included the first five–year marketing plan in their analysis. During this plan, domestic prices were held below world prices, resulting in substantial transfers to domestic consumers. If the first plan is omitted from the analysis, producers (consumers) become net gainers (losers), and the cumulative social costs are substantially lower at $A 159 million.

Myers, Piggott, and MacAulay (1985) used econometric simulation to analyze the effects of Australian wheat price policy on the level and stability of key market variables, such as producer prices and industry revenues (Table 2.9). They concluded that only modest effects were achieved. In part, this was because home-consumption pricing proved to be a relatively ineffective instrument, given proportionately small domestic sales. In addition, in some stabilization plans, the buffer fund rules limited the extent to which stabilization funds could be used to moderate producer prices (see, for example, Campbell and Fisher, 1982: 130).

Current Policy

Those who have been critical of the wheat policies which existed over the 1948–1989 period would have to claim that current policy is an improvement. Stabilization instruments have been removed and, while pooling is still a feature, there is much more flexibility available to the Wheat Board in terms of how it conducts its pooling operations.

Policy Issues

Because the Australian economy is so heavily dependent on commodity exports, there are many current policy issues which, in one way or another, impinge upon the grains sector. Hence, the choice of topics discussed in this

TABLE 2.9 Effects of Wheat Market Intervention on Real Levels and Stability of Some Key Industry Variables, 1953/54 to 1983/84

Variable	Percentage change in	
	Mean	Coefficient of variation
Price of food wheat	7.1	–32.1
Price of stockfeed wheat	2.5	–11.6
Export price	–0.6	2.7
Price to growers	2.0	–16.7
Quantity exported	2.5	–4.7
Production	1.7	–2.5
Revenue from exports	1.7	–2.5
Revenue to growers	3.6	–2.6

Source: Myers, Piggott, and MacAulay, 1985: Table 4.
Note: The effects were generated by simulating plans 2 to 6 over the period 1953/54 to 1978/79 and plan 7 over the period 1979 to 1983/84.

section is highly selective, being confined to international competitiveness and monopoly exporting arrangements.

International Competitiveness

While much attention has been focused on the pros and cons of deregulation of grain marketing in recent years, another issue which has received much attention is Australia's competitiveness in world markets, especially in light of EC and US trade policies. (The two issues are related to the extent that the highly regulated marketing arrangements in existence prior to 1989 amounted to a cost impost on grain producers.) The Industries Assistance Commission (1988) found that interventions by other countries in their wheat markets in 1986 reduced international prices by about 15 percent, causing a decline in average net farm income in the wheat–sheep zone of about 5 percent. Australia has been at the forefront of countries pushing for reform in international agricultural trade. However, it is also the case that Australia's international competitiveness is eroded by domestic economic policies. A range of domestic economic reform measures (including reform of waterfront work practices and reform of the centralised wage–fixing system) are in place which should improve the competitiveness of Australian grain producers in world markets.

Monopoly Exporting Arrangements

One of the powers which the Wheat Board retained under the 1989 Wheat Marketing Act was a monopoly on the export of Australian wheat. This is not to say that no Australian export wheat is sold through international grain traders; rather, the Wheat Board does utilise the skills and knowledge of private traders in accessing certain markets. However, in principle as well as in practice, the Board has complete control over the volume and the destinations of Australian wheat exports from a given wheat crop. The local terminology for this arrangement is 'single–desk selling.' It not only applies to wheat exports, but also to the export of some other grains.

The main arguments advanced in favour of the retention of monopoly grain exporting authorities are that such agencies enjoy a better relationship with government-controlled buying organizations in importing countries than do private firms; that the agencies can exercise some market power in export markets; and that competition in the grain markets would otherwise lead to the exploitation of producers. There is little empirical evidence to support any of these contentions. It is quite clear that other exporters allow private traders to sell grains to importing countries with government procurement agencies (e.g., the US sells wheat to China). It is unlikely that even the Australian Wheat Board, which is by far the largest Australian crop marketing agency, can significantly affect the market, given its share in most years. The danger with single–desk selling arrangements, of course, is the lack of a competitive discipline to ensure that Australian grain producers are obtaining the highest returns possible (considering both costs and revenues) from the export of their grain.

It is very difficult to determine empirically whether single–desk selling arrangements result in a price premium from particular markets. What might be construed as a price premium could be no more than a payment in return for certain services or conditions associated with a particular sale. Alternatively, it could be a premium associated with some quality attribute in which case there would be no need for monopoly selling in order to obtain the payment.

Summary

Grain production in Australia is a risky farm enterprise that is usually undertaken in conjunction with beef and/or wool production. Farms tend to be quite large and yields relatively low because of soil and climatic factors. The principal grain crop is wheat, the vast majority of which is sold on unstable export markets. Production and price risk together resulted in the institution of stabilization policies, which also contained elements of support as well.

Until 1989 wheat marketing was controlled through a series of five-year marketing or stabilization plans. Important instruments embodied in those plans included home-consumption pricing, buffer-fund arrangements, underwriting or deficiency payments, and wide-scale pooling of both costs and returns. The impact of home-consumption pricing diminished through time because of the relative unimportance of the domestic market, but, in early plans, it caused substantial transfer of income between producers and consumers. In general, "stability" of key industry variables did not greatly improve as a result of the marketing arrangements. The ubiquitous practice of pooling created distortions that would be absent in a deregulated environment.

Wheat marketing was largely deregulated in 1989 although the Australian Wheat Board retains a monopoly status with respect to exports. Results from deregulation include the emergence of a number of private companies trading grain on the domestic market and an increase in private storage capacity. The wheat industry is no longer characterized by five–year stabilization plans and, in general, growers and the Australian Wheat Board have much more flexibility in how they trade grain.

International competitiveness is the major policy issue confronting the grains sector. The issue arises in large part from grains policies in place in the US and EC. The monopoly status of the Australian Wheat Board in relation to the export of Australian wheat is likely to remain a contentious issue whilst ever it exists.

Note

The final drafting of this chapter took place when major changes in the Australian grain marketing system were being canvassed by Australial federal and state governments. For example, the new wheat marketing legislation, which takes effect in July 1989, may result in substantial deregulation of the domestic market.

References

Australian Bureau of Agricultural and Resource Economics. Commodity Statistical Bulletin, 1990. Australian Government Publishing Service, Canberra, 1990.
————. Quarterly Review of the Rural Economy 3/3(1991).
Bureau of Transport Economics. "Competition and Regulation in Grain Transport: Submission to Royal Commission." Occasional Paper 82, Australian Government Publishing Service, Canberra, 1987.
Campbell, Keith O. and Brian S. Fisher, Agricultural Marketing and Prices. Melbourne: Longman Cheshire, 1982.
Davidson, B. R. "The Economic Structure of Australian Farms." In D.B. Williams (ed.), *Agriculture in the Australian Economy*, 2nd ed., Sydney University Press, 1982.
Industries Assistance Commission. The Wheat Industry. Report No. 329, Australian Government Publishing Service, Canberra, 1983.
————. The Rice Industry. Report No. 407, Australian Government Publishing Service, Canberra, 1987.
————. The Wheat Industry. Report No. 411, Australian Government Publishing Service, Canberra, 1988.

Johnson, D. T., A. A. Powell and P. B. Dixon. "Changes in Supply of Agricultural Products". In D.B. Williams (ed.), Agriculture in the Australian Economy, 3rd ed, Sydney: Sydney University Press, 1990.

Longworth, J. W. "The Stabilization and Distribution Effects of the Australian Wheat Industry Stabilization Scheme." Australian Journal of Agricultural Economics 11(1967): 20-35.

Longworth, J. W. and Phillip Knopke. "Australian Wheat Policy, 1948-79: A Welfare Evaluation." American Journal of Agricultural Economics 64(1982): 642-54.

Myers, R. J. R. R. Piggott, and T. G. MacAulay. "Effects of Past Australian Wheat Price Policies on Key Industry Variables." Australian Journal of Agricultural Economics 29(1985): 1-15.

National Farmers' Federation. Australian Agricultural Yearbook 1986. Melbourne: Publishing and Marketing Australia, 1986.

———. Australian Agriculture: The Complete Reference on Rural Industry. Melbourne: Morescope Pty. Ltd., 1987.

Quiggin, J. and B. S. Fisher. "Pooling in the Australian Grain Handling and Transport Industries." In B. S. Fisher and J. Quiggin (eds), The Australian Grain Storage, Handling and Transport Industries: An Economic Analysis. Research Report No. 13, Department of Agricultural Economics, University of Sydney, 1988: 47-53.

Royal Commission into Grain Storage, Handling and Transport. Volume 1 Report. Canberra: Australian Government Publishing Service, February, 1988(a).

———. "Pricing Practices." In Volume 3 Supporting Papers. Canberra: Australian Government Publishing Service, February, 1988(b), Supporting Paper 6.

Scobie, G.M. and P.R. Johnson. "The Price Elasticity of Demand for Exports: A Comment on Throsby and Rutledge. Australian Journal of Agricultural Economics 23(1979): 62-66.

Spriggs, J., J. Geldard, W. Gerardi, and R. Treadwell. "Institutional Arrangements in the Wheat Distribution System in BAE Occasional Paper 99, Australian Government Publishing Service, Canberra, 1987.

3

Canada

Colin A. Carter

Introduction

The purpose of this chapter is to provide an overview of the Canadian cereal grain markets, focusing on the major crops. The domestic production and disposition of grains is covered, followed by a discussion of important government policies that pertain to grain and a description of what part Canadian grains play in international trade flows. This chapter also provides the reader with background information on the size, scope, and regulation of the Canadian grain industry.

Canada is the world's seventh largest industrial economy, behind the United States, Japan, West Germany, France, Italy, and the United Kingdom. Average per capita income in Canada in 1986 was $US 14,174, making its citizens some of the most affluent in the world. In addition, the Canadian economy has performed quite well in recent years compared to many other industrial nations. Its real GDP grew by an annual average of 4.2 percent between 1982 and 1987. This was almost a full percentage point faster than the Organization for Economic Cooperation and Development (OECD) average. Growth was export-led, with exports accounting for about 25 percent of Canada's GNP. Its balance of trade has been positive, at about $C 10-17 billion per year for the past several years. The unemployment rate is below 10 percent, which is a five-year low. Inflation in 1987 ran at 4-5 percent per year, which was slightly higher than the previous two years. The importance of international trade to Canada's economy cannot be overstated. Canada exports about $US 3,330 per person, which is over twice as much as Japan, Australia, and the United States. Major exports include agricultural products, forest products, metals, minerals, and automobiles.

At the primary production level, the farming sector in Canada employs approximately 4 percent of the nation's labor force and accounts for about 3 percent of the total gross domestic product (GDP). Based on these measures alone, the agricultural sector appears to be a relatively small component of the Canadian economy. However, other factors, in particular agriculture's share of export revenue, elevate its importance. In fact, agricultural exports are some of the major reasons that Canada continues to maintain a positive net merchandise trade balance. In the recent past, Canadian agricultural exports averaged $C 9.7 billion per year, which was 10 percent of total exports. At the same time, Canadian agricultural imports averaged about $C 5.5 billion per year, yielding an annual trade surplus in agricultural products of around $C 4.2 billion.

Grain Production

Wheat contributes a greater share to farm cash receipts in Canada than any other commodity, with beef a close second. Wheat completely dominates the Canadian cereal grain industry, largely for basic climatic and agronomic reasons. Normally, about 29 million ha are planted to grains each year in Canada, and close to 40 percent of this is sown to wheat. Most of the wheat is grown in the western prairie provinces of Alberta, Saskatchewan, and Manitoba (Figure 3.1). For climatic reasons, maize production is generally limited to the eastern province of Ontario. Almost all of the wheat is grown under dryland conditions, with a very short growing season. Farms are relatively large in western Canada (average size per producer is about 275 ha), and the trend is toward even larger and more mechanized operations. Annual precipitation in the prairie region ranges between 350 and 550 mm. The predominant crop is spring wheat rather than winter wheat. Spring wheat is planted in May, and harvest occurs from late August through early October.

Measured by output, the major grains/oilseeds produced are wheat, barley, maize, oats, and canola (rapeseed) (Table 3.1). In terms of the total value of production, the order of importance is wheat, canola, barley, and maize. Normally, about 75 percent of the wheat, 50 percent of the canola, and 50 percent of the barley are exported. Wheat exports are plotted in Figure 3.2.

There are four major classes of wheat grown in Canada: hard red spring, red winter, soft white spring, and amber durum. The red spring wheats are used around the world to blend with lower protein wheats (from other countries) to produce bread flour. All-purpose flour for rolls, cakes, and muffins is milled from the red winter wheat and the soft white wheat. Durum wheat is used for pasta products. However, the standard class of wheat in Canada is hard red spring wheat, which is high in protein "content" and protein "quality" (or strength), both desirable characteristics for pan bread. In contrast, the dominant wheat class in the United States is hard red winter, and in Australia it is white wheat. Canada has a reputation for producing a high quality, very uniform wheat.

About 85 percent of wheat produced is hard red spring wheat and 10 percent is durum. Output increased considerably between 1970 and 1990 (from 9 to 22 million metric tons). It is important to note that the production of wheat in Canada was unusually low in 1970 because of a one-year government program

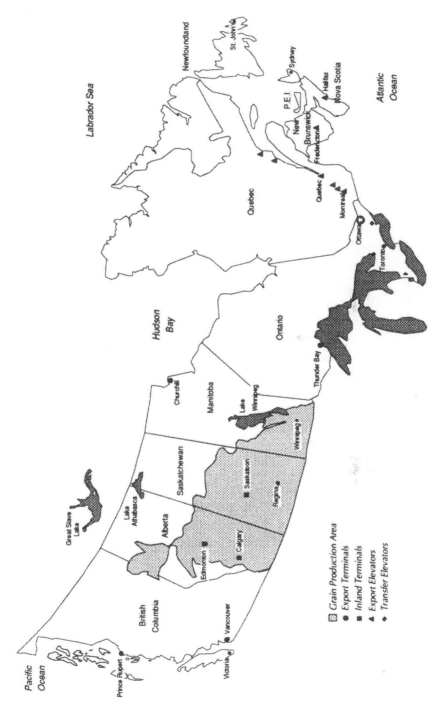

FIGURE 3.1 Western Canadian Grain Production Area. Source: Canadian Wheat Board, Winnipeg

TABLE 3.1   Canadian Grain Production and Exports (Kt)

Grain	1986/87	1987/88	1988/89	1985/86	1989/90	Average 1986-91
Wheat						
Production	31,378	25,992	15,996	24,575	32,709	26,130
Exports	20,783	23,519	12,419	17,425	22,104	19,250
Oats						
Production	3,251	2,995	2,993	3,546	2,851	3,127
Exports	257	286	728	737	383	478
Barley						
Production	14,569	13,957	10,216	11,674	13,925	12,868
Exports	6,719	4,954	2,879	4,506	4,635	4,667
Rye						
Production	609	493	268	873	713	591
Exports	201	221	115	295	342	235
Flaxseed						
Production	1,026	729	373	497	936	712
Exports	690	624	455	498	494	552
Rapeseed (canola)						
Production	3,787	3,846	4,288	3,096	3,281	3,660
Exports	2,126	1,750	1,949	2,048	1,888	1,952
Maize						
Production	5,911	7,015	5,369	7,157	5,912	6,366
Exports	(499)	188	(946)	534	(381)	(434)
Soybeans						
Production	960	1,270	1,153	1,219	1,292	1,179
Exports	(70)	35	135	(94)	24	6

Sources: Canada Grains Council, Statistical Handbook 1991. Winnipeg: Canada Grains Council.
Note: Crop year is August 1-July 31.

that paid farmers to set aside acreage. This was called the Lower Inventories For Tomorrow (LIFT) program. In 1988 there was a severe drought that reduced yields dramatically.

Much of the growth in wheat production that occurred since 1970 was caused by a rise in planted acreage rather than because of yield improvements. The increase in acreage resulted from changes in management practices, which gradually reduced the amount of fallowed area. Fallowing is commonly used to preserve soil moisture and control weeds. Hectares fallowed range from 20 to 25 percent of the total arable land in the prairie provinces.

There is very little supply response (either acreage or yield) to prices. The supply elasticity for wheat in Canada is estimated to be only 0.38 (Meyers, et al.,

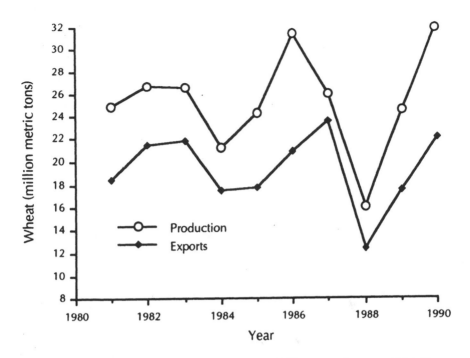

FIGURE 3.2 Canadian Wheat Production and Exports. *Source:* Canadian Grains Council, Statistical Handbook 1991, Winnipeg: Canada Grains Council.

1986). This is not surprising because there are few alternative crops that can be grown in the prairie region.

The data presented in Figure 3.3 clearly show that there are rather substantial year-to-year fluctuations in yields in important wheat-exporting countries, including Canada, the United States, the EEC, and Australia. However, the data also demonstrate that, on average, Canadian wheat yields have not significantly increased since the early 1970s. This is in sharp contrast to wheat yields in the United States and the European Community. The growth of wheat yields in Australia was similarly low. However, wheat yields grew by 121 kg/ha annually in the European Community and by 37 kg/ha/yr in the United States.

Disposition of Grains Through
Domestic Consumption and Exports

The Canadian wheat industry is driven by exports. Canada produces only about 5 percent of the world's wheat harvest in any given year, but with its relatively small population of 26 million people, most of the wheat is not domestically consumed.

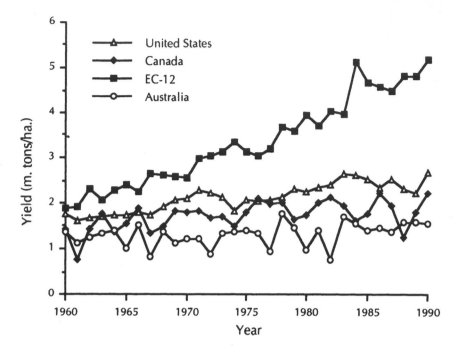

FIGURE 3.3 Wheat Yields in Major Exporting Countries. *Source:* USDA, PS&D View database, 1991.

In any given year, the domestic market absorbs only about one-fourth of all Canadian wheat sales (Table 3.2). Ninety-five percent of Canadian wheat is produced in western Canada, and prairie farmers are much more dependent on the export market than are eastern Canadian wheat farmers.

On average, about 10 percent of Canadian production is milled for home use, 10 percent is sold domestically for feed, and 5 percent is used locally for seed purposes. The market for domestically milled wheat has limited growth potential because the demand for flour and semolina has leveled off in Canada, and flour exports have fallen dramatically. Average flour exports over the 1977-81 period were 1.1 million metric tons, declining to only 0.41 million metric tons over 1980-90. Canada lost market share in the international flour market largely because of an increase in subsidized "sales" and food aid from the European Community and the United States.

The per capita consumption of wheat in Canada stands at about 80 kg/year compared to 75 kg in the United States and 96 kg in the European Community. Per capita consumption is gradually declining in Canada, but a low population growth acts as an offsetting force to maintain total consumption at a relatively constant level.

TABLE 3.2 Supply and Disposition of Canadian Wheat ('000' tonnes)

	1984/85	1985/86	1986/87	1987/88	1988/89	1989/90	1990/91
Supply							
Opening stocks	9,190	7,598	8,569	12,731	7,305	5,032	6,442
Production	21,188	24,252	31,378	25,992	15,996	24,565	32,709
Total	30,378	31,850	39,947	38,723	23,301	29,607	39,151
Disposition							
Exports	17,543	17,683	20,783	23,519	12,419	17,425	22,104
Human food	1,969	2,127	2,075	2,150	2,253	2,268	2,406
Feed	1,979	2,007	3,040	4,468	2,275	2,133	3,074
Other uses	1,288	1,464	1,318	1,281	1,322	1,339	1,353
Carry-over	7,598	8,569	12,731	7,305	5,032	6,442	10,214

Source: Canadian Grains Council, *Statistical Handbook*, various issues.

Unlike milled wheat, most of the feed wheat consumed on the prairies is either handled outside of the licensed elevator system or used on-farm. Livestock enterprises, in particular beef, are an important component of the Canadian agricultural sector. About 80 percent of the beef cow herd and 30 percent of the hog population is located in western Canada. In total, there are approximately 10 million hogs and 3.2 million beef cows in Canada. Both beef and pork are exported from Canada, primarily to the United States and Japan. Demand for wheat for animal feed in western Canada remains fairly constant at 2 to 2.5 million metric tons per annum and is not very responsive to changes in price. A trade model developed at Iowa State University estimated the feed demand elasticity in Canada to be -0.12 (Meyers, et al., 1986). Wheat feeding in the United States is much more responsive to changes in market conditions, with the price of wheat relative to maize acting as the major determinant. Meyers, et al. (1986) estimated the U.S. demand elasticity for feed wheat to be -3.01. For example, in 1983/84 when the relative price of maize rose dramatically, the use of wheat as a livestock feed nearly doubled in the United States. The use of wheat for feed remained high in the United States in 1984/85 at approximately 11.2 million metric tons (35 percent of total domestic use), but then declined to 7.7 million metric tons in 1985/86. A similar pattern of wheat feeding did not develop in Canada. As a percent of the total used, the domestic feeding rate is normally quite high in Canada (about 40 percent of total domestic use) compared to the United States. In Canada, wheat's portion of total domestic feedgrains is also relatively high, about 12 percent per year, while in the United States, 4-5 percent is the average. The feed market offers the best potential for increased wheat demand in Canada, but given the current relatively high feeding rates and the introduction of higher-yielding dwarf barley varieties, the volume of wheat used for feed is not expected to grow rapidly.

Government subsidization of freight rates encourages the movement of feedgrains away from the prairies (see Government Regulation in the Grain Sector). However, sales of prairie wheat to eastern Canada for feeding purposes have declined because of increased maize production in Ontario and Quebec provinces. A significant shift in the location of livestock production (from the east to the west) in response to changes in transportation rates for grains would alter this situation, but this is not likely to occur.

The domestic feedgrain market is analyzed in depth by the Canada Grains Council in a report, *The Structure of the Feed Grain Market in Western Canada* (1985a). Researchers studied the three major markets for prairie feedgrains: (a) the feed market in western Canada; (b) the feed market in eastern Canada; and (c) the export market. Between 1974 and 1983, the feed market in western Canada stagnated, demand from the eastern Canadian market declined, and the export market grew slowly.

Although wheat stocks fluctuate considerably from year to year in Canada, they averaged close to 10 million metric tons from 1971 to 1990. This represents about 50 percent of annual production (Figure 3.4). The average stocks-to-production ratio in Canada is about the same as in the United States over this time period. Furthermore, Canadian stocks are largely held by farmers (or farmer-owned grain cooperatives), while in the United States the government stores a large share of stocks.

International Trade in Grains

Historically, Canada has been the world's second largest wheat exporter, with a market share of between 18 and 26 percent over the past fifteen years. The other major exporters are the United States, the European Community, Australia, and Argentina. The United States and Canada together account for over 60 percent of world wheat trade.

Several different categories of wheat are traded internationally, but Canada specializes in high-quality and durum wheat. Canada's ability to compete in the international market is enhanced by the fact that it offers a high-quality, uniform product. High-quality wheats (No. 1 and No. 2 Canadian Western Red Spring (CWRS)) represent over one-half of Canadian exports, whereas medium-quality wheats account for almost two-thirds of U.S. exports. No. 1 CWRS is by far the dominant grade exported, accounting for an average of 45 percent of exports. The major importers of high-quality Canadian wheats are, in order of importance, Japan, the United Kingdom, the USSR, China, Cuba,

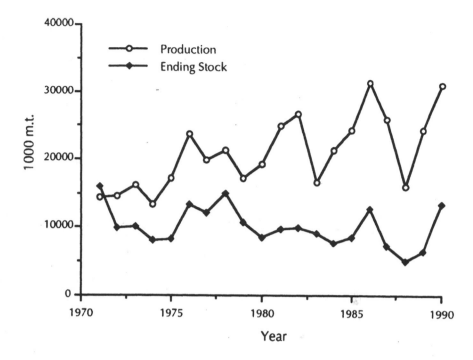

FIGURE 3.4 Wheat Stocks and Production in Canada, 1970-90. *Source:* USDA, PS&D View database, 1991.

and Brazil. The largest overall importers are the USSR, China, Japan, Brazil, and the United Kingdom (Table 3.3).

The high-quality wheat market has, however, been growing very slowly compared to that for cheaper, medium-quality wheats. Improvements in baking techniques worldwide permit flour with a lower protein content to be used without sacrificing bread quality, which in turn reduces the need for high-quality Canadian wheat.

Global trade in wheat increased from 54 million metric tons in 1970 to over 100 million metric tons in the mid 1980s. Large gains were made in the 1970s when grain trade grew approximately twice as fast as world production. During that decade, Canada's wheat exports increased by about 30 percent. There was also an important distributional shift in the pattern of world wheat trade. Wheat imports by the developed countries stagnated, while the centrally planned economies (CPEs) markedly increased their world market purchases. The Canadian Wheat Board (CWB) established a firm position in this market and now exports more than 50 percent of its wheat supplies to CPEs.

During the 1970s, Canada sold about 20 percent of its wheat to Western

TABLE 3.3 Canadian Wheat Exports by Country (in percentages)

Crop year	Brazil	China	Egypt	India	Italy	Japan	United Kingdom	USSR	West Germany	Others
1970/71	2.9	20.3	4.1	6.0	2.8	8.6	13.4	4.2	3.4	34.3
1971/72	2.6	20.8	0.6	4.5	3.7	10.1	10.3	19.4	1.7	26.3
1972/73	2.3	28.0	0.2	2.8	2.3	8.7	8.0	26.7	1.6	19.4
1973/74	6.8	11.8	—	3.0	5.9	14.4	10.2	13.6	2.8	31.5
1974/75	8.8	21.2	—	4.5	5.2	10.6	14.4	2.8	0.8	31.7
1975/76	4.3	9.9	—	3.9	5.6	13.2	10.0	26.0	1.0	26.1
1976/77	7.6	14.9	1.7	1.1	3.9	10.2	10.2	9.3	2.6	38.5
1977/78	5.4	20.9	3.4	—	5.9	8.5	10.1	10.6	0.4	34.8
1978/79	7.8	23.5	1.2	—	3.0	9.1	9.8	14.0	0.1	30.7
1979/80	6.9	17.6	0.2	—	4.6	8.6	9.2	13.7	—	39.2
1980/81	8.3	16.9	0.1	0.2	5.1	8.5	7.8	25.5	—	27.7
1981/82	7.3	17.3	1.8	0.5	2.9	7.6	7.6	28.0	0.1	27.1
1982/83	7.2	21.1	0.1	—	3.0	6.4	5.3	33.2	—	23.7
1983/84	6.4	16.1	3.1	2.4	3.5	6.2	4.5	31.8	0.1	25.9
1984/85	6.7	16.3	2.6	—	1.3	7.7	3.7	35.3	—	26.4
1985/86	5.7	14.8	2.7	—	2.1	7.3	4.0	30.1	—	33.3
1986/87	3.9	20.1	1.0	0.2	0.4	6.7	2.5	26.7	0.1	38.7
1987/88	1.9	32.7	0.4	—	—	6.4	1.8	19.4	—	37.1
1988/89	0.1	23.2	—	—	—	11.2	3.4	21.9	0.1	40.2
1989/90	1.3	26.6	—	0.1	0.1	8.5	1.6	20.3	—	41.7

Sources: Canada Grains Council, Wheat Grades for Canada, p. 77 (for 1970/71-1980/81). Canadian Wheat Board, Annual Report, 1989/90 XI, pp. 11-12. (for 1981/82-1990/91).

Europe, declining to 10 percent in the early 1980s because of decreased sales to the United Kingdom. Sales to Japan as a percent of total Canadian exports have also become less significant, while markets in Eastern Europe, the USSR, and Latin America have increased in importance. Sales strategies employed by the CWB have also changed over the last fifteen years. In the 1960s, most Canadian wheat sales were made to multinational grain companies, who in turn sold to the importer. In the 1970s, the CWB began to deal directly with the importers. This was facilitated by the growing importance of the CPEs in the market and their use of state trading agencies to import grains. The multinational companies now play only a limited role in marketing Canadian wheat to overseas customers.

Although international trade in feed wheat is relatively small, Canada is a major exporter, along with the European Community and Australia. The USSR is the largest feeder of wheat in the world, and in 1986-87, about 25 percent of its wheat imports (approximately 4 million metric tons) were of feed-grade quality. This was supplied primarily by Canada and the European Community.

Canada is the third largest contributor of world food aid. Of the wheat and flour exported each year, about 4 percent (over 700,000 metric tons) is part of food aid programs. Approximately 50 percent is donated under bilateral programs, and the remainder is distributed through the United Nations World Food Program.

In the early 1970s, Canadian exports of flour were about 5 percent of wheat exports, but this has fallen to less than 2.5 percent. Commercial markets for Canadian flour disappeared (e.g., the United Kingdom), and almost all of Canada's flour exports are now in the form of food aid shipments.

Marketing and Government Regulation in the Grain Sector

Marketing Channels

There are three major marketing channels—the Canadian Wheat Board (CWB), the dual CWB open-market system, and the open market—through which Canadian grains flow (Figure 3.5). Figure 3.5 refers to all grains, but the CWB handles wheat and barley only. Because there is an intricate mix of government agencies, farmer cooperatives, and private companies involved in marketing Canadian grain, the system is very complicated. For example, barley is handled by private traders and there is an active futures market, but at the same time, the CWB also purchases barley, and it regulates all producer sales through the imposition of delivery quotas. This means that the futures market does not operate as efficiently as it would without government interference.

In the case of wheat and barley marketing, the CWB is basically a government

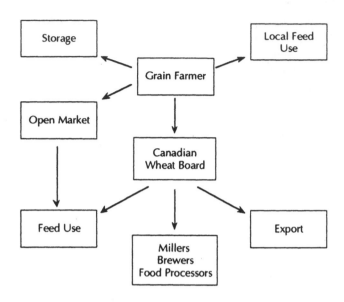

FIGURE 3.5 Major Marketing Channels in Canadian Grain Markets. *Note:* The Canadian Wheat Board handles only two grains; wheat and barley.

sales agency and owns no grain handling facilities. It employs the services of both private and cooperative elevator companies to physically move grain. Even though it is a government agency, the CWB's mandate is to bring the highest possible returns to producers and to give them equitable access to the export market. The CWB is the world's largest single grain marketing agency. It employs between three and five commissioners, who are appointed by the government, and includes a staff of about 525. The commissioners, which can be roughly equated with a board of directors, periodically seek advice from an advisory committee elected by farmers, but the committee has no legal control over the commissioners. Unlike the Australian Wheat Board, the Canadian Board only answers to the federal government and not to farmers.

Most of the wheat produced in western Canada is marketed through the CWB because it has monopoly rights over all wheat exports and all domestic sales for human consumption. Approximately 95 percent of the wheat produced enters the primary elevator system, while the remainder is used on-farm for feed or seed or sold locally. Of the wheat that does enter the elevator system, 97 percent is delivered to the CWB and 3 percent to private traders. Private traders are only permitted to buy feed wheat, which they subsequently sell on the Winnipeg cash and futures market. Unlike the Australian Wheat

Board, the CWB does not trade on the futures market. Wheat grown in Ontario is sold through the Ontario Wheat Producers' Marketing Board, which is a much smaller agency that is controlled by farmers. Ontario is different because it lies outside the jurisdiction of the Canadian Wheat Board.

At the beginning of each crop year (August 1), the Canadian government establishes initial producer prices for grain sold to the CWB. Prices are announced in advance, normally in April, to allow farmers to adjust their seeding plans. Separate prices are established for each grade of wheat. Receipts from CWB sales to the domestic and export markets are pooled, and producers receive an initial payment at delivery. In some years, they receive an interim payment during the crop year (if prices strengthen considerably), and a final payment once the crop year is over. At the end of the crop year (July 31), the pool is closed, and the CWB deducts its administrative expenses, interest costs, etc. Each producer receives the same price (before freight deductions) no matter when the wheat is sold to the CWB within a particular crop year. The CWB has separate pools for hard red spring wheat and durum wheat.

When selling to the CWB, producers' marketing costs are deducted in two stages. Freight costs and primary elevator handling costs are deducted from the initial payment at the time of delivery. Other costs, which include interest, insurance, storage, etc., and the Board's operating costs, are later charged against the pool before the final payment is advanced to the farmers.

Until 1988, domestic sales of wheat to millers by the CWB took place at prices that were partially insulated from world levels. This was referred to as the two-price wheat policy and was established in 1967. During the 1970s, the Canadian government fixed the domestic price to mills at relatively low levels and thus subsidized consumers (assuming the millers and bakers passed on this savings) when world prices were high. The two-price policy was very controversial over the years and was altered a number of times. The Canadian government discontinued the two-price system in 1989.

Government Regulation of the Grain Industry

The Canadian grain market is also heavily regulated in addition to the strictures imposed by CWB. Some regulations were developed to provide equity to producers, some to facilitate ease of marketing, and some to appease special interest groups. Major regulations and controls include grain freight rates, producer delivery quotas, rail car allocations, varietal licensing, and grading.

The wheat produced in western Canada must be moved over vast distances to reach a sea port. Most of the grain is moved to port by rail rather than by truck or barge. The farmers deliver their grain to primary elevators, which are located along rail lines. The rail freight rates for shipment to export destinations are regulated by the government, and they have not changed much in the last ninety years. Prior to the turn of the century, the federal government entered into an agreement (the Crow's Nest Pass agreement) with the Canadian Pacific Railway to fix the rates, and in return, the railway received a subsidy. These rates were generally considered sufficient to provide a return to the railways until the inflationary period of the 1960s and 1970s (Gilson, 1982). Inflation made profit impossible under the old rates, so the railways discontinued investment in the transportation system, which rapidly deteriorated. Farm stocks of wheat increased, and without proper transport facilities, the CWB could not meet the demands of all of its customers. After much study and negotiation, the federal government increased its subsidy to the railways, and farmers now pay a larger portion of shipping costs. Variable freight rates (e.g., a discount for unit trains) are now being increasingly employed. As a result, Canadian grain transport bottlenecks have almost disappeared.

Producer delivery quotas are another important aspect of regulation in the Canadian grain industry. Deliveries to primary elevators are governed by producer quotas. These quotas, which were originally imposed to provide equitable access to markets for all producers, have been in place since 1940. In the 1950s, the quotas were used to ration storage space in primary elevators. Then in the 1960s and 1970s, the sales strategies of the CWB and the capacity of the transportation system began to influence quota levels, justifying their continuation. For instance, delivery quotas were used to match farm sales with CWB sales. The overall capacity of the transportation system was also strained by the increase in wheat marketings caused by growth in production, and quotas were used to reduce pressure on the system.

The entire quota system was recently reviewed and restructured by the Canadian government in an attempt to reduce biases that had crept into the system. Under the new system, delivery quotas are based on the volume of grain each producer agrees to make available for sale to the CWB. Supply agreements must be signed between the farmer and the CWB soon after the harvest is completed.

The government regulatory agency that is responsible for quality control of Canadian grain and for the supervision of grain handling is the Canadian Grain Commission. Some of its most important functions include grain inspection and grading. Grain is visually inspected, and any new variety licensed for sale must be easily distinguishable from any existing variety and must also have superior milling and baking characteristics. Canadian regulators have historically stressed the high quality of Canadian wheat. This includes a high protein content, strong baking strength, and uniformity. Canada now has the reputation for supplying good quality wheat, but this has resulted in lower average yields than those of competitors. For example, many of the semi-dwarf spring wheats that are grown in the United States are higher yielding than Canadian varieties. However, because most of the U.S. varieties are visually indistinguishable from existing Canadian varieties, they cannot be licensed in Canada. Some farmers smuggled American seed into Canada in the early 1980s, selling the crop as "unprescribed" varieties, which meant that it sold for feed purposes. Because these types of wheat cannot be distinguished from CWRS varieties, graders were afraid that the unlicensed grain would get mixed up with CWRS varieties. However, the Canadian Grain Commission found that CWRS grades were seldom contaminated with unlicensed varieties (Committee on Unprescribed Varieties, 1986). Estimations of the economic costs of inspection and grading are very high, ranging between 5 and 17 percent of annual net farm income in Canada (Carter, Loyns, and Ahmadi-Esfahani, 1986).

By 1985, there were approximately 500,000 acres of wheat seeded to unprescribed varieties in Canada (Committee on Unprescribed Varieties, 1986). Census figures indicate that this figure approached 600,000 acres in 1986. In response to farmers' wishes to plant higher-yielding semi-dwarf wheats, the Canadian government followed the advice of the Committee on Unprescribed Varieties and licensed Oslo wheat in 1987. Oslo is visually distinguishable from the Canadian Neepawa variety, and it is eligible for the newly established "Prairie Spring" grade. The introduction of the Oslo variety will raise farm returns in some areas.

Grain Prices

Wheat prices rose dramatically in the 1973/74 crop year. While declining somewhat during the mid- and late 1970s, prices peaked in 1981 but then fell again until the late 1980s. A key factor in the downward trend in global wheat prices is increased yields worldwide. In the 1960s, yields grew approximately 2.5 percent per annum, slowing to an average of 2.2 percent per annum in the 1970s. Growth in the 1980s has averaged 3.6 percent thus far, largely because of spectacular achievements in the developing countries of China, India, and Argentina in particular. Yields also noticeably improved in the European Community and the United States (Figure 3.3).

Data in Table 3.4 show that import prices "landed" in Japan (basis c&f Japan) for premium Canadian wheat (No. 1 CWRS, 13.5 percent) dropped over 20 percent from $US 264 to $US 207/metric ton between 1980/81 and 1985/86, with prices for other classes of wheat showing similar declines. The premium for CWRS wheat held steady, while the discount on U.S. hard red winter wheat (as a percent of average import prices) increased. This is contrary to the conventional wisdom advanced in the early 1980s, which predicted that the spread between Canadian and U.S. wheat would narrow. The Canadian price premium was highest in 1974, 1976, and 1981 when there were temporary shortages of high-protein wheat. However, another reason the price for Canadian CWRS wheat appears to have held firm is that some importers maintain that the quality (with uniformity as the key factor) of U.S. wheat is

TABLE 3.4 Wheat Import Prices by Class, Basis C and F, Japan ($US/m. ton)

Crop year	Australian Standard White	Can. No. 1 CWRS 13.5%	US No. 2 Dark North Spring 14%	US No. 2 Hard Winter 13%	US No. 2 Hard Winter Ordinary	US No. 2 Western White
1970/71	68	76	73	73	68	69
1971/72	64	72	70	69	65	65
1972/73	83	105	98	100	99	99
1973/74	180	224	213	220	223	215
1974/75	183	223	215	206	192	187
1975/76	159	204	200	185	168	162
1976/77	126	154	148	140	128	125
1977/78	127	146	138	131	125	132
1978/79	161	179	168	166	154	159
1979/80	204	234	220	216	207	194
1980/81	216	264	243	230	220	201
1981/82	202	234	215	213	203	191
1982/83	196	225	211	215	201	202
1983/84	187	227	213	212	189	182
1984/85	174	212	196	188	177	167
1985/86	158	207	192	177	162	161

Sources: 1970/72-1980/81: Canada Grains Council, 1985: 127; 1981/82-1983/84: International Wheat Council, World Wheat Statistics, 1985: 72; 1984/85: International Wheat Council, Review of World Wheat Situation 1984/85: 42.

declining. This allowed the CWB to continue to charge a premium for high-quality Canadian wheat.

Canadian Policy Developments

Major Policy Elements

Agricultural policy in Canada takes a variety of forms at both the federal and provincial levels.[1] Over three-quarters of total federal government expenditures on agriculture ($C 2.5 billion in 1985) are direct transfer payments, while research and extension receives about 18 percent of these monies (Hedley and Groenewegen, 1984). The major elements of Canada's agricultural policy include the following: (a) Income and price stabilization programs (federal and provincial); (b) Grading, inspection, and quality control of agricultural products (federal); (c) Transportation and input subsidies (federal and provincial); (d) Research, education, and extension programs (federal and provincial); (e) Regional economic development programs (federal); (f) Producer marketing boards and federal marketing agencies (private or quasi-public); and (g) International trade agreements and market development (federal).

There are some who believe that the government has intervened too much in Canadian agriculture (Forbes, Hughes, and Warley, 1982). In their opinion, there has been an unnecessary escalation in the number of stabilization programs, marketing boards, formula prices, and border controls. A thorough discussion of the major instruments (Table 3.5) used to enforce Canadian grain policy is beyond the scope of this chapter; however, income stabilization programs and transportation subsidies are particularly important for grains, and they receive special attention in this chapter.

Stabilization Programs

Stabilization programs are pervasive in Canadian farm policy. Generally speaking, price stabilization is provided for commodities that are used domestically, while income stabilization is furnished for a number of goods that are exported. The major federal programs are the Agricultural Stabilization Act (ASA: enacted in 1958 and amended in 1975) and the Western Grain Stabiliza-

TABLE 3.5 Major Instruments of Grain Policy in Canada

	Commodity		
Instrument	Wheat	Maize	Barley
PRODUCTION/CONSUMPTION			
Producer-guaranteed price	X		X
Deficiency payments			
Government purchases			
Producer sales quota	X		X
Input subsidies			
- credit	X	X	X
- irrigation	X		
- seed			
Crop insurance	X	X	X
Controlled consumer price			
Income stabilization	X		X
Price stabilization	X		
TRADE			
Imports - tariff		X	
- quota			
- subsidies			
- licensing	X		X
- state trading			
Exports - taxes			
- state trading	X		X
- export credit	X		
- food aid	X		
OTHER			
Marketing subsidies			
- transport	X	X	X
State marketing	X		X

tion Act [WGSA: enacted in 1976 and replaced in 1991 with the Gross Revenue Insurance Program (GRIP) and the Net Income Stabilization Account (NISA)]. The ASA price stabilization program applies to maize, soybeans, oats, and barley grown outside of the region covered by the WGSA.

The ASA provides deficiency payments to producers, which are triggered when the market price (adjusted for changes in cash costs) falls below 90 percent of the previous five-year average. Payments to farmers under this program have been quite modest -- averaging only 2-3 percent of net farm income. Formerly, red meats were also under the ASA but are now part of tripartite stabilization programs, which are supported by contributions made by the federal government, the provinces, and farmers.

The WGSA was a voluntary income stabilization program jointly funded by farmers and the federal government. There are seven eligible crops: wheat, barley, oats, canola, flaxseed, rye, and mustard seed. The basic objective of the WGSA was to provide protection against variations in net cash flow from grain sales. It was designed so that the aggregate net cash flow in the prairie region will not fall below the previous five-year average. A participating producer received stabilization payments in proportion to levies paid into the program. There were payouts in 1978 ($C 115 m), 1979 ($C 253 m), 1984 ($C 223 m), 1985 ($C 522 m), 1986 ($C 856 m), 1987 ($C 1,395 m), 1988 ($C693 m), and 1989 ($C 176 m). The effectiveness of the WGSA as far as stabilizing farm income was widely recognized. In fact, a special study of the United States Congress in 1984 recommended that a cash flow stabilization program similar to the WGSA be considered as an option by U.S. farm policy planners. However, since the WGSA ran into deficit problems in the late 1980s, it was discontinued.

In response to the "subsidy war" between the European Community and the United States, a one billion ($C) dollar Special Grains Program (SGP) was announced by the Canadian government in late 1986. Individual crop producers received a maximum payment of $C 25,000 under the program. The payment for each crop was inversely related to the relative price decline attributable to the export subsidy war. A second set of payments, this time in excess of one billion ($C), was announced in December 1987.

The federal and provincial governments reached agreement on both the NISA and GRIP programs early in 1991. GRIP was designed as an income stabilization program. Under the program, individual farmers can "insure" a target revenue per acre for any major crop. The target level is based on historical yields and historical prices. Farmers pay 33% of the program costs, the federal government pays 42% and the provincial government pays 25%. A payout is made whenever actual revenue falls below the "insured" target revenue for an individual farmer.

NISA is more of an income "smoothing" program and designed to complement GRIP. Farmers participating in NISA are permitted to contribute 2% of their net sales to their NISA account and these contributions will be matched by the federal and provincial governments (shared equally). Farmers can withdraw from NISA whenever the farm's gross revenue falls below the previous five year average, or when net income falls below $10,000.

Transportation Policy

A development in Canadian policy that was important for international agricultural trade was the passage of the Western Grain Transportation Act (WGTA) in November 1983. This act removed the historically low grain freight rates in Canada and facilitated the transportation and handling of western grains destined for export. Under the act, the federal government makes annual contributions to the railways of $C 658.6 million, and farmers pay higher freight rates. At the present time, farmers pay about 22 percent of the actual freight rate for grain shipment, and the remaining 78 percent is effectively paid by the government. According to the WGTA, the government's share will decline over time. The goal of the WGTA is to ensure that Canada will more efficiently transport and handle grains in the future. This will be accomplished by continually upgrading the transportation and handling system.

Trade Policy

A major strength of Canadian agriculture is low production costs for crops such as wheat, barley, maize, and canola. This is because of a relatively extensive and fertile land base. With few alternative uses (except grain production), the opportunity cost of producing grain on the Canadian prairies is very low. Surplus feedgrains and a large and productive land base also contribute to economical livestock production. The major weaknesses of the agricultural sector in Canada are its inability to effectively diversify the production base, and its relatively high degree of dependence on the international marketplace. These problems have left Canada exposed to the risks of price and income instability, while in recent years protectionist trade actions in the European Community, the United States, and Japan in particular have depressed both prices and trade flows, damaging income potential for Canadian producers. Consequently, monetary gains from international trade are likely to remain at least temporarily depressed.

The response of Canadian farmers to depressed prices has been to continually improve agricultural productivity, while the government has searched for more liberal trade arrangements to dispose of the additional surplus. In the 1975-84 period, the volume of agricultural output in Canada increased an average of 2.2 percent per annum, and this growth is expected to continue. On the trade front, government policy is directed toward developing arrangements that will allow Canada to better compete in world markets. Two important policy thrusts are the Canadian-United States free trade agreement

TABLE 3.6 Grain Subsidies for the Five Most Important Grains in Canada, 1970-85 (percent*)

Crop year	Wheat	Barley	Commodity Oats	Rapeseed	Flaxseed
1970/71	30.2	9.9	2.5	3.8	4.3
1971/72	15.5	7.7	1.6	4.9	5.8
1972/73	19.8	3.2	0.8	2.8	2.1
1973/74	5.4	3.4	1.0	2.2	1.7
1974/75	7.0	5.6	2.1	3.8	3.4
1975/76	8.1	6.4	3.9	5.1	4.5
1976/77	8.5	11.3	5.1	6.9	5.8
1977/78	16.5	13.4	5.5	8.1	6.1
1978/79	11.4	14.5	2.4	8.3	8.2
1979/80	11.6	8.2	1.8	5.3	4.2
1980/81	12.3	8.2	2.3	6.0	5.5
1981/82	8.9	9.7	3.2	7.5	5.4
1982/83	11.5	13.0	3.0	7.3	6.7
1983/84	13.7	15.4	4.4	7.4	10.9
1984/85	20.5	15.7	5.0	11.0	12.4
Mean	13.4	9.7	3.0	6.0	5.8
Coefficient of variation	0.5	0.4	0.5	0.4	0.5

Source: Carter and Glenn, 1988.
*Ratio of total producer subsidy to total producer value (gross returns).

and the Uruguay round of GATT (General Agreement on Tariffs and Trade) negotiations. Unless the reduction of agricultural trade barriers is vigorously pursued under these agreements, the future commercial viability of Canadian agriculture is threatened.

Government subsidies to grain producers are shown in greater detail in Table 3.6. The figures are expressed in 1970 Canadian dollars, and the nominal dollar figures were all deflated by the consumer price index. Some of the policies directed at producers of the five major grains and oilseeds provide direct transfer payments from the federal government, but the majority furnish indirect support by influencing the prices that farmers receive (Carter and Glenn, 1988). For wheat, which is the major crop, the proportional subsidies varied between 30.2 and 5.4 percent. Over the period, major subsidies were also paid to the railways for the transport of export grain to seaports. Canadian wheat producers do not receive quite as much government support as do their U.S. counterparts, but the difference between the two countries is not very substantial (Carter and Glenn, 1988).

Summary

In this chapter, we sketched the production, distribution, and regulation of the Canadian grain industry. The production of wheat, the most important cereal grain, is characterized by large-scale, dryland, mechanized farms. Canada has a large land base, and, as a consequence, production practices are not as intensive as in many other countries. For this reason, the costs of production are relatively low. The types of wheats produced are primarily hard red spring and durum. The Canadian government, through regulations, has stressed the importance of high-quality wheat. Rapid increases in yields were largely foregone because lower-yielding varieties with excellent milling and baking qualities were grown instead.

Given distributional shifts in the world wheat market, Canada has become very dependent on continued large purchases from the CPEs, with the Soviet Union and China as its two largest customers. The Canadians purposefully developed this market with long-term agreements. The market in the United Kingdom has become less and less important over the years, and as a result, the Canadian grain marketing system is still undergoing adjustments.

The Canadian Wheat Board has been operating for more than fifty years, and it is presently the largest wheat merchant in the world. It has successfully penetrated markets controlled by government importing agencies. Canadian farmers are highly dependent on the export market, and they have fewer cropping/production alternatives for climatic and agronomic reasons than their counterparts in Australia and the United States. For this reason, Canadian grain policy will continue to both promote exports and to emphasize the high quality of their product. The Canadian economy stands to benefit from the liberalization of global agricultural trade, and the future viability of Canadian grain farmers is dependent on freer trade.

Notes

1. See Anderson and Gellner, 1985; Barichello, 1984; and Forbes, Hughes, and Warley, 1982 for a more complete discussion of Canada's agricultural policy.

References

Anderson, W.J. and J.A. Gellner. "Canadian Agricultural Policy in the Export Sector." *Canadian Journal of Agricultural Economics* (Proceedings Issue) 32(1985): 170-85.

Barichello, R.R. "Recent Canadian Agricultural Policy and Its Relevance for the United States." Working paper prepared for the American Enterprise Institute Agricultural Studies Project. Washington, D.C.: 1984.

Canada Grains Council. *The Structure of the Feed Grain Market in Western Canada.* Winnipeg: Canada Grains Council, 1985a.

_____. *Statistical Handbook.* Winnipeg: Canada Grains Council, various annual issues.

_____. *Wheat Grades for Canada – Maintaining Excellence.* Winnipeg: Canada Grains Council, 1985b.

Canadian Grain Commission. *Canadian Grain Exports.* Winnipeg: various annual issues.

Canadian Wheat Board. *Annual Report.* Winnipeg: various annual issues.

_____. *Grain Matters.* Winnipeg: various monthly issues.

Carter, C. "The System of Marketing Grain in Canada." In G.L. Cramer and W.G. Heid, Jr. (eds.), *Grain Marketing Economics.* New York: John Wiley and Sons, 1983.

Carter, C. and M. Glenn. "Government Transfers to North American Grain Producers: Levels and Implications." *Agribusiness* 4(1988): 285-97.

Carter, C., A. Loyns, and Z. Ahmadi-Esfahani. "Varietal Licensing Standards and Canadian Wheat Exports." *Canadian Journal of Agricultural Economics* 34 (1986): 361-. 77.

Committee on Unprescribed Varieties. "Report of the Committee on Unprescribed Varieties." Winnipeg: 1986.

Forbes, J.D., D.R. Hughes, and T.K. Warley. *Economic Intervention and Regulation in Canadian Agriculture.* A Study Prepared for the Economic Council of Canada and the Institute for Research on Public Policy. Ottawa: Minister of Supply and Services Canada, 1982.

Gilson, C. *Western Grain Transportation Report on Consultations and Recommendations.* Ottawa: Minister of Supply and Services Canada, June 1982.

Hedley, D.D. and J. Groenewegen. "Canadian Agricultural Policy and Trade." In G. Lee (ed.), *World Agricultural Policies and Trade.* Saskatoon: University of Saskatchewan, 1984.

International Wheat Council. *World Wheat Statistics.* London: various annual issues.

_____. *Review of World Wheat Situation.* London: various annual issues.

Meyers, W.H., S. Devadoss, and M. Helmar. "Baseline Projections, Yield Impacts and Trade Liberalization Impacts for Soybeans, Wheat, and Feed Grains: A FAPRI Trade Model Analysis." Working Paper No. 86-WPZ, Center for Agricultural and Rural Development (CARD), Ames, Iowa, 1986.

4

The European Community

David Blandford and Madeleine Gauthier

The European Community (EC) is an economic union of twelve countries (Belgium, Denmark, France, Greece, Ireland, Italy, Luxembourg, the Netherlands, Portugal, Spain, the United Kingdom, and West Germany) with a combined area of 2.3 million km^2 and a population of over 320 million (Commission of the European Communities (CEC), 1987a). In terms of gross domestic product, the Community is second only to the United States; nevertheless, the economies of the member countries are quite different. Per capita income ranges from less than $US 4,000 in Greece to over $US 12,000 in Germany (World Bank, 1988). Economic diversity is accompanied by social and political diversity. There are nine national languages in the Community, and country governments range across the political spectrum.

The agricultural sector also contains a wide range of participants. The Community is the second largest producer and trader of grains in the world after the United States; yet its farms range from small traditional farms in Greece and Portugal, to large industrialized units in France and the United Kingdom. Such contrasts make it difficult to generalize about the grain market in the Community. What does link the countries together is a set of structures and regulations that comprise the "Common Agricultural Policy" (CAP). However, national grain markets do differ significantly, and Community policies are often supplemented with national policies. For this reason, in this chapter particular emphasis is placed on the CAP as the major unifying force in the grain market, while less attention is devoted to national market structures. Even though Community regulations apply across all the countries, there are few centralized grain market operations, and each country maintains its own distinct marketing system and institutions.

The Community: A Loose Federalism

The European Economic Community was founded in 1957 with the signing of the Treaty of Rome by six countries (Belgium, France, Netherlands, Italy, Luxembourg, and West Germany). The original Community of six, or EC6, expanded to the EC9 in 1973 with the addition of Denmark, Ireland, and the United Kingdom. Greece joined in 1981 to makè the EC10; finally, Spain and Portugal acceded in 1986 to give the current EC12 (Figure 4.1).[1]

Conceived originally as a customs union to promote trade among its members, the Community developed into a loose political confederation, which is viewed by some as the transitional stage prior to the creation of a true political federation of states. Under the current arrangements, the balance of power is equally shared by national governments, who still retain substantial flexibility over policy making at the national level. Community policy decisions are made by the Council of Ministers. In agricultural matters, for example, decisions are made by the agricultural ministers of the member states. There is a European Parliament composed of elected representatives from the member countries. This has limited but increasing powers and acts as a forum for discussion and debate on Community policies. The main administrative body is the European Commission. Located in Brussels, this is the civil service of the Community and is charged with helping to design EC policies and with overseeing their implementation. Finally, there is a European Court of Justice whose function is to ensure consistency between Community regulations and national laws.

Historical circumstances after the Second World War were particularly conducive to the formation of an economic union in Europe. The EC provided a vehicle for economic cooperation between France and Germany and a way to politically rehabilitate Germany after two disastrous European wars. The European economies also needed freer movement of goods and services to stimulate recovery. However, the agricultural sector proved to be a major problem for the economic union because all of the member countries intervened in some fashion in their agricultural economies. France, the most competitive agricultural producer, demanded access to Germany's agricultural market in exchange for German access to France's industrial market. Because the six founding states had complex and extensive national agricultural policies, it was necessary to devise a common policy that would provide protection for domestic producers from cheap imports from non-member countries, but would allow members' agricultural products to move freely within the Community. The Common Agricultural Policy (CAP), whose principles and operational mechanisms were established after the signing of the Treaty of Rome, was the result of these imperatives.

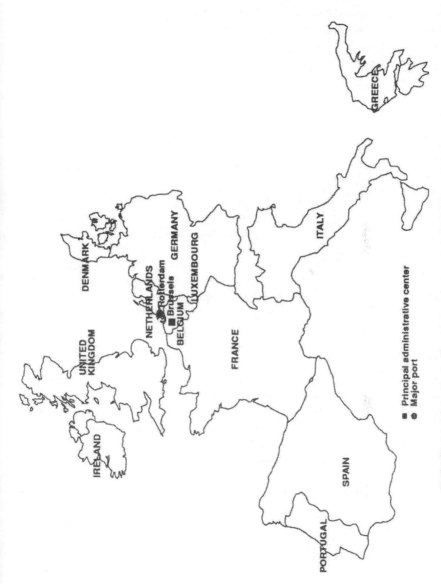

FIGURE 4.1 Main Grain-Producing Areas of the European Community

Agriculture in the Community: Diversity and Structural Change

Agriculture occupies a significant economic and political position in the Community.[2] Although its relative importance has been declining, agriculture still accounts for almost 6 percent of gross domestic product (GDP) and over 7 percent of total employment (Table 4.1). Behind these averages lie important national and regional differences. In 1989, agriculture accounted for almost 27 percent of total employment in Greece, 15 percent in Ireland, but less than 3 percent in the United Kingdom. Farms in the Community range from large-scale, capital-intensive units to small-scale, labor-intensive farms. These differences are reflected in national averages for farm size and productivity. Average farm sizes range from under 9 hectares in Greece, Italy and Portugal, to over 30 hectares in Denmark, and almost 70 hectares in the United Kingdom (Table 4.1). Productivity, measured by value added, is less than 4 thousand ECUs in Portugal and more than 57 thousand ECUs in the Netherlands.[3] Such differences are economically and politically important and have affected the evolution of the Community's agricultural policy.

In 1989, there were roughly 127 million hectares of agricultural land in the EC12 (including arable land, permanent meadows and grasslands, permanent crops, and gardens) (CECa, 1990). The use of land for agricultural purposes has fallen in all member states since the early 1970s, with the notable exception of Ireland, where it expanded by more than 1 percent annually, partly in response to the increase in commodity prices following the country's membership in the Community in 1973. Agricultural land area in Greece also increased slightly since the early 1970s. Despite the overall reduction in cultivated area and in agricultural employment in the Community, production grew dramatically. Over the period 1987-88, the annual rate of growth in the value of final production was almost 0.8 percent in the EC12 (CECa, 1990).

In the EC10, approximately 55 percent of the value of agricultural output is derived from livestock and livestock products and 45 percent from crop production (CECa, 1990). Livestock farms throughout the Community produce beef, veal, pork, lamb, chevon, chicken, eggs, and dairy products. The value of output increased rapidly since the early 1970s, particularly for poultry and dairy farms where the annual rate of growth has ranged between 5 and 6 percent annually. Cereal production (excluding rice) accounts for about 12 percent of the total value of output in the EC10 (CECa, 1987a).

Because of differences in climatic and soil conditions, the proportion of total agricultural area devoted to cereals varies substantially across the member countries, ranging from a low of 6 percent in Ireland to a high of 57 percent in Denmark in 1988 (CECa, 1990). For the EC12 as a whole, 27 percent of the total agricultural area was planted to cereals in 1988 (Table 4.1). This represents a slight decrease since 1975. In 1987, however, less than 9 percent of the farms in the EC12 were classified as grain farms (CECa, 1990).[4]

TABLE 4.1. Economic Indicators for the European Community

	EC-12	Belgium	Denmark	France	Germany	Greece	Ireland	Italy	Luxembourg	Netherlands	Portugal	Spain	UK
Population 1989 (millions)	326.1	9.9	5.1	56.1	62.1	10.0	3.5	57.5	0.4	14.8	10.3	38.9	57.2
GDP per capita 1989 (1,000 ECU)	17.1	17.5	18.2	18.6	19.3	9.3	11.5	17.8	20.8	17.6	9.4	13.0	18.2
Ag share of GDP 1989 (%)	3.0	2.2	3.8	3.2	1.6	16.4	10.9	4.1	2.3	4.2	5.2	5.1	1.4
Ag share of employment 1989 (%)	7.0	2.8	6.0	6.4	3.9	26.6	15.1	9.3	3.4	4.7	18.9	13.0	2.2
Ag area 1989 (million ha)	127.3	1.4	2.8	30.7	11.9	5.7	5.7	17.3	0.1	2.0	4.5	27.1	18.0
Av farm size 1989 (ha)	17.3	32.5	30.7	17.6	5.3	5.3	22.7	7.7	33.2	17.2	8.3	16.0	68.9
Value added per farm 1988/89 (1,000 ECU)	17.1	40.5	24.9	24.6	23.0	8.8	16.0	12.5	28.4	57.5	3.7	9.0	45.8
Share of ag area under cereals 1988 (%)	27.3	25.0	56.9	29.9	39.7	23.7	6.1	24.6	26.9	9.8	20.4	28.8	21.6

Sources: CEC, The Agricultural Situation in the Community, 1990. The World Development Report, The World Bank, 1991.

The Community uses a farm accounting system (the Farm Accountancy Data Network -- FADN) that identifies farm types based on the relative share of production of each commodity in the gross margin (receipts minus production costs) of the holding. Cereal farms are generally more specialized than other farm types. On average, cereal production accounts for over 70 percent of the gross margin of farms classified as cereal-based (CEC, 1990). By comparison, dairy farms derive less than 70 percent of their gross margin from milk, with the remainder coming from other products.

Cereal Production: Decreasing Area — Increasing Yields and Capital Intensity

Most of the major cereals are grown in the European Community. The geographical pattern of production is largely a reflection of differences in climate (Figure 4.1). The two most important grains are wheat, both "common" wheat (all bread and feed wheat varieties) and durum wheat, and barley (Table 4.2). Common wheat (dominated by low-protein or soft wheats) and barley are used mainly for livestock feed, particularly for poultry and hogs. Collectively, wheat and barley accounted for 78 percent of total cereal production in 1989 (Table 4.2), while maize, which is also used for feed, accounted for an additional 16 percent. Other minor cereals include oats, rye, and various summer cereals (for example, sorghum and millet). Rye and oats are more significant crops in the northern countries, such as Germany, where rye flour is a major ingredient in bread. Over the years, the production of rye and oats has become increasingly regionally concentrated as these grains have been displaced by higher-yielding, more profitable wheat and maize. Rice is produced in the south of the Community (particularly Italy and Spain), but on a relatively small scale. Because of its limited importance, rice will not receive any further attention in this chapter.

France is by far the largest cereal producer, accounting for over 30 percent of the EC12 total in 1989 (Table 4.2) and is the leading producer of wheat, barley, and maize. Other important producers include Germany with 17 percent, the United Kingdom with 13 percent, Italy with 10 percent, and Spain with 14 percent. In Germany, the United Kingdom and Spain, the major grains are common wheat and barley. Durum and common wheats and maize are the major grains in Italy.

Between 1973 and 1988, EC12 cereal production expanded from 123 million metric tons to 165 million metric tons, an increase of over 30 percent (USDA, Commodity Supply and Utilization Database). With the exception of Luxembourg, production grew in all member countries. Wheat accounted for most of the increase, both common and durum, with production rising roughly 4 percent per year between 1973 and 1988. The production of coarse grains, primarily barley and maize, grew by about 1 percent per year. Increases in

TABLE 4.2 Cereal Production in the European Community, 1989

	Area ('000' ha)	Yield (m. tons/ha)	Production ('000'm. tons)	Share of EC production (percent)
Total cereals[a]				
EC12	35,000	4.68	163,800	100.0
France	9,375	6.00	56,219	34.3
Germany	4,657	5.82	27,112	16.6
U.K.	3,900	5.42	21,121	12.9
Italy	4,437	3.68	16,307	10.0
Spain	7,741	3.01	23,317	14.2
Denmark	1,576	5.12	8,067	4.9
Greece	1,366	3.36	4,586	2.8
Ireland	340	6.45	2,194	1.3
Belgium	345	6.40	2,208	1.3
Portugal	1,027	1.28	1,318	0.8
Netherlands	202	6.04	1,220	0.7
Luxembourg	34	3.85	131	0.1
Common wheat				
EC12	13,453	5.40	72,664	100.0
France	4,725	6.48	30,612	42.1
U.K.	2,098	6.59	13,819	19.0
Germany	1,769	6.21	10,993	15.1
Spain	2,166	2.36	5,122	7.0
Italy	1,146	3.80	4,351	6.0
Durum wheat				
EC12	2,802	2.15	6,013	100.0
Italy	1,806	1.69	3,056	50.8
France	305	4.40	1,341	22.3
Greece	515	2.19	1,130	18.8
Barley				
EC12	11,717	3.97	46,554	100.0
France	1,832	5.36	9,816	21.1
Spain	4,257	2.19	9,308	20.0
Germany	1,753	5.57	9,758	21.0
U.K.	1,657	4.80	7,960	17.1
Denmark	1,001	4.98	4,982	10.7
Maize				
EC12	3,889	6.74	26,204	100.0
France	1,883	6.69	12,606	48.1
Italy	805	7.93	6,382	24.4
Spain	516	6.25	3,224	12.3
Greece	200	8.50	1,700	6.5

Source: CEC, The Agricultural Situation in the Community, 1990.
[a]Excluding rice.

coarse grain production are completely attributable to yield improvements. Harvested cereal area in the EC12 actually fell from 37 million hectares in 1973 to 35 million hectares in 1988, with all of the reduction occurring in coarse grain area (Figure 4.2). However, the decline was not uniform throughout the Community. The UK, Ireland, and Spain increased cereal area slightly, while area remained more or less constant in France and fell in other member countries. Overall, the amount of land planted to wheat, barley, and maize expanded at the expense of rye and oat area, resulting in decreased rye and oat production in spite of improvements in yields. Changes in grain production in the Community can mainly be traced to various technological innovations and to the impact of pricing policies. These factors are discussed in more detail below.

Between 1973 and 1988, the amount of land devoted to wheat in the EC12 remained roughly constant. However, wheat area in France, the Community's largest producer, expanded by 800 thousand hectares (21 percent), as common wheats (mostly high-yielding soft wheats) displaced pasture and other less-profitable crops. Wheat area also expanded substantially in the United Kingdom, which is now the second largest producer in the Community. This increase was due to a shift in cropping patterns in response to higher Community prices after Britain joined the Community. Arable land previously used as pasture for livestock was plowed to produce more profitable wheat crops. Italy is the largest producer of durum wheat. The area devoted to durum increased by over 300 thousand hectares, or 19 percent, between 1973 and 1988 (CEC, 1990).

As indicated above, since the early 1970s, a striking feature of the Community's grain industry has been the remarkable increase in yields. Yields of wheat, for example, grew at an average annual rate of over 3 percent in the EC12 since the early 1970s, and the Community average is rapidly approaching 5 metric tons per hectare (Figure 4.2). Coarse grain yields grew only slightly less rapidly. Nevertheless, significant yield differences exist among member countries. Cereal yields in Spain and Portugal are approximately one-half the average for the EC12 (Table 4.2). In Spain, for example, the yield for common wheat in 1989 was about 3 metric tons per hectare, compared to between 6 and 5 metric tons in France, the UK, and Germany. Yields in Spain are likely to grow rapidly with the intensification of production and the adoption of new technologies.

Since the early 1970s, the number of cereal holdings declined, existing units grew larger, and production became increasingly capital-intensive. According to the European Commission, grains are now more capital-intensive than any other major crop in the Community (CECa, 1986). However, despite the trend toward larger units, individual farms are usually operated by a single family. Each farm employs only a few, generally related individuals. Most of the land in the Community is transferred through inheritance rather than sold through market channels. Consequently, land rental is becoming increasingly impor-

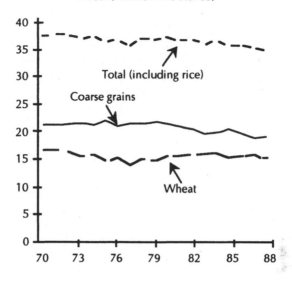

Area (million hectares)

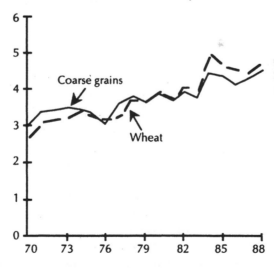

Yield (metric tons per hectare)

FIGURE 4.2 EC12 Grain Area and Yields. *Source*: Based on data from USDA, ERS. PS&D View Database. Washington, D.C.: 1989.

tant in enlarging the scale of operations. Cereal farms, particularly those in France and the United Kingdom, are highly profitable. In 1987/88, the net value added by cereal farms in these two countries was 65-70 percent above the average for all farms (CECa, 1990).

Between 1973 and 1986, input use (energy, fertilizers, pesticides, and machinery) in the agricultural sector almost doubled in the EC12 (CECa, 1987). Although a breakdown by farm enterprise is not available, substantially more inputs were likely used on cereal crops, accounting for a large portion of yield increases. Farmers now plant higher-yielding winter wheats in place of traditional spring wheats and apply large amounts of fertilizers and chemicals to maintain soil fertility and to control weeds and pests. As discussed below, pricing policies in the Community have encouraged the extensive use of inputs and the adoption of new technology.

In summary, since the early 1970s, major structural changes have occurred on cereal farms in the European Community. First, cereal production is more geographically concentrated within countries, and the mix of cereals has shifted toward wheat, particularly high-yielding, low-protein varieties. Second, there are fewer cereal farms, and they are larger and more capital-intensive. Third, major growth in yields have more than offset decreases in area. Collectively, these changes have contributed to increases in output and exports, which created problems for the Community's agricultural policy and for other cereal exporters.

There have been relatively few attempts to estimate aggregate supply elasticities for grains in the European Community because of technical and institutional problems. There are statistical difficulties in calculating an aggregate response for countries with different agricultural sectors. Further, time series data have only been available since 1967, the year that Community-wide support prices were implemented. Finally, the support program led to relatively stable prices and limited short-run variation in output. Researchers concluded that acreage is not responsive to changes in price, although in many instances they found that significant adjustments were made in yield. The elasticity of response of wheat yields to price ranges from 0.4 in Meilke and de Gorter (1986) to 0.7 in Devadoss, et al. (1986). Total output elasticity ranges from a low of 0.5 in Gardiner, et al. (1989), to 1.4 in Meilke and de Gorter (1986). The data collected in most empirical studies make it difficult to explain the substantial growth in Community grain production, and in particular, the impressive improvement in yields. Researchers, using a comprehensive data set for farms in eastern England, recently concluded that increases in yield were related at least as much to long-term changes in varieties and improvements in disease prevention methods as to changes in relative prices (Melencovitch, 1988). The Community's grain policies clearly helped to establish an environment favorable to the adoption of new technology, resulting in intensified production. These policies are discussed in greater detail below.

Consumption: The Importance of Animal Feed

Approximately 60 percent of all cereals used in the Community are fed to livestock, 25 percent are directly consumed as food, and the remaining 15 percent are used for seed or industrial purposes. The demand for grains has been virtually stagnant since the early 1970s, while the feeding of coarse grains has actually declined since the end of the 1970s (Figure 4.3), despite a rise in livestock production. The real value of livestock output in the EC10 grew at an average annual rate of over 5 percent between 1973 and 1986. More wheat was fed to livestock, primarily poultry and hogs, partly because of subsidies to feed compounders, although nonfeed uses of wheat slumped. Sluggish consumption of grain in the food industry reflects slow population growth and a low income elasticity for bread and other cereal products. Overall, total cereal usage in 1988, at 139 million metric tons, was virtually the same as in 1974 (USDA, Commodity Supply and Utilization Database). Because the community's population increased over the period, both per capita feed and nonfeed consumption of grain has actually declined since the early 1970s.

The stagnant demand for feedgrains in the face of growing livestock output was caused by improvements in feeding efficiency and the availability of cheaper feed substitutes, such as maize gluten meal, manioc, citrus pulp, molasses, and soybean meal, most of which are imported. Although soybean meal is primarily a protein source, livestock feeders sometimes use it as a grain substitute (OECD, 1981). Imported cereal substitutes enter the EC duty-free or with low tariffs fixed under the General Agreement on Tariffs and Trade (GATT). Since 1982, the Community has restricted manioc imports from its principal supplier, Thailand, through a "voluntary" export restraint agreement. The imposition of similar restrictions or tariffs on other products has been discussed, but the Community's ability to take action is limited by its obligations under the GATT and by strong resistance from the United States, one of its principal suppliers of substitute feeds. Imports of energy-rich cereal substitutes by the EC10 (primarily manioc, maize gluten, and molasses) increased from roughly 11 million metric tons in 1976 to 19 million metric tons in 1982. In 1985, largely because of restrictions on manioc, imports totalled 18 million metric tons. Imports of protein-rich products (primarily soybeans and soybean meal) increased from 18 million metric tons in 1976 to 25 million metric tons in 1985.

Substitutes have become particularly important in the production of mixed feeds. In 1974, cereals made up roughly 45 percent of the total volume of ingredients used in mixed feeds in the EC10; but, by 1984, the proportion had fallen to 35 percent, helping to contribute to the general decline in demand for cereals. Whether mixing their own feed rations or purchasing complete feeds, farmers in the Community are feeding less cereals to animals. The high support prices for grains are directly responsible for this trend. Very little manioc is

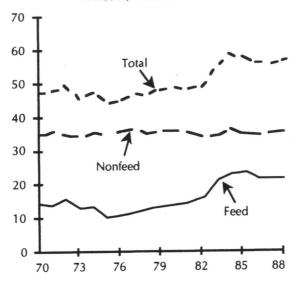

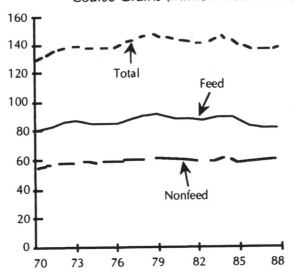

FIGURE 4.3 EC12 Wheat and Coarse Grain Use. *Source*: Based on Data from USDA, ERS. PS&D View Database. Washington, D.C.: 1989.

traded outside of the Community, and maize gluten is a minor feedstuff in other countries. The role of high grain prices in creating the demand for substitutes is acknowledged by the Commission, but the Council of Ministers is unwilling to reduce grain prices to solve the problem.

Because of stagnant demand, a declining proportion of wheat produced in the EC is destined for direct human consumption. In 1988/89, roughly 50 percent of common wheat production in the EC was consumed by humans, compared to over 60 percent in 1973/74, while animal use grew from roughly 30 percent in 1973/74 to 40 percent in 1988/89. Additional protein required in bread flour is obtained from imported high-protein wheats, which are blended with domestic soft wheat. Over 80 percent of the Community's durum wheat is consumed directly by humans, primarily in pasta and related products. Total consumption of durum wheat declined by roughly 9 percent between 1973/74 and 1988/89 because of reduced intake of starchy foods. Almost 80 percent of barley produced in the Community is fed to livestock. The remainder is used in the brewing and distilling industries, while direct human consumption is minimal. Maize is principally used for livestock feed (roughly 80 percent of the total), although the amount consumed by people in breakfast cereals and other products has increased.

Cereal use in the Community is responsive to market conditions, and most of the empirical studies of demand indicate high price elasticities, particularly for feed. Meilke and de Gorter (1986) and Bahrenian, et al. (1986), estimate that the price elasticity of demand for feed wheat is between -1 and -3. The price elasticity for food use is low, and the income elasticity is small or negative. The price sensitivity of feed use and low income elasticities of demand, combined with rapid technological change and the effects of pricing policies, have contributed to a growing structural imbalance between supply and demand, which is reflected in the Community's international trade.

Trade: The Community as a Growing World Force

The Community is the most important trading bloc in the world, and its trade in agricultural and food products is no exception. In 1989, the EC12 imported $US 64 billion worth of agricultural products from the rest of the world and exported $US 40 billion, representing about 13 percent of world agricultural exports and 23 percent of world imports (CEC, 1990). Germany is the largest importer of agricultural products, and France is the largest exporter. The United States is the Community's most important agricultural trade partner, accounting for almost 18 percent of total EC exports and 15 percent of imports (CEC, 1987b). The Community has a negative agricultural trade balance with the United States and, in 1986, imported products valued at over 3 million ECUs more than exports.

The Community is the world's second largest exporter of cereals, after the

TABLE 4.3 Cereal Trade in the European Community

	1973/74 (orig. data)	1987/88 (Kt)	1988/89 (Kt)
All cereals			
Imports	24,407	7,838	7,332
Exports	9,287	26,963	34,868
Net exports	-15,120	19,125	27,536
Common wheat			
Imports	4,084	1,775	2,103
Exports	5,165	14,400	18,040
Net exports	1,081	12,625	15,937
Durum wheat			
Imports	1,410	809	431
Exports	108	1,850	3,212
Net exports	-1,302	1,041	2,781
Barley			
Imports	1,968	452	734
Exports	2,957	8,307	11,392
Net exports	989	7,855	10,659
Maize			
Imports	14,696	3,601	3,257
Exports	654	1,864	1,957
Net exports	-14,042	-1,737	-1,300

Source: CEC, The Agricultural Situation in the Community, various years.
Notes: Data for 1973/74 are for the EC9, those for other years are for the EC12. All cereal totals exclude rice.

United States. In 1985, cereal exports from the EC10 accounted for 17 percent of total agricultural exports and 5 percent of imports. The major grains traded by the Community are wheat, barley, and maize (Table 4.3). The EC is a leading exporter of wheat, particularly soft wheat, which is produced in France. In 1988/89, the Community had net exports of more than 15 million metric tons of common wheat, close to 15 percent of the world total. It also had net exports of over 2 million tons of durum wheat and just over 10 million tons of barley. In addition, the Community is a major maize trader, with net imports totalling 1 million metric tons in 1988/89.

The largest grain export market for the EC is the collection of countries which comprised the former Soviet Union (Table 4.4). The Community has captured a large share of the fast-growing demand for grains in these countries, and exports of wheat, which were negligible in the 1970s, averaged 5 million metric tons in 1986-87. These countries also purchased roughly 2 million metric tons of barley. Other important recent wheat customers are Poland and China, and Saudi Arabia has made major barley purchases. Collectively, other countries in Asia and Africa, particularly North Africa, are important wheat customers,

TABLE 4.4 The EC's Important Cereal Trade Partners by Grain Type (Kt)

	Average 1981/82[a]	Average 1986/87[a]
EXPORTS		
Common wheat		
USSR	1,214	5,073
Poland	1,530	829
Morocco	1,205	58
China	669	353
Barley		
USSR	397	1,782
Poland	406	116
Saudi Arabia	1,609	2,892
Switzerland	315	220
IMPORTS		
Common wheat		
USA	1,670	820
Canada	1,578	907
Durum wheat		
USA	580	103
Canada	565	400
Barley		
Canada	402	57
Maize		
USA	7,656	2,353
Argentina	391	618

Source: EUROSTAT, External Trade: Analytical Tables, Series C, Luxembourg, various issues.
[a] Calendar years.

although their annual purchases are variable. In recent years, the Community has expanded its cross-Atlantic wheat exports by trading with Cuba and other Latin American countries.

The Community obtains almost all of its maize imports from North America (Table 4.4). The second largest supplier is Argentina. The Community continues to buy some hard, bread-making wheat from North America. In 1986-87, these imports averaged almost 2 million metric tons. Until recently, the Community was a net importer of durum wheat, but since 1982/83, it has exported a small surplus. Most of the Community durum wheat requirements also come from Canada and the United States. With the growth in Community production, however, these imports are declining.

The degree of self-sufficiency in food and agricultural products varies significantly between individual countries, but net exports of grain have grown in the Community as a whole (Figure 4.4). Since 1974, the EC12 has been a significant net exporter of wheat in most years. It was a net importer of coarse grains until 1985, when burgeoning barley exports made it a net exporter. Most

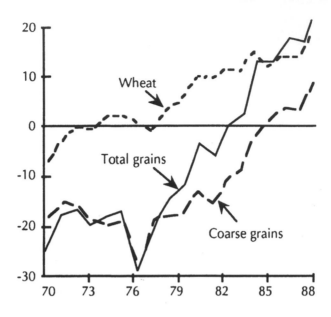

FIGURE 4.4 EC12 Net Grain Exports. *Source*: Based on data from USDA, ERS. PS&D View Database. Washington, D.C.: 1989.

dramatic is the shift in the Community's position from that of a net importer of 20 million metric tons of grains in 1973 to that of a producer of major world surpluses with net exports of 27 million metric tons in 1988. As indicated above, this was caused by sharply higher yields, coupled with sluggish domestic demand. The recent membership of Portugal and Spain, which are both net importers of cereals, has reduced the expansion in net exports, but the rising trend seems likely to resume if current policies are maintained.

The Common Agricultural Policy:
The Centerpiece of the System

The centerpiece of the Community's grain market is the Common Agricultural Policy, which is based on three fundamental principles (Harris et al., 1983; Fennell, 1987). First, the Community functions as a single market for agricultural commodities. Given the history of agricultural protectionism in the original member countries, this implied the replacement of national price support policies with a common price support system. Second, preference is always given to domestic producers over foreign competitors. This requires the use of import measures, such as duties and levies, to keep the price of imported grain above domestically produced grain and community prices above world prices. The third principle states that community members jointly finance CAP costs. This led to the creation of the European Agricultural Guidance and Guarantee Fund (EAGGF) to administer Community agricultural expenditures.

These principles were developed after the signing of the Treaty of Rome in 1957. They were embodied in the price support system for cereals, which was introduced as the first commodity support program under the CAP in 1962. The program covers all agricultural products and includes whole grains (common wheat, durum wheat, barley, maize, rye, oats, buckwheat, canary seed, grain sorghum, and other cereals); wheat and rye flour, groats, and meals; and first stage processed cereal products (e.g., flour), starch, and glucose. More details on the major instruments used in Community grain policies are listed in Table 4.5.

After a transition period when support prices differed by country, common support prices for grains were introduced throughout the EC6 by 1 July 1967. However, these were short-lived because of the devaluation of the French and German currencies in 1969, and the subsequent move to floating exchange rates following the devaluation of the US dollar in 1971. To cope with fluctuating currency values, the Community created a set of special exchange rates for agricultural prices called green rates. These rates were used to determine the prices of agricultural products in the currencies of member countries. A system of border taxes and subsidies called Monetary Compensatory Amounts (MCAs) was established to adjust for differences between the green rates and market rates. Thus, for example, if the value of the British pound declined against the French franc on foreign exchange markets, the effect of the decline on agricultural prices was offset by imposing a tax on exports of agricultural commodities from Britain to France and a subsidy on exports from France to Britain. Without this system, British producers would find it profitable to ship products to France, be paid the support price in francs, and then sell the francs for pounds, rather than selling products in Britain at the support price. Conversely, French farmers would find it less profitable to ship products to Britain for payment in depreciated pounds. The MCAs in effect created a system of tariffs and subsidies on intra-Community agricultural trade and, to some extent, are a contradiction to the principle of free trade in agricultural products within the Community.[5]

The Price Support System

The price support mechanism is founded on target, intervention, and threshold prices. The target price is the cornerstone because it determines the levels of the intervention and threshold prices. The target price is the farm-gate price that all farmers are entitled to receive for their products, regardless of circumstances. As explained below, free-market prices equal target prices only when the Community is a net importer. In the case of the Community cereal

TABLE 4.5 Major Grain Policy Instruments Used in the European Community

Instrument	Coarse grains	Wheat
PRODUCTION/CONSUMPTION		
Producer guaranteed price	X	X
Deficiency payments		
Government purchases	X	X
Producer taxes	X[a]	X[a]
Production quota		
Input subsidies[b]		
- credit		
- fertilizer/pesticide		
- irrigation		
- machinery/fuel		
- seed		
Crop insurance		
Controlled consumer price		
TRADE		
Imports - tariff		
- variable levy	X	X
- quota		
- subsidies		
- licensing		
- state trading		
Exports - taxes		
- restrictions		
- subsidies	X	X
- licensing		
- state trading	X[c]	X[c]
OTHER		
Marketing subsidies[b]		
- storage		
- transport		
- processing		
State marketing		

[a]Co-responsibility levies.
[b]Some national governments maintain programs that provide production and marketing subsidies.
[c]Tender system for exports of surpluses.

market where supplies are in excess of domestic requirements, free-market prices are always lower than target prices. The Commission sets three target prices for standard qualities of cereals: one for common wheat, one for durum wheat, and one for rye, barley, maize, and sorghum. Target prices differ geographically, but Duisburg in Germany, which is located in the main cereal deficit area, is used as the basing point.

The threshold price is the minimum price at which grain imports are permitted to enter the Community. It is calculated such that imported cereals transported to Duisburg will sell at no lower than the target price. However, domestically produced grains may sell below the target price in Duisburg, hence Community grains are protected from import competition. The basing point for the threshold price is Rotterdam (the largest port in the Community), but the same price applies to all Community ports. The threshold price acts as an effective ceiling price in domestic markets because domestic grains are price-competitive with imports up to the threshold price (plus transportation costs) but not above. A variable levy paid by importers is charged to ensure that all imports enter the Community at the threshold price. The levy, calculated daily, is the difference between the lowest representative world price—referred to as an "offer" price—and the threshold price.

The intervention price, the third cornerstone of CAP price policy, is a support price at which the Community will purchase grain from farmers if they cannot obtain a higher price on the open market. The basing point for this price is Ormes in France (the main surplus area). The Commission sets two intervention prices for grains: a lower price for feed wheat, feed rye, barley, and sorghum, and a higher one for maize, breadmaking wheat, and breadmaking rye. The same prices are used throughout the Community with the exception of Spain and Portugal, which are in the process of adopting the system. There is a range between the target price and the intervention price within which internal market prices for domestically produced grains are expected to remain. If the Community is a net importer, the market price in Duisburg will approximate the price of imports, which will, as explained, equal the target price. If the Community is a net exporter, domestic prices will drop below the target price and, with a sufficient surplus, will approximate the intervention price. Community grain prices are not supposed to fall below the intervention price, but market prices typically lie at the lower end of the price band because domestic supply most often exceeds demand (except for maize). Monthly adjustments are made to all support prices to account for storage costs. These "monthly increments" are meant to promote the provision of a smooth and continuous grain supply throughout the year by removing pressure to sell immediately after harvest (Harris, et al., 1983: 73). To promote private stockholding, any stocks remaining at the end of a season are eligible to be purchased by the Community at a price that includes a "compensation allowance," which equals the difference between the target price for the last month of the old season and the first month of the new season.

In Figure 4.5, wheat target, threshold, and intervention prices and a representative offer (import) price are graphed to make it easier to see the various relationships among the prices. The threshold price is substantially higher than both the offer and intervention prices, and the gap has widened considerably since the mid-1970s. Typically, the intervention price is above the offer price, although there have been periods, most recently in 1983-84, when the reverse

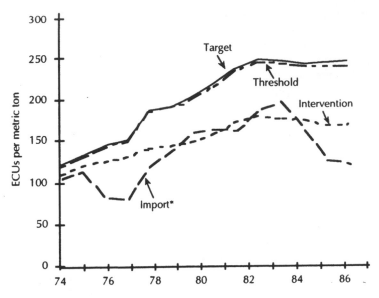

FIGURE 4.5 Wheat Prices in the European Community. *Sources:* CEC. Agricultural Situation in the Community. Various issues; USDA, FAS. Grains Situation. Various issues. *Offer price for Grade 2 soft red wheat in Rotterdam.

was true. Since the offer price is the lowest import price for wheat including transportation and other charges, the corresponding export price for Community wheat is below this price and consequently below the intervention price. Exports are subsidized through an export restitution equal to the difference between the export price and the intervention price.

The cereal policy regime is revised annually and includes levels of support prices, monthly increments, and quality criteria. Proposals for next year's prices are made by the Commission to the Council of Ministers based on market and budgetary trends, but the Council makes final decisions on prices. When the system was originally introduced, prices were announced prior to planting to allow farmers to adjust production. The original regulations specified that prices be fixed by August 1 for the marketing year that begins on this date. However, protracted political wrangling characterized the price-setting process, and support prices were never announced on time. The Commission tended to delay submission of its price proposal to the Council to allow an opportunity to procure up-to-date information on domestic and world market conditions (Fennell, 1987). In the mid-1970s, the timing was officially changed. The Commission now submits proposals at the end of the calendar year, and the Council (after deliberation and negotiation) votes on these the following spring. Thus, farmers have already made planting plans before prices are announced.

In addition to prices that are decided by the Council, the revision of the cereal regime may also include changes in rules and regulations that deal with particular problems in the marketplace. In recent years, a number of measures were introduced to curb what are referred to as "structural surpluses." These measures, which are discussed below, were designed to reduce the budgetary costs of the policy and to restrain growing exports in response to pressure from other exporters.

To dispose of production in excess of what can be absorbed by the domestic market at Community prices, exporters receive export restitutions equal to the difference between domestic and international prices. These are paid by the Commission from the Community's budget. Exports are closely monitored by the Commission, and the refunds paid to exporters are calculated based on market trends, including quality and quantity of cereals and internal trade costs. To determine the actual restitution, the highest world market price at the time of shipment is subtracted from the per unit value of the shipment at domestic prices. Transport costs to importing countries are not covered, and the precise destination of shipments is not monitored. Other exporters label the restitutions as export subsidies, and as EC grain exports have grown, the payments have become an increasingly contentious issue. It is argued that Community refunds depress world prices, displacing other suppliers from "traditional" markets. These allegations are a major factor in the debate on farm legislation in the United States and resulted in the US use of counter-subsidies.

Grain Marketing and Distribution Systems

Although the Community maintains a common pricing policy for grains, cereal marketing is not centralized. Prior to the signing of the Treaty of Rome, each country had well-established (and often very old) marketing channels. The common policy was designed to function within rather than replace existing systems.

The Commission itself does not normally undertake any marketing activities. Apart from its role in influencing the volume of exports through restitutions, the Commission monitors private and national grain handlers. The administration and coordination of Community policies at the national level is managed by national governments. Their most important area of responsibility is the intervention system.

The Intervention System

National governments grant intervention agencies, which include producer cooperatives, grain traders, and occasionally processors, licenses to operate collection centers in areas likely to have surplus grains. The range of commodities delegated to the agencies varies. In France and the Netherlands, for instance, separate intervention centers are usually in place for each commodity subject to EC market organization, while in the United Kingdom, a single organization deals with all commodities (Fennell, 1987). Intervention agencies are required to purchase grains offered to them, provided they meet minimum standards set by the Commission for quality and quantity (10 metric tons for durum and 80 metric tons for other grains). Countries can set minimum quantities at higher levels, and most do to encourage private grain marketing and to control costs (Toepfer International, 1988). Sellers receive the intervention price plus the appropriate monthly adjustment, but must pay costs to transport the grain to storage. The costs of operating intervention centers are met by national governments, but are reimbursed by the Commission.

Intervention stocks are offered for sale through tender or public auction. By law, the domestic sales price cannot be lower than the intervention price. The Commission can recommend that stocks be sold on the domestic market but cannot require this without the approval of member countries (Fennell, 1987). Alternatively, stocks can be marketed outside the Community. The decision to export usually is made by the cereals management committee in the Commission composed of representatives of all member countries. The majority of export certificates in the Community are awarded by way of tenders. An invitation to tender, issued by the Commission, states the nature, quantity, quality, port of shipment (determined by the location of the grain,) and geographical area of destination (e.g., North Africa). Companies interested in submitting a bid must do so within a specified time period. The bid must include the export refund and any compensatory amount requested. The trader with the lowest bid is awarded an export certificate. It then pays a monetary deposit, refunded when it can be certified that the grain has been shipped outside the Community. Trading companies buying on the open market can request an export certificate at any time by applying to an intervention center. At the time of sale, they immediately receive the applicable export refund from the Commission.

Similar procedures exist for importing cereals into the Community. Companies, typically multinational corporations, must apply for an import licence at an intervention center. Licenses are issued for a specific amount, which can be imported anywhere in the Community. A license can be transferred once and is valid only for a specified period. The import levy is not fixed in advance, but is calculated on the day of importation.

Beyond these general characteristics, there is considerable diversity in the structure and organization of national grain markets. In the remainder of this section, we describe the French market to illustrate how pricing and marketing policies function in a major grain-producing country in the Community. France is a particularly good example because it is the EC's leading producer and exporter of grains.

The Cereal Marketing System in France

The French government has traditionally played a significant role in grain production, which is a major component of French agriculture. The Office National Interprofessionnel des Cereales (National Cereal Industry Organization – ONIC), established in 1940, is responsible for implementing Community cereal regulations. ONIC is administered by a council composed of participants in the domestic cereals market, but dominated by producers. Before France's accession to the EC, ONIC controlled cereal marketing in France and was responsible for setting support prices and supervising grain assembly and storage. It also had a monopoly over external trade. With the implementation of the common regime for cereals, ONIC became the executive body of the CAP for grains in France. It grants licenses to intervention centers, monitors the operation of these centers, and issues import and export licenses. It is, however, not involved in any commercial activities (Pivot, 1983).

The demand for cereals in France is divided between primary (processors) and secondary (end users) users (Figure 4.6). Primary users are flour mills and feed mills. Secondary users are bakeries, pasta makers, seed merchants, and industrial processors. Primary users are more directly affected by Community regulations than secondary users because they deal with intervention centers and merchants that act as intermediaries for intervention centers, while secondary users usually obtain supplies from millers. In France, as in Italy and Germany, family enterprises traditionally dominate the flour trade. In contrast, in the UK and the Netherlands, large corporations control much of the market (Pivot, 1983).

Producer cooperatives play a major role in French agriculture. In 1986, 75 percent of all cereals were sold through cooperatives, compared with, for instance, 20 percent in the UK (CEC, 1987a). Cooperatives are typically farmer-owned and are sanctioned by the government because they protect producers' interests. They monopolized cereal assembly between 1936 and 1952 and still remain an important market force today. Cooperatives also transport, store, and market grains.

When farmers market grain, the first step is to transport it to an assembler, which may be a producers' cooperative or a private company. Assemblers hold stocks for later delivery to users, or are processing firms with storage facilities. Some assemblers do not own storage capacity and immediately sell the grain to other firms. In France, the same firm usually assembles and stores the grain, but seldom has processing capacity. ONIC grants licenses to assemblers and supervises their activities under Community regulations. Over 70 percent of the assembly licenses (to undertake intervention activities) are held by producer cooperatives, which also own more than two-thirds of the country's storage capacity (Pivot, 1983).

Further down the marketing chain, the role of cooperatives declines and that

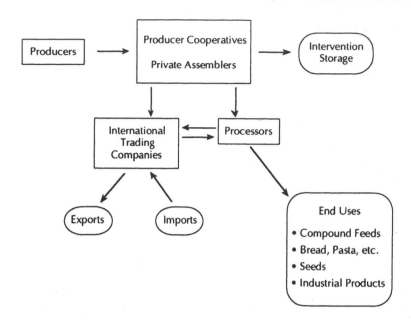

FIGURE 4.6 The Marketing System for Grains in the European Community.

of private firms rises. Most processing, as well as grain imports and exports, are controlled by private companies, although in recent years a few cooperatives have engaged in exporting. Private merchants, traders, and brokers act as intermediate distributors, providing spatial and temporal links between grain producers and final users. Private merchants handle most of the trade in the Community within national boundaries and among member states, while multinational corporations manage trade outside of the Community. After grain is assembled, it is transported by rail or boat. Grain traded among member countries in particular is often transported via the Community's extensive network of internal waterways (Debatisse, 1982).

Under CAP rules, grain processors and manufacturers can apply for a license to hold stocks. However, their storage costs are usually higher than grain handlers who own specialized storage facilities. Therefore, end users seldom hold significant stocks. To ensure a continuous supply, users generally enter into one of two types of contractual arrangements with producers. The first, a "short circuit," is a two-party contract that includes a producer (or a cooperative) and a user. The contract stipulates delivery conditions for a certain amount of grain on a particular date or dates. The second type, the "long-circuit," is becoming increasingly popular. It involves a series of two-party contracts between producers and intermediaries, and the user. The merchandise specified in the contract can be moved from silo to silo, but usually the title is transferred without any physical transport of the grain. This system allows the trade of forward contracts to spread price risk (Pivot, 1983).

The price supports provided by the CAP have removed a considerable amount of the price uncertainty related to production for farmers but have not eliminated all of the risk borne by stockholders, distributors, and users. Monthly price adjustments for storage, monetary compensatory amounts, and export restitutions do not necessarily fully make up for variations in the interest charges that affect handling and storage costs, or for movements in exchange rates. The risks created by these fluctuations, in addition to the instability of world market prices, helped to make the long-circuit market popular. This form of forward marketing spreads price risk among the various actors in the marketing process, but is less transparent and competitive than the open trading of futures contracts (Debatisse, 1982).

Financing and the Reform of Community Grain Policies

As indicated earlier, the EAGGF administers the financial portion of the Community's agricultural programs. CAP expenditures are divided into two components or sections. The Guarantee section includes price supports, while the Guidance section includes monies spent on structural policies (such as farm consolidation and the conversion of land to alternative crops or uses). The Guarantee section accounts for the majority of expenditures. In 1986, for example, more than 22 billion ECUs were spent on price supports, while structural policies cost less than 1 billion ECUs.

Since its introduction, the CAP has accounted for the majority of all Community expenditures. Agricultural price supports have increased steadily and now represent between 60 and 70 percent of total Community expenses (Figure 4.7). Of these, cereal price supports typically account for 10-16 percent of total price supports. The funds for EAGGF operations are provided from the budgetary resources of the Community. These are obtained primarily from the revenues collected by member states from value added taxes (VAT). Currently, transfers equivalent to a VAT rate of 1.4 percent are required to meet Community budget requirements. Import levies also provide the Community with financial resources, but these have become less important as the volume of imports, particularly agricultural imports, has declined.

In the late 1960s and early 1970s, cereal price supports accounted for 30 to 40 percent of all support expenditures (Harris, et al., 1983). This share dropped during the 1970s when the gap between world and EC market prices narrowed and expenditures on other commodities, for example milk, rose. Cereal price supports have averaged 10-15 percent of EAGGF's total support expenditures since the mid-1970s. In 1973, roughly 1 billion ECUs were spent on price

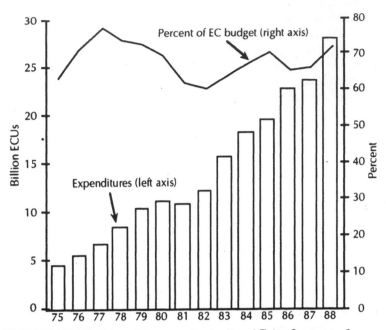

FIGURE 4.7   EC Expenditures on Agricultural Price Support. *Source:*
Commission of the European Communities. Agricultural Situation in the
Community. Various issues.

supports for cereals, of which approximately 50 percent was for export
restitutions (Figure 4.8). The expenditures for 1987 totalled over 3.5 billion
ECUs, of which more than 70 percent were devoted to export refunds. During
the early 1980s, in response to falling world prices and rising surpluses, the
Community increased spending on longer term cereal storage in place of
exporting. In 1987, for example, over one-quarter of total expenditures were
spent on long-term storage (Figure 4.8).[6] However, export subsidies recently
increased, and the use of storage as a supply management tool is declining. The
subsidy war between the Community and the United States, triggered by the
1986 US farm legislation, contributed to this change.

The growing cost of agricultural support and the increase in commodity
surpluses place a substantial strain on the CAP. Pressures intensified after the
1984 harvest reached a record high of 173 million metric tons (Toepfer International, 1988). Production has since declined slightly, but not enough to
measurably reduce structural surpluses (excess supply over domestic demand). Problems in the cereal market helped result in an increasingly
acrimonious debate about the level of support prices and sources of funding for
the agricultural support system. The United Kingdom, which is a major net

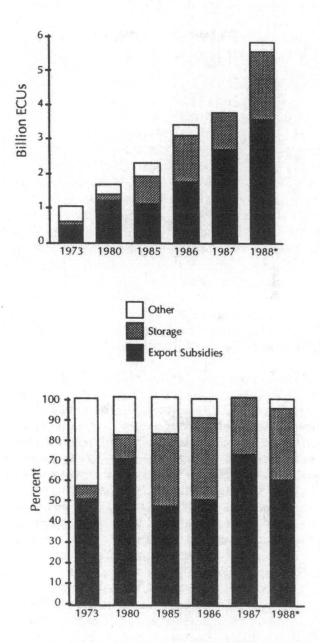

FIGURE 4.8 EC Price Support Expenditures on Cereals. *Source:* Commission of the European Communities. *Agricultural Situation in the Community.* Various issues.

contributor to the Community's budget, is particularly resistant to increases in support prices. In 1980, the UK obtained a refund of 1.2 billion ECUs from the Community's budget. The system of refunds has, however, continued, as has British lobbying for a more permanent solution to the budgetary costs of the CAP.

Problems with the price support system for cereals are not new, and the rules and regulations have been revised numerous times since their inception in 1967. The original policy specified an entire set of regional intervention prices, differentiated by transport costs. It was thought that markets were not sufficiently integrated to guarantee that deficit areas could obtain consistent deliveries from surplus areas at the desired price. The complexity and rigidity of this system made it difficult to operate. In 1976 a revised program, the silo or cathedral hierarchy of prices, was introduced that abolished regional differentiation and instituted a single set of support prices for the entire Community (for more details, see Harris, et al., 1983 and Debatisse, 1981). This system introduced greater flexibility by allowing market prices to reflect quality differentials and seasonal variation. These reforms did not, however, address the problem of structural surpluses in the Community, nor did they anticipate the increasing competition in the feed industry between cereals and substitutes. Subsidies were later introduced to make wheat more attractive to feed manufacturers, but these proved costly, and it is difficult to ensure that feed wheat receiving payments is not resold on the domestic market for other purposes.

In the 1980s in response to the growing costs of support, several production-related measures were adopted. The most important is the co-responsibility levy, introduced in 1986/87. This is a fixed levy of 5.38 ECUs per ton assessed on cereals sold on domestic markets, into intervention, or exported. The objective of the levy is to make producers pay for the cost of surpluses and to discourage excess production. The proceeds from the levy are put back into the guarantee section of the EAGGF. In its first full year of operation, the levy raised 412 million ECUs (Toepfer International, 1988). Small producers who market less than 25 metric tons of cereals per year receive an offsetting subsidy equal to the levy. An additional co-responsibility levy for cereals, the guarantee threshold, was introduced in 1988/89. This operates in a way similar to the basic co-responsibility levy but is applied only if total Community production exceeds 160 million metric tons.

In addition to reducing producer revenues with levies, a program to encourage the withdrawal of land from agricultural production, similar to the "set-aside" program in the United States, was also adopted in 1988/89. National governments administer the program, and farmer participation is voluntary. Participants receive a subsidy (of 100 to 600 ECU per hectare per year), paid by national governments but partially refunded by the Commission, if at least 20 percent of their area previously sown to cereals is removed from cultivation.

The land may be used for other purposes; pasture for instance. Farmers who agree to set aside 30 percent or more of their land are also exempt from the co-responsibility levy on the first 20 metric tons of cereals sold.

In an additional attempt to discourage surpluses, the procedures for selling cereals to intervention centers were tightened in 1987. The buying-in system allows intervention purchases to occur only after market prices at specified export ports fall below the intervention price. Cereals can then be brought to intervention centers, but sellers receive only 94 percent of the intervention price.

Conclusions

The European Community is an increasingly important actor on world grain markets. Originally an economic union of six countries, its membership now includes most of western Europe. Sustained by a system of high and stable support prices, the Community's grain industry underwent major structural changes since the formation of the EC in the late 1950s. Yields of cereals increased dramatically and are among the highest in the world. Cereal production is a very capital-intensive activity, and in the major producing areas, cereal farms are large compared to other farms, specialized, and highly profitable.

Grain consumption in the Community is stagnant because of high prices and competition from livestock feed substitutes, such as manioc. Growth in production, combined with stagnation in consumption, has changed the Community from a substantial net importer of grains to a leading net exporter. In order for Community grains to compete on international markets, exports must be heavily subsidized.

Various adjustments have been made to grain policies in the Community in response to the problem of surpluses, primarily because of growing budgetary pressures. There is, however, great reluctance, to directly reduce support prices. A variety of indirect measures have been used to discourage producers from selling surplus grains into intervention, but these have not solved the problems of high production and stagnant consumption. These measures are designed to circumvent the inflexibility of the price support system. As a result, Community grain policies tend to be complex and are largely driven by internal political considerations. Subsidized grain exports from the EC generate substantial friction with other exporting countries, particularly the United States. With the expansion of the Community to include Portugal and Spain, which have yet to achieve grain yields typical in the rest of the EC, another phase of production growth appears likely. Although temporary increases in world prices may help to reduce the burden of the cereals policy on the Community's budget, it seems unlikely that internal and external pressures on the Community's grain policies will diminish in the foreseeable future.

Notes

1. The European Economic Community (EEC), the customs union, was only one of several Community organizations established in the 1950s. To reflect the totality of these organizations, the more general term, European Community, is used in this chapter.

2. Because of successive additions to the Community, it is sometimes difficult to assemble comparable statistics. When available, figures are provided for the EC12. Unless otherwise stated, the years used in this chapter are calendar years. Split years are August-July cereal marketing years.

3. The European Currency Unit (ECU), a system for stabilizing currency values in the Community, is the centerpiece of the European Monetary System, and is also used for budgetary purposes. Its value is determined by a "basket" of currencies whose weights are determined by a number of factors, including the relative size of the economies of member countries and their trade volume. The value of the ECU has fluctuated considerably against the US trade dollar. In 1980, it was worth $1.39, but fell to $0.76 in 1985. In the late 1980s, it ranged between $1.10 and $1.20.

4. This only includes "commercial farms," which are defined as those that market the bulk of their output and have a minimum gross margin (value of production less costs) of 2 thousand ECUs.

5. The Community has announced its intention to eliminate remaining barriers to the free movement of goods and services across national frontiers by 1992. This will probably require the elimination of Green Rates and MCAs.

6. Total storage expenditures exceed the amount shown in Figure 4.8 because short-term storage costs are incurred on grains brought into intervention. The storage figure quoted is for long-term storage or "storage proper" as defined by the Commission.

References

Bahrenian, A., et al. *FAPRI Trade Model for Feed Grains: Specification, Estimation, and Validation.* CARD Staff Report #86-SR1, Iowa State University, Ames, Iowa, December 1986.

Commission of the European Communities (a). *The Agricultural Situation in the Community.* Brussels. Various issues.

_____(b). *External Trade, Analytical Tables.* Brussels. Various issues.

Debatisse, M.L. *EEC Organization of the Cereal Market: Principles and Consequences.* The Center for European Agricultural Studies Occasional Paper No. 10, Wye College, Ashford, Kent, 1981.

_____. *Cereal Export.* Paris: ATYA Edition, 1982.

Devadoss, S., et al. *FAPRI Trade Model for the Wheat Sector: Specification, Estimation, and Validation.* CARD Staff Report No. 86-SR3, Iowa State University, Ames, Iowa, October 1986.

Eurostat. *External Trade: Analytical Tables.* Luxembourg: various issues.

_____. *National Accounts.* ESA Aggregates 1960-1986. Luxembourg: 1987.

Fennell, R. *The Common Agricultural Policy of the European Community.* 2nd Ed. Oxford: PSP Professional Books, 1987.

Gardiner, W.H., V.O. Roningen, and K. Liu. *Elasticities in the Trade Liberalization Database*. Washington, D.C.: USDA, ERS, 1989.

Harris, S., et al. *The Food and Farm Policies of the European Community*. Chichester: J. Wiley & Sons, 1983.

Meilke, K.D. and H. de Gorter. "An Econometric Model of the European Economic Community's Wheat Sector." International Agricultural Trade Research Consortium (IATRC) Working Paper No. 86, Minneapolis, Minnesota. February 1986.

Melencovitch, A. "Changes in United Kingdom Cereal Production: An Econometric Investigation into Area and Yield Response in the Eastern Counties of England, 1970-85." M.S. thesis, Cornell University, Ithaca, NY, 1988.

OECD. *Animal Feeding and Production: New Technical and Economic Developments*. Paris: 1981.

Pivot, C. *Le Ble Francais Face a l'Environnement International*. Paris: Economica, 1983.

Toepfer International. *The EEC Grain Market Regulation 1988/89*. Hamburg: October 1988.

World Bank. *World Development Report, 1988*. New York: Oxford University Press, 1988.

5

The United States

David Blandford and Harry de Gorter

As in most industrialized countries, the contribution of agriculture to employment and national income has been declining in the United States for many years. In 1970, agricultural output accounted for 2.9 percent of gross national product (GNP) and 4.4 percent of national employment.[1] By 1985, agriculture's share fell to 2.3 percent of GNP and 2.7 percent of total employment. Nevertheless, agriculture continues to account for a significant portion of U.S. exports. During the mid-1970s, agriculture's share of the value of exports surpassed 20 percent. Following a depression in agricultural trade in the mid-1980s, caused primarily by slower world income growth and a strong dollar, agriculture's share fell to roughly 12 percent in the late 1980s.

The agricultural sector in the United States is very diverse. Livestock and livestock products typically account for 50 percent or more of the total value of farm marketings. Depending on prices, grains account for 30-40 percent of the value of crops marketed. In 1988-89, grains averaged 50 percent of the harvested area of crops and contributed one-third of the total value of U.S. agricultural exports.

Grain Production: Growth and Diversity

The United States produces most grains in significant quantities. Area planted and harvested, total production, and yield for the major grains in recent years are shown in Table 5.1. Wheat and maize are the two most important, collectively accounting for 80 percent of the grain area harvested. Total production averaged 250 million metric tons in 1987-89, representing a significant decline from the peak of almost 350 million metric tons in 1985. Changes in U.S. grain production are closely related to changes in U.S. grain policies; an issue explored in detail later in this chapter.

TABLE 5.1 U.S. Grain Production (average 1987-89)

	Area planted (million hectares)	Area harvested for grain (million hectares)	Production (million m. tons)	Yield (m. tons/ha)
Maize	27.82	24.62	165.84	6.73
Wheat	28.06	23.11	54.00	2.34
Sorghum	4.69	4.14	16.30	3.93
Barley	4.04	3.49	8.83	2.53
Oats	5.92	2.60	4.66	1.79
Rice	1.08	1.07	3.02	2.83
Rye	0.92	0.24	0.40	1.72

Source: USDA, National Agricultural Statistics Service. Crop Production 1989 Summary. Washington, D.C., 1990.
Notes: Area planted includes all uses, including cover crops, while area harvested is for grain only. Rice production and yield are for rough rice (unmilled basis). Years beginning Sept. 1 for maize and sorghum; June 1 for oats, barley, wheat and rye; and Aug. 1 for rice.

Major Grains Grown in the United States

Maize is the most important grain crop in the United States. In 1987-89, it accounted for about 40 percent of harvested area and 65 percent of total production. Some maize is harvested as silage for dairy and beef cattle feed, which accounts for the difference between harvested and planted area (Table 5.1). Approximately 95 percent of the maize produced in the United States is dent maize, so called because of the indentation in the kernel caused by the shrinkage of starch. The Federal government sets grain standards in the United States. It classifies dent maize into yellow, white, and mixed categories. Six levels of quality are identified on the basis of such factors as test weight, moisture content, and the proportion of broken or damaged kernels and foreign material.

Maize is grown in almost every state and occupies roughly 20 percent of total U.S. cropland. However, the bulk of grain production is located in the Midwest (Figure 5.1). The major maize-growing area, known as the "Corn Belt," includes the states of Ohio, Indiana, Illinois, Iowa, Minnesota, Missouri, and Nebraska. Two of these states, Iowa and Illinois, accounted for 35 percent of total production in 1987-89 (Table 5.2). These states are particularly suited to maize production because of their rich, well-drained soils, warm summer climate (average temperature 20^0-25^0C.), and well-distributed precipitation. Maize is sown as early as March in the southern states, and planting for grain is completed in the north by late May. In drier areas, such as central Nebraska, irrigation is used. Harvest begins in early October in the Corn Belt and finishes by mid-November. Most of the maize is field-shelled using a combine with a corn head. The shelled corn is then dried, either on farm or at commercial drying facilities.

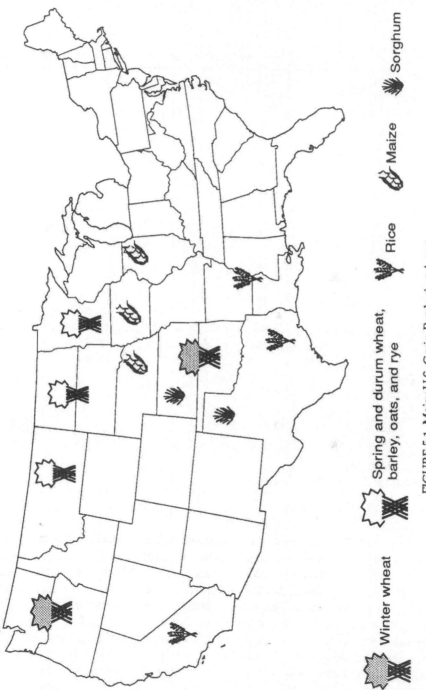

FIGURE 5.1 Major U.S. Grain-Producing Areas

Winter wheat

Spring and durum wheat, barley, oats, and rye

Rice

Maize

Sorghum

TABLE 5.2 Leading Grain-Producing States (1987-89)

	Percent of Total Production		Percent of Total Production
Maize		*Wheat*	
Iowa	18.7	Kansas	15.2
Illinois	16.5	North Dakota	10.3
Nebraska	12.7	Oklahoma	7.7
Minnesota·	8.5	Montana	6.0
Total these states	56.4	Washington	5.9
		Total these states	45.0
Sorghum		*Barley*	
Kansas	35.2	North Dakota	23.2
Texas	24.2	Idaho	14.2
Nebraska	16.2	Minnesota	9.9
Total these states	75.6	Washington	8.1
		South Dakota	5.0
		Total these states	60.5
Oats		*Rice*	
Minnesota	12.1	Arkansas	40.9
South Dakota	12.1	California	20.3
Iowa	12.0	Louisiana	14.6
Wisconsin	11.4	Texas	13.2
North Dakota	6.6	Total these states	88.9
Total these states	54.3		
Rye			
South Dakota	22.1		
North Dakota	15.8		
Georgia	10.6		
Minnesota	6.7		
Nebraska	6.6		
Total these states	61.7		

Source: USDA, National Agricultural Statistics Service. Crop Production 1989 Summary. Washington, D.C., 1990.

Wheat is raised in many areas of the country, particularly where precipitation is too low to reliably produce maize. Five states (Kansas, North Dakota, Oklahoma, Montana, and Washington) accounted for 45 percent of total wheat output in 1987-89 (Table 5.2). Wheat is divided by the U.S. federal grading system into hard red winter, soft red winter, hard red spring, durum, red durum, white, and mixed. Federal standards specify six quality grades for wheat, corresponding to those for maize. Hard wheats are generally more vitreous and have more protein than soft wheats, although there is some overlap in the protein ranges. Typically, hard red spring has the highest protein content (up to 18 percent); and white has the lowest at 8-10 percent. United

States wheat production is dominated by winter wheat, which is used as a cover crop in many areas to protect soils from wind erosion, or as a source of early spring grazing for livestock. In 1987-89, winter varieties accounted for over 75 percent of total production.

The leading class of wheat grown in the United States is hard red winter, which makes up 40-45 percent of total production. It is a medium protein wheat, typically ranging from 10-12 percent. Production is centered in the Great Plains states of Kansas, Nebraska, Oklahoma, and Texas (Figure 5.1). In these areas, precipitation averages less than 650 millimeters per year, with frequent dry periods. Winter temperatures are sub-zero, but there is usually adequate snow cover to protect the young plants during their winter dormancy. Hard red winter wheat is planted between late August in the north to late October in the south, and is usually harvested during mid-June in the south and mid-July in the north. Soft red winter wheat (9-12 percent protein) accounts for 20-25 percent of total wheat production and is grown primarily in more humid areas close to the Great Lakes (primarily the states of Illinois, Indiana, and Ohio) and the Atlantic coast. Planting dates are similar to those for hard red winter, but harvest can extend into August in the more northern states.

Hard red spring accounts for 15-20 percent of the total U.S. wheat crop and is grown primarily in the upper Plains states, such as North Dakota and Minnesota, where extreme winters make the climate unsuitable for winter wheat production (Figure 5.1). Planting takes place from mid-April to the end of May, and the crop is harvested during August or early September. Durum wheat (10-16 percent protein) makes up less than 5 percent of U.S. production and is planted principally in North Dakota, but also in Minnesota, Montana, South Dakota, and Arizona. Planting and harvest periods are the same as those for hard red spring wheat.

White wheat, accounting for 10-15 percent of U.S. wheat production, is mainly grown in the northwest in Washington, Oregon, and Idaho (Figure 5.1). The planting period varies considerably depending on the state. In Oregon, for example, white wheat can be planted as early as mid August or as late as the end of January. Harvest typically extends from mid-July to late August. Both white wheat and soft red winter wheat are produced east of the Mississippi river in Illinois, Indiana, Ohio, and also in the East Coast states.

Sorghum has become an important grain crop in the United States. In 1987-89, it accounted for close to 7 percent of total grain area harvested and 6 percent of total grain production (Table 5.1). Sorghum is planted under a wide range of soil and climatic conditions because it tolerates limited moisture and high temperatures. A large portion of U.S. production is located in the Southern Plains states; with almost 60 percent of the total in Kansas and Texas (Table 5.2). Sorghum is divided into four classes for grading purposes: yellow, white, brown, and mixed. These classes are graded 1 through 4 depending on test weight, moisture content, and the proportion of damaged or broken kernels.

Barley accounted for 6 percent of the total harvested grain area in the United States in 1987-89 and 3 percent of total production (Table 5.1). Like wheat, barley is well adapted to the cooler and drier sections of the country. Most U.S. barley is grown in the Northern Plains, where spring wheat production is also concentrated (Figure 5.1). The leading producers are North Dakota and Idaho, which collectively accounted for 37 percent of barley output in 1987-89 (Table 5.2). Typically, 30 percent of the barley grown in the United States is low-protein malting barley, which is used in the brewing industry; the rest is fed to livestock. Barley varieties are classified as either malting or feed, and the malting varieties are further subdivided into two-row and six-row, each having unique properties and malting characteristics. Number grades are employed for each subclass of barley, in addition to descriptive terms such as bleached, smutty, or stained (Heid and Leath, 1978).

Most barley is seeded in the spring, although winter varieties, which are planted in the fall, are also grown. In the arid parts of the Northern Great Plains, barley is spring planted on land fallowed the previous year to gather moisture. Malting varieties are often seeded on irrigated land in this area because the protein level can be controlled more easily by increasing fertilizer applications (particularly nitrogen) which requires a secure water source. Malting quality is also improved when more phosphates are applied. In the more humid parts of the Northern Great Plains and in eastern areas, spring barley is sown as part of a continuous crop rotation. Barley is frequently planted on land that is plowed in the fall and fallowed over the winter. Winter barley is most common in the south, where it is used for winter pasture as well as grain production.

Oats are grown throughout the United States, but the northern states of Minnesota, South Dakota, Iowa, and Wisconsin are the principal areas of production. In 1987-89, these four states accounted for over 45 percent of total U.S. oat production. Oats made up roughly 4 percent of the national harvested area of grains and 2 percent of grain production (Table 5.1). Spring or northern white oats are the most common species because they are most suited to the cool climate areas of the northern United States. Red oats are grown in areas that are too warm for white oats, such as the South. Grey oats are grown on the West Coast. Oats also double as pasture, silage, nurse, and cover crops, thus explaining their widespread cultivation throughout the United States. Spring oats (white oats) are usually planted in April and May and harvested in July or August. In dairy areas, oats are harvested green for silage or "oatlage". Winter oats (red oats) are planted in the early fall and are used as winter pasture for livestock. The livestock are subsequently removed, and the grain is usually harvested in May or June. A second crop, for example, soybeans, is then planted. Alternatively, in the spring the oat stubble may be plowed under, and soybeans or sorghum planted.

Rice accounted for less than 2 percent of total grain area in 1987-89 and roughly 1 percent of total grain production (Table 5.1). Rice production is largely confined to Arkansas, California, Louisiana, and Texas, which accounted for almost 90 percent of output in 1987-89 (Table 5.2). Roughly 70 percent of production is long-grain rice, with medium-grain rice making up the remainder. Cultural practices vary slightly between areas, but all are highly mechanized. Most rice land is leveled to facilitate irrigation and drainage. Seeding, typically by airplane, takes place in April and May, except in the Gulf Coast area where planting occurs from early March to late June. Irrigation water is supplied by wells, on-farm water storage structures, and rivers. The crop is harvested with self-propelled combines.

Rye is the least important grain produced in the United States. In 1987-89, it accounted for less than one-half of one percent of harvested grain area and one-fifth of one percent of total grain production (Table 5.1). In recent years, three states, South Dakota, North Dakota, and Georgia, produced about 50 percent of the total rye grain crop (Table 5.2). As evident from Table 5.1, a substantial portion of the area planted is not harvested. In many areas, rye is used as a cover crop or green manure, for straw, or as winter or early spring forage for livestock.

Changes in Area and Yields

Yields in the United States vary substantially by grain type. In 1987-89, production per hectare ranged from a low of 1.7 metric tons for rye, to 6.7 metric tons for maize (Table 5.1). New varieties and innovations and improvements in production technologies have had a major effect on grain yields in the United States since the 1930s. Most dramatic is maize, whose yields rose from the 1930-34 average of 1.4 metric tons per hectare (shelled basis), to over 7 metric tons per hectare during the 1980s because of the introduction of high-yielding, fertilizer responsive hybrids and the increased use of fertilizer and other inputs. The yields of other grains also grew substantially. In the years following the Second World War (1945-49), U.S. wheat yields averaged just over 1 metric ton per hectare, while in recent years, the average exceeded 2.3 metric tons per hectare (Table 5.1).

Trends in yields of coarse grains (maize, sorghum, barley, oats, rye, millet, and mixed grains), wheat, and rice from 1960-89 are shown in Figure 5.2. During this period, yields improved steadily, particularly for coarse grains. As a whole, the rate of increase was 2 percent per year, with a 2.4 percent growth rate for coarse grains (Table 5.3). All yields are quite variable, however, with an average annual deviation from trend of roughly 9 percent. Coarse grain yields fluctuate most, with an average annual variation of 11 percent per annum. Periodic droughts affect coarse grain production, the most recent being in 1988 (Figure 5.2). The use of hybrids and large quantities of fertilizer led to rapid growth in maize production, but also made yields more susceptible to the effects of drought because high-yielding varieties are less tolerant of adverse conditions. Rice yields are most stable since the crop is produced with irrigation.

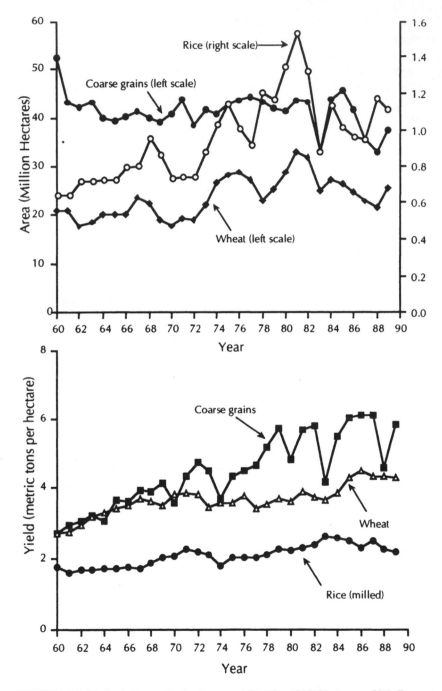

FIGURE 5.2 United States Grain Area and Yields, 1960-89. *Source:* U.S. Department of Agriculture, Economic Research Service. PS&D View Database, 1989.

TABLE 5.3 Growth and Variability of U.S. Grain Production, 1960-89 (percent)

	Area	Yield	Production
Annual growth rates			
Total grains	0.2	2.0	2.2
Coarse grains	-0.4	2.4	2.0
Rice	2.0	1.2	3.2
Wheat	1.2	1.4	2.6
Variability[a]			
Total grains	8.9	9.2	14.8
Coarse grains	8.2	10.9	16.3
Rice	16.9	6.3	16.4
Wheat	13.7	7.2	16.1

Source: U.S. Department of Agriculture, Economic Research Service. PS&D View Database, 1989.
[a]Coefficient of variation net of trend.

When changes in harvested area are taken into account, the rates of increase in output were highest for rice and lowest for coarse grains. Rice area expanded rapidly until the early 1980s in response to strong demand in international markets (see the section on trade below). Generally, coarse grain area declined since the early 1960s (Figure 5.2), largely because of government programs (see the section on policies). Wheat area expanded during the 1970s, again because of the effects of government programs and export demand. It declined during the 1980s as export markets weakened, and land was removed from production under government programs to support producer prices. Changes in grain area are often closely associated with each other (Figure 5.1). For example, in 1983, the harvested area of all three grain categories fell simultaneously. This occurrence can be attributed to government programs, which have a major impact on grain production.

Estimates of supply elasticities for grains in the United States vary widely. The main sources of disagreement are the extent of effects of changing technology on yields and of government policy on planting decisions. Isolating the effects of technological change and government policy from traditional factors that affect supply decisions is difficult. Nevertheless, it is generally agreed that grain yields respond more to price than does area in the United States. Johnson (1973) argues that yield elasticities are in the 1-2 range, while area elasticities hover around 0.1. This is because total crop area in the United States is largely fixed in the short run. In the medium term, the rate of responsiveness rises, but area is still inelastic in relation to variations in price. Government programs have a very important impact on area and yield response, through direct effects on area planted and indirectly through price. These factors are discussed in more detail below.

Grain Use in the United States

The United States uses over 200 million metric tons of grain each year (Table 5.4). Roughly 70 percent is fed to livestock, and only 10 percent is consumed directly as food. The remainder is used industrially, primarily in brewing and distilling (for beverages and fuel), and in the production of sweeteners, starch, and related products. In 1988-90, each person in the United States consumed approximately 88 kg of grain as food, but total per capita disappearance was over 840 kilograms (Table 5.5).

Wheat is the primary foodgrain in the United States. Most wheat is consumed directly as bread, baked goods, or in other processed foods, but the proportion fed to livestock has tended to increase since the early 1960s (Figure 5.3). The use of wheat for feed is highly variable and is largely a function of prices. Simply put, when wheat prices are depressed, more is fed to animals. Conversely, when prices are high, feed use declines. Strong export markets during the late 1970s meant that only 10-15 percent of the wheat used in the United States was fed to livestock. Weak export markets during the early 1980s increased this proportion to 20-25 percent. Typically, 80 percent or more of coarse grain consumed domestically is fed to animals, although industrial uses also account for a sizeable portion of the total, particularly for maize and barley. The demand for grain for industrial purposes is growing rapidly. Over 20 percent of maize consumed in the United States goes to manufacture industrial products, primarily maize sweetener and alcohol fuel. This compares to 10 percent in the early 1960s. The production of maize sweetener increased fourfold since the early 1970s because of high sugar prices. Sweeteners from maize now make up almost 45 percent of total U.S. consumption of caloric sweeteners.

In per capita terms, the food use of wheat has grown slightly, particularly since the mid-1980s. This is probably because of the greater sensitivity among Americans about the possible health consequences of high consumption of meat and livestock products, particularly feed-intensive products such as beef. These concerns have led to an increase in the popularity of vegetable and grain products and a switch by consumers to white (poultry) meat, which requires less grain per unit of output than red meat. These changes in demand for livestock products are reflected in a gradual decline in per capita use of coarse grains for livestock feed (Figure 5.3). Should these trends continue, they could have increasingly significant implications for U.S. grain farmers. Either production will have to adjust to the change in domestic demand or additional export markets will have to be found.

Empirical studies of the demand for grain in the United States suggest that short-run changes in response to price fluctuations occur primarily in the use of grain for feed. Grain for food and industrial uses is relatively unresponsive to short-term variations in prices. The high level of consumer income in the United States and the small proportion of this income spent on food (less than

TABLE 5.4 U.S. Grain Balances

	Area harvested (mill. ha.)	Yield (m.tons/ha.)	Production (m. m. tons)	Beginning stocks (m. m. tons)	Imports (m. m. tons)	Exports (m. m. tons)	Feed use (m. m. tons)	Apparent consumption (m. m. tons)	Ending stocks (m. m. tons)
Total grain									
1965-69	61.8	3.2	195.4	64.1	0.4	41.3	123.6	155.2	63.4
1970-74	62.7	3.5	217.7	56.4	0.4	58.9	133.5	167.2	48.4
1975-79	70.5	3.8	268.8	62.6	0.4	91.0	127.0	167.0	73.8
1980-84	71.1	4.0	289.2	100.8	0.7	103.4	134.6	184.6	102.7
1985-90	63.9	4.5	288.6	132.9	1.8	88.6	143.6	205.7	128.9
Coarse grains									
1960-64	44.1	3.0	132.4	67.0	0.3	15.6	107.7	120.4	63.7
1965-69	40.0	3.8	153.8	44.0	0.3	20.4	120.2	134.8	42.9
1970-74	41.0	4.2	171.1	35.7	0.4	31.6	129.0	144.8	30.7
1975-79	43.0	4.9	209.2	38.1	0.3	57.1	124.0	143.9	46.7
1980-84	40.9	5.2	214.1	67.0	0.5	59.4	128.3	156.4	65.8
1985-90	38.2	5.8	223.9	96.8	9.4	53.0	135.8	173.2	95.6
Rice									
1960-64	0.7	3.0	2.1	0.3	a	1.1		1.0	0.3
1965-69	0.8	3.6	2.9	0.3	a	1.7		1.2	0.4
1970-74	0.8	3.7	3.0	0.4	a	1.8		1.4	0.3
1975-79	1.1	3.6	3.9	0.9	a	2.2		1.6	1.0
1980-84	1.2	3.8	4.7	1.4	a	2.4		2.0	1.6
1985-90	1.0	4.4	4.6	1.5	0.1	2.4		2.6	1.2
Wheat									
1960-64	19.6	1.7	33.3	35.8	0.1	19.5	1.0	16.4	33.2
1965-69	21.0	1.8	38.7	19.7	a	19.2	3.4	19.1	20.1
1970-74	20.9	2.1	43.6	20.3	a	25.5	4.5	21.1	17.3
1975-79	26.4	2.1	55.7	23.5	0.1	31.7	3.0	21.5	26.0
1980-84	29.0	2.4	70.5	32.5	0.1	41.6	6.4	26.1	35.3
1985-90	24.8	2.5	60.0	34.7	0.5	33.3	7.8	29.8	32.1

Source: U.S. Department of Agriculture, Economic Research Service. PS&D View Database, 1991.
Note: Figures are averages for the years shown. Rice is milled rice.
[a]Less than 0.05. Detail may not add because of rounding.

TABLE 5.5 U.S. Domestic Use of Major Grains (1986-90) (million metric tons)

	Maize	Wheat	Sorghum	Barley	Oats	Rice	Total Major Grains
Food	a	20.1	a	a	a	1.6	21.7
Industrial	34.0	b	0.5	4.5	2.2	b	41.2
Seed	0.5	2.8	c	0.4	0.7	0.1	4.5
Feed	114.7	3.9	13.4	4.8	6.2	c	143.1
Total	149.3	26.7	13.9	9.7	9.2	2.1	210.9
Per capita (kg)	600.0	107.5	55.8	39.0	36.8	8.5	847.7

Sources: USDA, Situation and Outlook Reports for Feed, Rice and Wheat. Various issues.
*Included with industrial use.
bIncludes food use.
cLess than 0.05.

20 percent) means that final demand for grain responds little to price variations. Industrial demand is also unresponsive to price changes, particularly in the short run, because the cost of grain is only a fraction of the final cost of processed products, and substantial fixed capital is already committed to processing. Because feed costs form the major part of the total cost of fed livestock production, feed use does respond to price. If the relative prices of grains change, substitution among grains is common. Thus, feed-based livestock enterprises help bring about short-run adjustments to price changes in the U.S. grain economy.

Grain Trade: The World's Largest Exporter

The United States imports very little grain and is the world's largest exporter. Grain exports rose steadily during the 1970s from 40 million metric tons to a peak of 115 million metric tons in 1980 (Figure 5.4). Strongest growth was in coarse grains, which rose from 46 percent of total grain exports in 1970 to 62 percent in 1980. Expanding world income and competitive U.S. export prices stimulated global demand for U.S. grains, particularly for livestock feed. Exports declined sharply during the early 1980s because of slower world economic growth, foreign debt problems in some importing countries, the high value of the U.S. dollar, and high government support prices for grains (Blandford, Meyers and Schwartz, 1988). In 1985, grain exports fell to 63 million metric tons, almost half that exported in 1980. Since 1985, changes in U.S. support prices, a weakening of the dollar's value, the use of export subsidies, and higher rates of economic growth in some importing countries contributed to a partial recovery in export volumes (Figure 5.4).

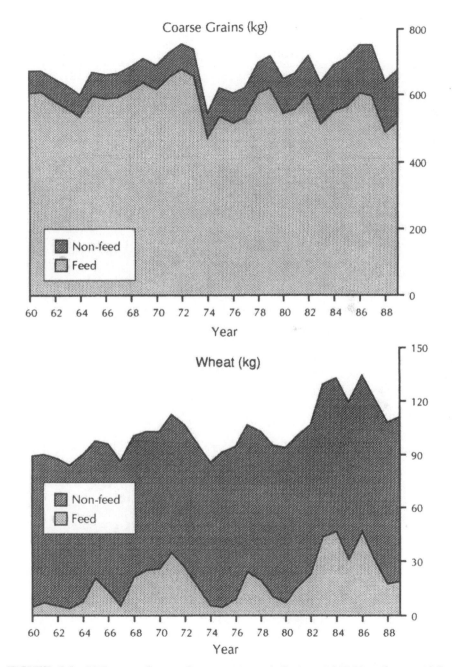

FIGURE 5.3   U.S. per Capita Consumption of Grains, 1960-89.   *Source:* U.S. Department of Agriculture, Economic Research Service.   PS&D View Database, 1989.

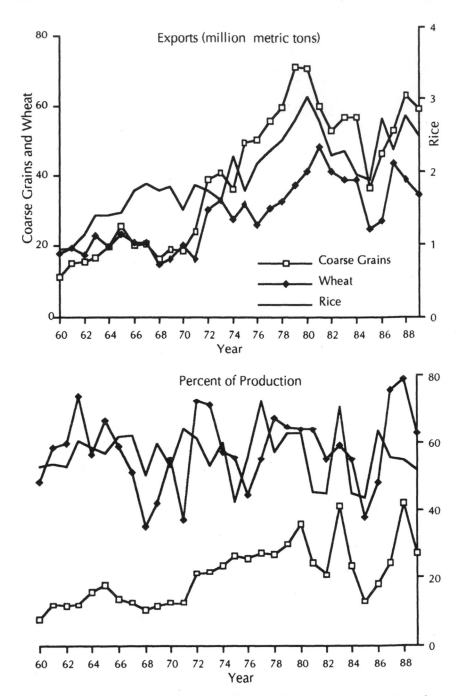

FIGURE 5.4 U.S. Grain Exports, 1960-89. *Source:* U.S. Department of Agriculture, Economic Research Service. PS&D View Database, 1989.

Relative to production, exports of wheat and rice are much more significant than coarse grain exports. The United States has a huge domestic market for coarse grains, particularly for feeding livestock. Apart from the high ratio of exports to production in 1983 (caused by a sharp decline in production attributable to government programs) and in 1988 (because of drought) (Figure 5.4), less than 35 percent of domestic coarse grain production is generally exported. This can be compared to ratios as high as 70 to 80 percent for wheat and rice. When market prices fell in 1985 because of continued oversupply, less than 14 percent of coarse grain production was exported and only 40-45 percent of wheat and rice output. The level of U.S. grain exports and their ratio to production has been extremely variable since the early 1970s. This is because of the interaction of macroeconomic forces in the United States and overseas, weather conditions, and the impact of U.S. farm programs.

Asia, in particular Japan, is the most important destination for U.S. grains (Figure 5.5). Japan is the single largest and most consistent purchaser of American grains, accounting for roughly 20 percent of the total in 1988-89. The Soviet Union and China are important but often erratic importers of U.S. grains. Between 1982 and 1988, for example, the USSR's grain purchases from the United States ranged from less than 3 million to over 18 million metric tons. China's purchases varied from 56,000 metric tons to over 8 million metric tons. As the European Community has become a large net exporter, Western Europe, which used to be an extremely important market for U.S. grains, is accounting for a declining proportion of grain exports. The developing countries in Asia, Africa, and Latin America now take a growing share of U.S. exports. The financial position of U.S. grain farmers is increasingly linked to developments in the markets of developing and centrally planned economies, such as those discussed in other chapters in this book.

Grain Marketing Systems and Institutions

Grain is harvested during a relatively short period. After harvest and drying (if necessary), producers store grain or sell it through a variety of outlets. These include other farmers, particularly livestock farmers, grain elevator companies, processors, and export marketing companies (Figure 5.6).

The most common point of sale is the country elevator. These elevators are located in small towns in production areas and primarily receive grain directly from farms by truck. They vary in capacity from as little as one thousand metric tons to as much as 50 thousand metric tons. Country elevators may be owned privately and independently, by a farm cooperative, or by a commercial grain company (line elevators). Country elevator firms frequently offer a variety of services, such as drying and conditioning of grain, the sale of farm supplies, and storage facilities for producers, processors, and the U.S. government. Grain is obtained from farmers by spot buying at the time of delivery, through forward

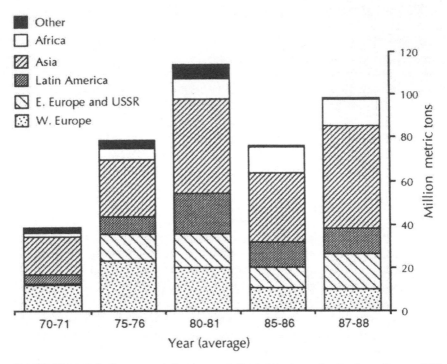

FIGURE 5.5 U.S. Exports of Grains and Grain Products by Region. *Source:* U.S. Department of Agriculture, Foreign Agricultural Trade of the United States, various issues.

contracts, and by accepting grain for storage and sale at a later date. Prices for spot sales are based on futures prices in major grain exchanges (see Chapter 1) or cash bids from dealers or terminal markets. The most common outlets for grain from country elevators are sub-terminal or river elevators and terminal markets.

Sub-terminal elevators are inland train-loading stations, typically located near metropolitan areas that buy grain from country elevators. River elevators act as an assembly point for grain transported by barge to terminal markets, exporters, or other consumers. Both sub-terminal and river elevators provide various services, such as grain blending to ensure uniformity, and storage facilities. Terminal elevators are usually large structures located near important marketing and transportation hubs, such as Chicago, Kansas City, and Minneapolis. These elevators receive grain from cash grain merchants and country and sub-terminal elevators by truck, rail, barge, or boat, depending on their location and handling facilities. As in the case of sub-terminal elevators,

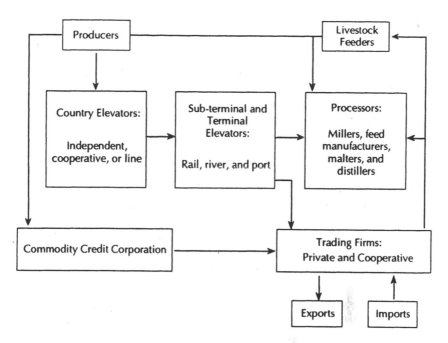

FIGURE 5.6 The U.S. Grain Marketing System

terminal elevators may be owned and operated independently, by cooperatives, or by an integrated grain company. The major outlets for grain from terminal elevators are millers or grain processors, feed manufacturers, and exporters.

Most transactions at the local elevator level and above are made by contract for deferred delivery. The contract specifies price, quantity, time of shipment, grade, and terms of sale. The delivery date may be set as little as five days in the future to up to 120 days. Legally, grain shipped within state borders need not be federally inspected, but, in practice, shipments are often inspected by state agencies, or private inspectors operating on behalf of grain exchanges or commercial organizations. When grain is shipped by rail, prices are usually based on official grades established by federal inspection at the point of shipment. Grain shipped by truck is graded but not necessarily federally inspected at the point of destination.

Grain merchandisers (sub-terminal elevators, terminal elevators, processors, feeders, exporters, and foreign buyers) typically use the futures market to hedge the price risk of purchasing grain at a contract price from country elevators or other sources for delivery and sale at a later date. The futures markets play an important role in providing price information to all levels of the grain marketing system, in reducing the price risks of participants, and in creating the potential for profits (or losses) for speculators. Standardized contracts calling for the delivery of grain in a designated future month are traded on four futures markets: the Chicago Board of Trade, the Kansas City Board of Trade, the Minneapolis Grain Exchange, and the Mid-America Commodity Exchange (also in Chicago). Contracts exist for wheat, maize, oats, and rye on the Board of Trade, wheat and grain sorghum in Kansas City, spring wheat and durum in Minneapolis, and wheat, maize, and oats on the Mid-America Exchange. Trading on the commodity exchanges is an important part of the price formation process for grain in the United States and on the world market (see Chapter 1).

Through the growth of cooperatives and private firms, the U.S. grain marketing system has become more economically integrated. The cooperative segment of the grain market is generally more oriented toward domestic rather than international sales. Cooperatives currently account for 40 percent of all off-farm grain sales at the first-handler level in the United States (Gilmore, 1982).[2] However, cooperative forays into the international market have met with limited success because of the difficulty in competing effectively with well-established private firms with multinational operations. One of the major export cooperatives (Far-Mar-Co), a subsidiary of Farmland Industries, experienced financial difficulties because of its export activities. Cooperatives rarely provide a complete range of export services, such as transportation and handling, and they typically use private firms for actual shipments, or enter into formal joint ventures with competing private companies.

Large, multinational grain trading companies occupy an important place in the U.S. grain marketing system and are dominant exporters of U.S. grains.[3] These firms own most of the sub-terminal space in the United States, but a minority of the country elevators. They are heavily involved in the operation and leasing of port elevators. They also hold milling, feed manufacturing, and grain processing facilities, both in the United States and overseas.

Government Policies: A Key Element

Until the 1930s, there was little direct government intervention in the pricing of agricultural products, except for the levying of tariffs on some imports. Agricultural policy in the United States was primarily oriented towards improving agricultural productivity through the development and spread of new technologies. This goal was pursued at the federal level by the U.S. Department of Agriculture, and at the state level through land grant colleges and universities, established as a result of the 1862 Morrill Act. The situation changed dramatically with the passage of the Agricultural Adjustment Act of 1933, enacted in response to the Great Depression during which farmers' incomes fell precipitously, especially in comparison with their urban counterparts. The price support programs introduced for "basic" commodities, including grains, still remain in force today in a modified and expanded form.

Farm legislation is revised every four or five years in the United States. The Administration, headed by the President, drafts proposals for the Farm Bill, which are considered by both the House and the Senate of the U.S. Congress. Each of these work out their own versions. After passage, differences are reconciled by a joint committee. Both houses then vote on a final version of the legislation, which must be approved by the President before being signed into law. Thus, the framing of farm legislation in the United States is surrounded by intense political activity. General farm and commodity organizations represent the interests of producers, and processors, marketers, consumers, and a variety of other special interest groups lobby representatives in the Congress in an attempt to influence the legislation to their advantage. The Administration also actively lobbies Congress to ensure that its wishes are met in the final bill.

It is difficult to identify exactly who frames farm policy in the United States. Farm legislation is the result of shifting coalitions within and among Congress, the Administration, and lobbying groups. Farm programs are influenced by historical precedent, inertia, and compromise. Ex post, policies may appear to display a measure of rationality, but it is misleading to view the agricultural policy process in the United States as one in which a set of specific, well-defined objectives are pursued through the choice of appropriate instruments by a single decision-making group.

The legislation that finally emerges from the bargaining process is often substantially different from that originally proposed by the Administration. The Food Security Act of 1985 is an excellent example. Originally, the Administration proposed major changes in U.S. farm programs. The major objective was to reduce the government's role in agriculture and to make the sector more market oriented. It was proposed to eliminate most government payments to producers and to make support prices more responsive to changes in world market prices. The legislation that finally emerged from Congress was radically different. Support prices were reduced, but large government payments to farmers were maintained. A variety of expensive export subsidy schemes were introduced, aimed primarily at increasing U.S. agricultural exports.

Agricultural policy in the United States is commodity-oriented, rather than driven by the overall interests of the agricultural sector. Only a limited range of crops are covered by legislation (primarily, grains, cotton, and some oilseeds). Apart from the dairy industry, which has a price support program for milk, livestock producers receive little consideration in the framing of commodity programs. Despite considerable debate among consumer and farm groups about the efficiency of domestic agricultural programs, and growing concern about the equity of the distribution of their benefits, the essential elements of the farm programs have remained intact for more than fifty years. Overall, the international implications of domestic policies are assigned little weight, despite the fact that the United States is a major player in world markets. The formation of U.S. agricultural policy is largely driven by domestic concerns and politics, with significant attendant implications for countries relying on the United States for grain supplies.

Public Policy and the U.S. Grain Sector

As a major component of agricultural production, grains have historically dominated farm policy in the United States. Price supports for grains were first introduced in the 1933 Agricultural Adjustment Act. Direct income support payments were initiated in 1963. An important feature of U.S. programs is that producer participation is voluntary rather than mandatory. If producers choose to enter the programs, they may have to meet certain conditions on the use of their land, such as, for example, not planting a portion of their acreage. Current farm policies in the United States for the major grains fall into three broad categories (Table 5.6): (a) price and income supports; (b) public stockholding; and (c) trade policies.

Price and Income Supports

A key element in U.S. government programs for grains is a support price called the *loan rate*, whose level is set annually through the political process. Participants in government programs have the option to place some or all of their production under loan with a public corporation called the Commodity Credit Corporation (CCC). Farmers receive a payment equal to the loan rate for each unit of production pledged as collateral to the CCC. The commodity is held in storage on behalf of the CCC on the farm or at some other location. Farmers then have the option of reselling the stored product on the open market during a one-year period. Typically, they will choose this route if market prices are above the loan rate plus the costs of storage, which they reimburse to the CCC. Otherwise, farmers can default on their loan and the commodity becomes the property of the CCC. In the short term, this system provides farmers with access to cash. In the longer term, it guarantees program participants a minimum price for their product. Farmers outside the program cannot access CCC loans and must pay to store their product or sell it on the open market. They benefit from the program indirectly only if CCC loan purchases result in higher market prices. Whether this occurs depends on many factors and is discussed further below.

Direct payments or subsidies, called *deficiency payments*, are made to participating grain farmers based on the difference between a *target price* and the higher of either the market price or the loan rate. The target price, which is generally above both the market price and the loan rate, only partially determines the total subsidy received. To limit government expenditures, payments are calculated using a hypothetical yield called the *program yield*[4] rather than a farmer's actual yield. Furthermore, the area on which the payment is based is limited by *base acreage limitations*. Each farm is allocated a base area for each crop., e.g. wheat, which is eligible for deficiency payments. The method by which base area is determined has varied, but typically takes into account the area on the farm historically used to produce the supported crop. Participating farmers may not plant more than their base area in any given year if they wish to be eligible for support payments. Deficiency payments are calculated using the eligible area and the program yield fixed by the government. There is a limit on support payments (for all crops) that can be made to any one farmer (typically $50,000). This limitation is not very restrictive because it is difficult to establish legally what constitutes a farm and a farmer. Through the use of partnerships, for example, the number of "eligible" farmers can be increased by including both the principal operator and the members of his/her family.

The last major provision of the deficiency payment scheme is an *acreage diversion* requirement, which is also set annually. To be eligible for price and income supports, farmers may be required to "set aside" or take out of production a minimum proportion of their base. The diverted land can sometimes be used for certain other specified crops, but not for program crops (which include most grains). In some years, farmers may be offered the choice of taking a further proportion of their base out of production in exchange for special *diversion payments*. The purpose of diversion is to raise the market price by reducing supply, and to limit the amount placed under loan and in CCC stocks. Set aside requirements can cause substantial variations in the amount of cropland used to produce grains. Increased set-aside requirements and other acreage reduction measures (particularly the Conservation Reserve Program described below) were the major cause of the fall in the area of grain harvested during the late 1980s (Figure 5.2).

In addition to deficiency payments, grain farmers may be eligible for *disaster payments* when their crops are affected by drought or other natural disasters. Some farmers purchase federally subsidized crop insurance as protection against crop losses due to weather. But even a farmer who has no such insurance may be eligible for disaster payments. The provisions that govern disaster payments have changed frequently, but typically, if the farmer's yield is less than a reference level, a payment is made to offset the loss. Under the Disaster Assistance Act of 1989, this payment is calculated as a fraction of the target price and applied to the difference between actual yields and program yields. Under the act, payments to farmers not participating in government programs were authorized, but the payment rate was based on a fraction of the loan rate rather than the target price (USDA, ERS, 1990).

As indicated earlier, farmer participation in U.S. price and income support

TABLE 5.6 Major U.S. Grain Policy Instruments

Instrument	Commodity		
	Coarse grains	Rice	Wheat
PRODUCTION/CONSUMPTION			
Producer guaranteed price	X^a	X^a	X^a
Deficiency payments	X^a	X^a	X^a
Government purchases	X^a	X^a	X^a
Production quota	X^a	X^a	X^a
Input subsidies			
- credit	X	X	X
- fertilizer/pesticides			
- irrigation		X	
- machinery/fuel	X^b	X^b	X^b
- seed			
Crop insurance	X	X	X
Controlled consumer price			
TRADE			
Imports			
- tariff			
- quota			
- subsidies			
- licensing	X	X	X
- state trading			
Exports			
- taxes			
- restrictions	c	c	
- subsidies	X^d	X^d	X^d
- licensing			
- state trading			
OTHER			
Marketing subsidies			
- storage	X	X	X
- transport	e		e
- processing			
State marketing			
Margin control			

[a]Only for farmers who participate in government programs.
[b]Tax exemptions for fuel.
[c]Periodic export embargoes.
[d]Direct subsidies, subsidized credit, and food aid.
[e]Construction and maintenance of inland water transportation facilities.

programs for grains is voluntary. Some farmers may choose not to take part because they are unwilling to accept the restrictions entailed in the programs. Others avoid outside the programs for financial reasons. The decision to join is made annually, and farmers may choose to participate in some years but not in others. The number of grain farmers in the programs has varied over the years, but relatively low market prices in the mid to late 1980s kept the participation rate at about 80 percent.

The net effects of the U.S. system of price and income supports on supply is controversial. One view is that base area limitations and area diversion reduce the supply of grain below the equilibrium level at existing prices, even though target prices are set well above market prices. An alternative view is that there is substantial "slippage," which results in a limited reduction in output. The most important potential source of "slippage" is the effect of the program on the amount of land farmers choose to plant. For example, base area under the 1985 Farm Bill was determined as a five-year moving average of area planted plus that diverted. Forward-looking farmers may have found it in their best interests to quit the programs temporarily and overplant to expand their base (Erickson and Collins, 1985), earning higher subsidy payments in the future. Even if farmers do not withdraw from the programs, they often plant more than they would otherwise for fear of losing base area and future subsidy payments. Consequently, total area planted is higher than if the provisions truly limited output.

A second major source of "slippage" occurs when farmers remove their worst land from production to fulfill diversion requirements, causing higher average yields per hectare. General consensus is that average yields increase 20 to 40 percent when poorer quality land is taken out of production (Gardner, 1987). Gains in output can also occur on the land remaining in production when farmers use fixed resources more intensively on a smaller area. A uniform target price across crop qualities also induces production of lower-quality, higher-yielding varieties. Further, in drier areas, diverted land planted the following year retains more moisture and yields more grain. All of these factors raise output and partially offset the effects of land diversion. The combination of diversion and base area limitations tend to reduce the responsiveness of supply to market prices, partially explaining why many studies report very inelastic supply response elasticities for grains in the United States.

An additional mechanism for long-term land retirement, the Conservation Reserve Program, was introduced in the 1985 Farm Bill. This program applies to land classified as highly erodible by the Soil Conservation Service of the U.S. Department of Agriculture. Farmer participants submit a bid stating the annual payment they would accept over a ten-year period to retire highly erodible land from crop production. If the bid is accepted, farmers convert the land to vegetative cover for ten years. By the end of 1989, over 12 million hectares of cropland had been removed from production under this program. Roughly 60 percent of this land was previously planted to grains; and 30 percent was used to produce wheat. In the mid-1990s, when the CRP contracts begin to expire, the government will have to decide whether to continue to keep land retired under the program out of production or to allow it to be used to produce crops again.

Stockholding

The various measures that the government uses to promote the movement of land into and out of production have a major impact on the supply of U.S. grains and ultimately on world market prices. Another instrument used to manage the market is public stocks. Stocks accumulated by the CCC under the loan rate program can only be sold domestically when market prices rise significantly above the loan rate. Otherwise, CCC stocks are used in food aid programs or are sold commercially by companies at a discount overseas. In recent years, sales of CCC stocks have been an important part of programs designed to counter the effects of subsidized exports from the European Community (see the discussion of generic certificates and trade policies below).

There is also a Farmer Owned Reserve (FOR) stockpile that operates on the same basic principle as the CCC loan program. Stocks are accumulated (up to a maximum level) when market prices fall below loan rates and are released when market prices exceed a threshold. Participating farmers are paid interest rate subsidies for the construction of storage facilities and additional subsidies to cover storage. The stated purpose of the reserve is to maintain a buffer stock for food security purposes. Because of high trigger prices, stocks are rarely released from the FOR, and the program is basically another means used to support producer prices.

Two important modifications affecting public stock management were introduced into a U.S. farm program in the 1980s. The first was the *commodity* or *generic certificate*, issued to farmers as a noncash payment or "payment in kind" (PIK) to partially replace deficiency payments. Instead of cash, farmers receive a transferable certificate (valid for eight months) with a fixed dollar value that can be redeemed for commodities in CCC or FOR stocks at current market prices. An active certificate market has developed between farmers and grain marketing firms. The aim of the program is to cut government cash outlays on deficiency payments and to reduce government-held stocks. When certificates are redeemed, market supply increases and prices drop, but farmers are protected from the effects of the decline in market prices on their own production by the loan rate program. If farmers do not use or sell the certificates, they can be exchanged for cash from the CCC (Hanthorn and Glauber, 1987). Commodity certificates (issued to exporters) have been a major part of the Export Enhancement Program, which allows exporters to market CCC grain overseas at competitive prices and helps to counter export subsidies used by other countries (see trade policies section below).

The second program modification affecting public stocks is the *marketing loan*. Under this provision, farmers can repay the CCC for a commodity under loan at the "loan repayment rate," which is set at the prevailing market price. This price is generally below the original loan rate that farmers received for their crop. This policy tends to reduce CCC stocks and further lower market prices, allowing U.S. grain to compete internationally. Marketing loans have only been used for rice. The Secretary of Agriculture has the discretion to extend their use to feedgrains and wheat, but commodity certificates have been used instead to dispose of stocks of these grains, particularly wheat, and to target the disposal of these stocks to particular export markets.

The net effects of the accumulation, maintenance, and disbursement of CCC stocks are controversial. Because the government spent large sums of money in the past to acquire stocks, one view is that these policies and the resulting stocks provide a floor under world market prices. That is to say, the net effect of the loan rate and attendant CCC activities is to increase world prices above competitive levels. Although true in the short run, CCC stocks inevitably end up on the market at a later date. Market prices reflect this, and the price-depressing effects of the sale of public stocks over time are likely to offset any initial jumps in price brought about by the acquisition of stocks. Furthermore, several studies show that private storers respond to the presence of government stocks by changing their behavior, in particular by storing less (Wright, 1985). The reduction is even more pronounced if the response of other countries to U.S. stockholding activities is considered. Exporters, such as Canada and Australia, do not maintain large stocks when the CCC is storing significant quantities of grain. Importers also react by reducing stocks when their import needs can be guaranteed through large inventories in the United States.

The major potential effect of CCC stockholding then is to *stabilize* market prices, rather than to affect substantially the *level* of market prices. Even then, it is possible that U.S. public stockholding destabilizes markets because of the rules under which stocks are accumulated and released. As indicated earlier, the rules for operating the buffer stock provided by the Farmer Owned Reserve were not designed to reduce short-term fluctuations in prices, but to provide long-term price support.

The relationship between wheat prices and inventories since 1950 is illustrated in Figure 5.7. This relationship is representative of those for other grains. Until the early 1970s, farm prices for wheat were close to the loan rate, and the CCC held large inventories, reflected in the stocks-to-use ratio. Market prices increased dramatically between 1971 and 1975, moving substantially above the loan rate, and the inventory ratio declined significantly. In the late 1970s, market prices again fell, approaching the 1977 loan rate. As a result, inventories accumulated rapidly. Much of this grain resided in farmer-owned reserves, which were subsidized by the government.

After a brief upturn at the end of the 1970s and the early 1980s, market prices again dropped. Both target prices and loan rates were boosted significantly by the 1981 Agriculture Act. Softening of demand in world markets, attributable to lower income growth and a higher dollar, caused market prices to dip below the target price to the loan rate. Total inventories -- both those owned by the CCC and those subsidized by the government through the farmer-owned reserve -- rose. The 1985 Food Security Act resulted in a substantial reduction

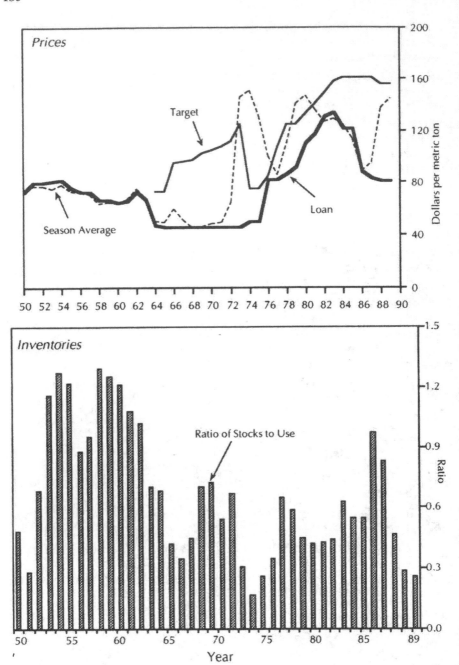

FIGURE 5.7 U.S. Wheat Prices and Inventories. *Source:* Data provided by the U.S. Department of Agriculture.

in loan rates, but not in target prices. Market prices followed the decline in the loan rate until drought in 1988 led to a large jump in prices. Only in periods when U.S. grain stocks are low have market prices been above the loan rate, and market prices are rarely above target prices.

The commodity programs for grains directly affect production, trade, and prices of grains in the United States. However, programs for nongrain commodities can have important indirect effects. One of the most notable cases is the sugar program. Sugar producers in the United States are protected from low-cost imports by import quotas. In recent years, the domestic sugar price has been two to three times that of the world price. High sugar support prices gave added impetus to the development of sugar substitutes, particularly maize sweetener. As indicated earlier, expansion of the sweetener market accounted for much of the increase in domestic use of maize in the United States, resulting in growth in production of an important feed by-product, maize gluten meal. This by-product is marketed in the European Community, where domestic feedgrain prices are kept high with price supports. The growth of maize sweetener/gluten meal production created tension between the United States and the Community. It has also meant that maize producers and sweetener manufacturers support domestic sugar producers in their lobbying efforts for the continuation of high support prices for sugar.

Trade Policies

Measures that directly affect grain trade in the United States are limited, but important. A licensing system is used to control imports. Such control is necessary to prevent imported grains from displacing domestic supplies when loan rates are above world market prices, and to limit the accumulation of public stocks under the loan rate program. Since the United States is a major grain exporter, most of its trade policies focus on exports. These include export subsidies, food aid, and subsidized credit.

With the exception of the CCC, the United States does not maintain a public institution similar to the producer marketing boards in Canada and Australia. However, during the 1970s and 1980s, the U.S. government directly intervened in trade by imposing a number of export embargoes. In 1974 and again in 1975, moratoria were placed on shipments of grain to the Soviet Union and, in the latter year, on shipments to Poland. These moratoria, each of which lasted only a few weeks, were instituted to stabilize the domestic grain market because of uncertainty about the adequacy of supplies in the face of production shortfalls in the United States and abroad. In January 1980, the United States imposed an embargo for sixteen months on the export of a number of products, including grain, to the Soviet Union because of its invasion of Afghanistan. During the embargo, the United States honored a prior commitment under a long-term grains agreement to supply 8 million metric tons of grain to the USSR. What impact these embargoes had is controversial. Evidence suggests that they had little long-term effect on world supply, demand and international trade, but possibly an important psychological effect on importers who no longer view the United States as a reliable trading partner (USDA, 1986).

Apart from these limited cases of direct intervention, the government's role in export sales is largely limited to monitoring and inspecting overseas shipments, funding of various export promotion activities, and providing food aid. It also influences trade through the rules and regulations that manage the accumulation and disposal of government stocks.

The United States instituted a food aid program with the passage of Public Law 480 in 1954, of which grains are a major component. Primary goals include disposal of excess CCC stocks, creation of new commercial markets for grains by changing consumer preferences, and achievement of humanitarian or political objectives. During the 1960s, shipments of grain under the PL 480 program were an important part of total U.S. grain exports, while in recent years, they have constituted only a minor portion. Nevertheless, the United States remains the world's major food aid donor (see Chapter 1). In addition, the United States offers credit under concessional terms to selected importers who purchase grain commercially from the United States. Short-term credit is supplied under the General Sales Manager (GSM) 102 program, longer-term credit is granted under GSM 103. In the 1980s, up to $U.S. 5 billion of credit was provided under GSM 102 and $U.S. 1 billion under GSM 103.

Changes in U.S. grain programs can have major implications for U.S. grain exports and international grain prices. Under the legislation in force during the early 1980s, production controls and the accumulation of public stocks were used to support high domestic prices that priced U.S. exports out of world markets. Because of the dominant role the U.S. plays in the world market, U.S. production controls and support purchases inevitably sustain world prices to a certain extent. A costly payment-in-kind (PIK) program in 1983-84 was introduced to reduce stocks. Under this program, producers were given government-owned commodities in exchange for controlling production. The program reduced production and stocks significantly in 1983, but had only a temporary effect on the domestic supply/demand balance and none on the uncompetitive position of the United States in international markets. It was not until the Food Security Act of 1985, under which loan rates were reduced, that exports began to rise and stocks to fall.

A variety of export subsidies were introduced under the 1985 Act, of which the most important were enacted through the Export Enhancement Program (EEP). Under this program, public stocks are used to subsidize exports of agricultural commodities, including grains, to specific markets. These stocks are released through the use of commodity certificates issued to exporters. The U.S. Department of Agriculture draws up a proposal for the sale of a commodity or commodities to a particular country through EEP. A Trade Policy Review

Group, made up of representatives from seven Federal agencies, including the USDA and the Department of State, reviews the proposal. If approved, private exporters submit a request to the USDA for the subsidy they need to make the sale. If the USDA accepts the request, the subsidy or "bonus" is paid to the exporter in the form of commodity certificates, which can be redeemed for CCC stocks to complete the export sale.

The Export Enhancement Program was developed because policy-makers felt that other exporters, primarily the European Community, were displacing the United States from its traditional markets with subsidies. EEP is expensive, costing over $2.3 billion by the end of 1989 (USDA, ERS, 1990). The principal commodity exported under EEP is wheat, and how much the program has contributed to boosting U.S. exports is controversial. The program did lower import prices for some countries, but may not have altered their total purchases. Whatever its cost effectiveness, the program is politically popular because it is viewed as a direct counter to EC grain subsidies, and one of the few ways that the United States has to try to force the Community to reduce these subsidies. U.S. agricultural trade policy is increasingly dominated by disputes with the European Community over its agricultural policies. U.S. responses to these policies affect world markets for grains, with attendant significant implications for importing countries and other grain exporters.

Other Policies

The major policies that influence the U.S. grain sector are the price and income support programs, which are part of farm legislation. There are, however, other measures that affect the competitive position of U.S. agricultural and grain exports. The government provides a variety of additional direct or indirect agricultural subsidies, ranging from preferential tax treatment to public funding of irrigation systems and waterways used to transport grain to ports. Other subsidies, such as those designed to promote the production of fuel alcohol from grain, are also important. The United States maintains an extensive, publicly supported research and extension system that provides information and technical assistance to farmers and others engaged in agriculture at little or no cost. Over the years, this system has contributed to steady improvements in agricultural productivity and to the maintenance of agriculture's competitive position. As gains in productivity continue, the real price of agricultural commodities declines, and it is becoming increasingly difficult to maintain support prices with public purchases and production controls alone.

As the agricultural economy and the grain sector in particular have become more capital-intensive and dependent on overseas markets, they have become increasingly sensitive to changes in domestic macroeconomic policies and to the state of the world economy in general. The level of interest rates, the value of the dollar, and the rate of economic growth in the rest of the world are key determinants of the financial health of U.S. agriculture. During the 1970s, low real interest rates, a low dollar value, and high income growth in the rest of the world contributed to a strong worldwide demand for U.S. agricultural exports, especially grains. During the early 1980s, a combination of restrictive monetary and expansionary fiscal policies caused real interest rates and the value of the dollar to rise at the same time as expansion real incomes in the rest of the world fell. This caused a substantial reduction in demand for U.S. agricultural exports, resulting in financial stress in the agricultural sector (Blandford, Meyers, and Schwartz, 1988). Macroeconomic policies in general place significant constraints on U.S. farm policies and have a major influence on the effect that farm policies have on domestic farm incomes and international trade. In addition, because the world's economies have become so interdependent, the domestic grain economy is also extensively influenced by overseas demand for U.S. grain products.

Program Costs

The costs of supporting grain prices and farmers' incomes through U.S. farm programs can be large in some years. When target prices are high relative to market prices or the loan rate, substantial deficiency payments are made. When the loan rate is high in comparison to market prices, costly stocks are acquired. Most of these expenses are borne directly by U.S. taxpayers. In the 1980s, farm programs were especially costly. As world market prices softened during the early 1980s, support expenditures rose (Figure 5.8). In 1981, they cost roughly $2 billion, but by 1983, expenses jumped to almost $12 billion. The decline in 1984 is apparent rather than real because expenses incurred by the Payment in Kind program in 1983 and 1984 are not included in the figures. PIK expenditures, which were not charged to the CCC's budget, totaled almost $10 billion (USDA, ERS, 1988). As indicated earlier, the passage of the Food Security Act at the end of 1985 led to a substantial rise in support costs. In fiscal year 1987, these totalled more than $18 billion. Feedgrains, primarily maize, are the largest source of total expenditures.

Government support payments became increasingly important to farmers during the 1980s. The producer subsidy equivalent (PSE) measures the gross transfers to producers from all types of government policies (Figure 5.9), including costs borne by consumers and taxpayers. In the case of U.S. grains, virtually all transfers to producers are from taxpayers. The estimates also include such items as transportation subsidies from the construction and maintenance of inland waterways and fuel tax exemptions for farmers (Table 5.6). In the early 1970s, the PSE for grains averaged 8 percent, but in 1986, the PSE stood at 53 percent. Subsidies to rice farmers in that year were equivalent to over 70 percent of the value of rice production. The relative assistance to grains has exceeded that for other products in recent years (Figure 5.9). Transfers to grain producers as a proportion of the total have risen from less

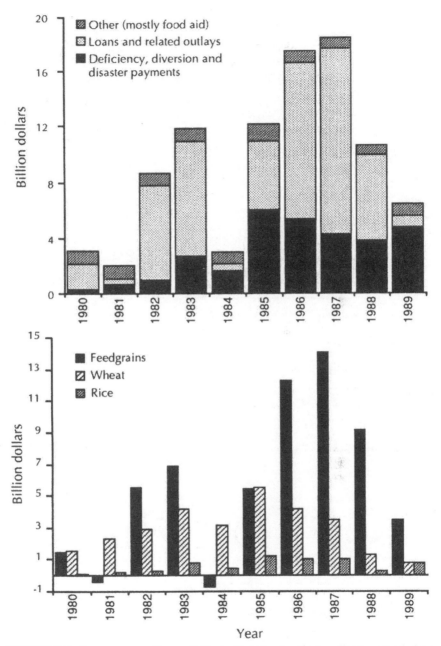

FIGURE 5.8 Government Costs of Grain Programs. *Source:* Data provided by the U.S. Dept. of Agriculture. *Note:* Years are Oct.-Sept. fiscal years. Figures exclude the costs of the PIK program in 1984-85, which were not included in the CCC's budget.

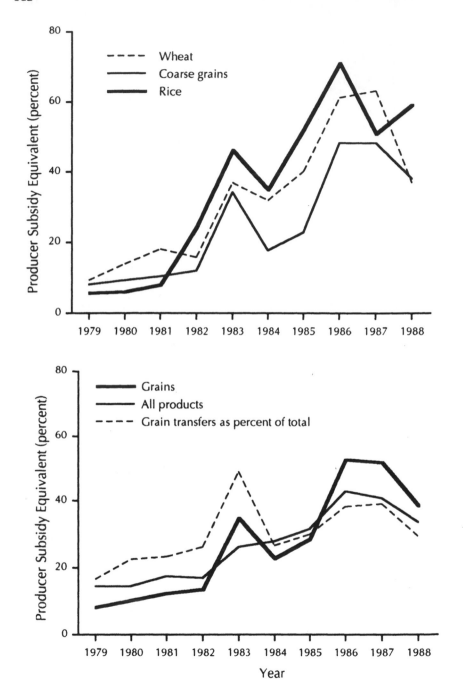

FIGURE 5.9 Producer Subsidy Equivalents for Grains. *Source:* OECD. Agricultural Policies, Markets and Trade, Monitoring and Outlook, 1988 and 1989.

than 20 percent in 1979 to almost 40 percent in 1986-87. Although expenditures dipped in the late 1980s, drought was the cause, and a return to large crops would again cause expenditures to rise. There have been several attempts to place a permanent cap on these expenditures as part of efforts to reduce the U.S. government's budget deficit, but the strength of the farm lobby has so far prevented this from occurring.

Conclusions

The United States is a major grain producing and consuming nation with a highly developed system for the production, processing, and marketing of grain. Private institutions, particularly the futures markets, provide an important vehicle for price formation and the spread of price information, both domestically and abroad. Public policies, particularly government price and income support programs, have a significant effect on the volume of U.S. trade and on international prices.

Since 1970, as continued improvements in productivity boosted output, the U.S. grain industry has come to depend on international markets as the primary outlet for growing surpluses. At the same time, domestic per capita use of grain, which was already very high, grew only modestly, with industrial uses for maize, including alcoholic beverages, maize sweeteners, and fuels, accounting for the bulk of the growth. The increasing reliance on international markets has not been without problems. Although traditional importers of U.S. grains, such as Japan, remain regular buyers, many newer foreign markets purchase irregularly. Fluctuations in the overseas demand for grain are caused by weather, changes in the policies of other countries, and an uncertain macroeconomic environment.

The United States' domestic grain price and income support programs are ill-equipped to deal with the new economic environment created by the growth of international trade. Designed originally for a closed economy and based on the assumption of price inelastic demand, the development of an unstable open economy with price and income elastic demand imposed major stresses on the price support system. Since the 1970s, large swings in prices and inventories have become a regular occurrence, and massive public stocks were accumulated in an attempt to support domestic prices. When the programs were initially designed, these developments were unimaginable. Rigid support prices have proved costly to maintain in a world of fluctuating exchange rates.

The American economy has not yet adjusted to the growing integration of its grain industry with international markets. The old price and income support policies continue to function, despite their apparent inefficiency in the changed economic environment. Agricultural policy is still driven by domestic political concerns, in spite of the overwhelming and obvious importance of international markets. Policy formation and implementation remain insensitive to international concerns and to the constraints imposed by international integration. Despite much rhetoric about reducing government involvement in agriculture, the government continues to be the key player in grain trade. Although free trade is widely espoused, U.S. grain exports are increasingly subsidized by public funds.

Policy-makers in the United States face difficult decisions as they try to revamp domestic grain policies to bring them more in time with today's economic situation. On the one hand, farmers need access to an expanding world market to maintain and increase their earnings. On the other, farmers are reluctant to be subjected to the international disciplines that guaranteeing this access would imply. A general reduction in the level of trade-distorting subsidies would help bring price stability to international markets, but would mean that U.S. domestic subsidies would have to be decreased or the payment method changed to avoid distortions in production. An improvement in U.S. access to overseas markets through the reduction of trade barriers would help to boost international trade and prices, but would require parallel concessions on the part of the United States. Agricultural trade barriers in the United States would have to be diminished, and steps might have to be taken to meet the security of supply concerns of trading partners. It seems likely that the U.S. grain market will become ever more closely integrated into international markets and correspondingly influenced by events in these markets. Thus, domestic grain policy-makers will inevitably be forced to adjust to increasing market integration in the future.

Notes

1. These percentages do not include the contribution of agriculturally related industries, such as fertilizer and machinery manufacturers, to national income and employment.
2. First handlers are the initial purchasers of grain from producers.
3. Major players include Cargill Inc., Continental Grain Company, Louis Dreyfus Corporation, Bunge and Born Corporation, Mistsui and Company/Cook Industries Inc., and Garnac Grain Company. For more details on these companies and other contenders in the grains trade, see Gilmore (1982).
4. The program yield was determined by averaging past yields on each farm. However, the yield was frozen in 1985.

References

Blandford, D., W. Meyers and N. Schwartz. "The Macroeconomy and the Limits to U.S. Farm Policy." *Food Policy* 13 (1988): 134-39.

Erickson, M.H. and K. Collins. "Effectiveness of Acreage Reduction Programs." In *Agricultural-Food Policy Review: Commodity Program Perspectives.* U.S. Department of Agriculture, Economic Research Service. Agricultural Economic Report

No. 530. Washington D.C., 1985.

Gardner, B.L. *The Economics of Agricultural Policies.* New York: Macmillan Publishing Company, 1987.

Gilmore, R. *A Poor Harvest; The Clash of Policies and Interests in the Grain Trade.* New York: Longman, 1982.

Hanthorn, M. and J.W. Glauber. *An Assessment of Marketing Loan Program Options.* Agricultural Economic Report No. 581. U.S. Department of Agriculture, Economic Research Service. Washington, D.C., 1987.

Heid, W.G. Jr. and M.N. Leath. *U.S. Barley Industry.* AER-432. U.S. Department of Agriculture, Economics, Statistics, and Cooperatives Service. Washington D.C., 1978.

Johnson, D.G. *World Agriculture in Disarray.* London: Trade Policy Research Center, 1973.

Organization for Economic Cooperation and Development. *Agricultural Policies, Markets and Trade: Monitoring and Outlook.* Paris, 1988 and 1989.

United States Department of Agriculture, Economic Research Service. *Embargoes, Surplus Disposal and U.S. Agriculture.* Agricultural Economic Report No. 564. Washington D.C., 1986.

_____. *Estimates of Producer Subsidy Equivalents: Government Intervention in Agriculture, 1982-86.* ERS Staff Report AGES880127. Washington, D.C., 1988.

_____. *Feed: Outlook and Situation Report.* Washington, D.C., Various issues.

_____. *Rice: Outlook and Situation Report.* Washington, D.C., Various issues.

_____. *Wheat: Outlook and Situation Report.* Washington, D.C., Various issues.

_____. *The Basic Mechanisms of U.S. Farm Policy: How They Work, with Examples and Illustrations.* Miscellaneous Publication No. 1479. Washington D.C., 1990.

United States Department of Agriculture, National Agricultural Statistics Service. *Crop Production: 1989 Summary.* Washington D.C., 1990.

Wright, B.D. "Commodity Market Stabilization in Farm Programs." In B.L. Gardner (ed.), *U.S. Agricultural Policy: the 1985 Farm Legislation.* Washington D.C.: American Enterprise Institute, 1985.

6

Cameroon

David Blandford and Sarah Lynch

The Republic of Cameroon is a coastal country roughly 475 thousand km² in size located in west-central Africa (Figure 6.1). It is bordered by Equatorial Guinea, Gabon and the Congo to the south, Chad and the Central African Republic to the north and east, and Nigeria to the west. The climate and local ecologies of the country are very diverse. As rainfall declines from south to north, the landscape changes from tropical rain forest, to savannah, and finally to semi-desert.

The population of Cameroon stands at roughly 11 million and is increasing at over 3 percent per year. The country is officially bilingual (French/English), but only 20 percent of its inhabitants live in the English-speaking Northwest and Southwest provinces (Figure 6.1). There are more than 200 ethnic groups in Cameroon, each with their own language, culture, and customs. Cameroon is rapidly urbanizing. Between 35 and 40 percent of the population now lives in urban areas, and this will reach 50 percent by the turn of the century if current trends continue. Douala, the main coastal commercial city, already has a population of over 1 million, while Yaoundé, the administrative capital 270 kilometers to the east of Douala, has approximately 750 thousand inhabitants. Rural population densities vary considerably. In the agriculturally rich areas of western Cameroon and the fertile river valleys in the extreme north, population densities are high, reaching 60 to 95 persons per km². In the southern forest zone and northern arid zones, population densities are as low as 4 persons per km². Urbanization and rapid population growth in neighboring countries has generated a growth in demand for food. Poor environmental conditions, war, and natural disasters have limited the ability of Cameroon's neighbors to meet domestic food demand. This increase in demand is exerting considerable pressure on Cameroon's food supplies in general and, in particular, on the country's grain system.

Cameroon's economy is dominated by agricultural production, both of food

FIGURE 6.1 Map of Cameroon

and export products, although the discovery of oil in the early 1970s had a major impact on the economic structure. Prior to the exploitation of oil reserves in the late 1970s, the economy grew at an average annual rate of roughly 5 percent. Agricultural exports, particularly cocoa and coffee plus forestry products, accounted for 85-90 percent of the country's exports. As the flow of oil rose in the early 1980s, the rate of growth accelerated to over 10 percent per year. In 1984/85, the peak year for oil revenues, petroleum accounted for 66 percent of exports and 20 percent of GDP.[1] The oil boom has now ended because of the decline in world oil prices and the depreciation of the U.S. dollar (in which exports are priced) against the Cameroonian currency, the CFA franc (CFAF).[2] The value of oil exports fell by almost 70 percent between 1984/85 and 1987/88. Real GDP has been falling since 1986/87, and the government's budget deficit increased from a modest 1 percent of GDP in 1985/86 to over 5 percent in 1986/87 and 1987/88. There is a liquidity crisis in the banking system because of the decline in oil revenues. Cameroon, aided by the International Monetary Fund and the World Bank, is engaged in a structural adjustment program to restore economic equilibrium.

Agriculture still accounts for 70 percent of domestic employment and 33 percent of gross domestic product. Despite its recent economic setbacks, Cameroon is one of the wealthier countries in Sub-Saharan Africa. Relative political stability under a one-party system of government has helped to maintain economic stability. Monetary discipline, generated by the country's membership in the Franc zone, has been complimented by fiscally conservative government policies. Efforts were made to invest the windfall income generated by the oil industry in the development of a more broadly based economy, rather than simply allowing consumption to increase. As a result of these policies, in 1986, per capita GNP was over 900 US dollars, placing Cameroon above countries such as Thailand, Nigeria, and Egypt (World Bank, 1988).

Because of its climatic diversity, Cameroon has a broad agricultural base. Animal health problems, such as the prevalence of the tsetse fly in the wetter parts of the country, limit the production of cattle and goats to the north and certain parts of the northwest. Poultry are found throughout the country, especially in the south and west. As incomes have grown, livestock products, particularly poultry meat, have become an increasingly important part of the diet, especially in urban areas. Export crops, such as cocoa, coffee, rubber and oil palm, are grown in the south and west by smallholders and on plantations, and cotton is planted on smallholdings in the north. Grains—maize, rice, sorghum, and millet—are staple foodcrops, but nongrain crops, particularly root crops, are also important, especially in the south. At present, about 10 percent of the total cropped area and 20 percent of the foodcrop area (including root crops) is planted to grains. These figures indicate that although grains are an important part of the food production system, other staples such as root crops (cassava, yams and cocoyams), banana, and plantain are also significant.

Production: Diversity and Dynamism

The main grains produced in Cameroon are maize, millet, rice, and sorghum. Attempts have been made to grow wheat commercially but with little success because of the hot climate. Foodcrop statistics are limited and unreliable, although an agricultural census in 1984 (not yet published officially) made more information available. Sample surveys using the census methodology have been conducted each year since the 1984 census, but much of the data are not yet processed. Prior to 1984, foodcrop production figures were approximated using area estimates taken from the 1972 census. The recent census figures suggest that these estimates are inaccurate. As a result, there are no reliable time series data on cereal production.

Cameroon's farms are typically divided into "traditional" (small holder) and "modern" (plantation) sectors. The traditional sector produces roughly 90 percent of Cameroon's agricultural output on 93 percent of the cultivated land. It is responsible for virtually all foodcrop production (except for rice). Most traditional farms also produce export crops: cocoa or tobacco in the south, coffee in the west, and cotton in the north. Foodcrops are typically produced for home use, and any surplus is sold locally. In 1984, only 15 percent of Cameroon's farms had no off-farm sales of agricultural products. Historically, the production of foodcrops was the preserve of women, while men were responsible for export crops, although this strict division is eroding. In most areas, traditional farms are highly diversified, growing numerous crops in small plots and practicing intercropping. Average cultivated area per farm is approximately 1.7 hectares. There is some regional variability because of land productivity and population densities, with farms in the South and Southwest provinces averaging over 2 hectares, and farms in Adamoua and the Littoral averaging 1.1 to 1.3 hectares.

Maize is produced in every province in Cameroon (Table 6.1). However, from 1984 to 1987, about 70 percent of the total was produced in the West and Northwest provinces (Tables 6.1 and 6.2, Figure 6.1). The climate in these areas is most suited to maize production, and the crop is the major staple of the indigenous people. Millet and sorghum are the dominant grains in the north, while little grain is produced in the high rainfall areas of the south, particularly in the Southwest, Littoral, and South provinces. Southern farming systems are based on other crops, such as cassava, yams, and plantain, in place of foodgrains.

In the northern provinces (Extreme North, North, and Adamoua), where most of the sorghum/millet is produced, the principal crop is planted in May-June and harvested in November through January. A second, lower-yielding crop is sometimes transplanted from nursery beds in October and harvested in January-February. In 1983/84, about 80 percent of the sorghum/millet was produced in the first crop cycle. Average yields are approximately 700 kg per hectare in the first crop cycle (Table 6.2), and roughly half this in the second production cycle. The census data suggest that millet and sorghum production

TABLE 6.1 Maize and Sorghum/Millet Production and Sales by Province, 1984

	Farms harvesting (thousand)	Percentage of farms harvesting (percent)	Percentage of farms harvesting with sales (percent)	Average quantity sold per selling farm (kg)	Average sales revenue per selling farm (1,000 CFA francs)	Share of regional farm product revenues (percent)	Ratio of regional sales to production (percent)
MAIZE							
Extreme North	27.2	10	16	491	63.7	2	32
North	37.3	38	28	310	33.1	3	25
Adamaoua	35.0	63	50	1,327	104.8	20	54
East	53.7	81	53	194	23.0	4	21
Central	139.0	85	34	63	7.4	1	19
South	44.3	80	32	53	7.6	1	20
Littoral	54.8	84	28	79	9.3	1	18
Southwest	62.2	83	47	128	11.1	1	33
Northwest	128.1	97	44	646	60.1	12	21
West	150.7	95	31	355	22.0	5	15
Cameroon	732.3	63	37	354	32.2	4	23
SORGHUM/MILLET							
Extreme North	231.4	81.1	9	231	29.2	5	4
North	84.0	85.1	14	221	26.1	3	6
Adamaoua	14.7	26.4	60	643	51.5	5	25
East	a	a	a	a	a	a	a
Central	a	a	a	a	a	a	a
South	0	0	0	0	0	0	0
Littoral	0	0	0	0	0	0	0
Southwest	0	0	0	0	0	0	0
Northwest	2.7	2.0	26	1,486	50.6	0	75
West	a	a	a	a	a	a	a
Cameroon	334.9	29.0	13	334	33.3	1	7

Source: Unpublished Agricultural Census Data, Ministry of Agriculture, Cameroon.
aIncluded in national figures.

TABLE 6.2 Grain Production in Cameroon

	Farms planting (thousand)	Area planted ('000 ha)	Area planted per farm (ha)	Farms harvested (thousand)	Total production ('000 tons)	Average production per farm harvested (kg)	Average yield per hectare (tons)
Maize							
1984	811.8	205.7	0.25	732.3	408.7	558	2.0
1985	776.3	186.1	0.24	733.9	313.4	427	1.7
1986	797.9	194.9	0.24	745.5	367.2	493	1.9
1987	762.1	203.0	0.27	718.1	461.5	643	2.3
Sorghum/ millet							
1984	365.4	373.5	1.02	334.9	207.7	620	0.6
1985	372.1	487.1	1.31	353.3	369.8	1,047	0.8
1986	375.4	533.8	1.42	368.6	562.7	1,527	1.1
1987	366.0	357.4	0.98	342.4	252.5	737	0.7
Rice (traditional)							
1984	17.9	5.1	0.28	17.0	7.3	431	1.4
1985	23.5	9.9	0.42	21.9	18.9	863	1.9
1986	19.2	6.2	0.32	18.1	13.5	748	2.2
1987	12.0	4.8	0.40	10.1	9.2	912	1.9
	(number)	('000 ha)	('000 ha)	(number)	('000 tons)	('000 tons)	tons
Rice (modern)							
1984	2	18.3	9.2	2	59.4	29.7	3.2

Source: Unpublished Agricultural Census data, Ministry of Agriculture, Cameroon.
Note: Rice production figures are for hulled rice.

is relatively static in the north, but more area is being planted to maize. Livestock production, particularly cattle grazing, is also important in this region. Approximately 75 percent of all farms in the three northern provinces keep livestock, compared to an average of 50 percent for the country as a whole. Because of the reduced incidence of animal pests, in particular the tsetse fly, 60 percent of Cameroon's cattle, 55 percent of the goats, and 60 percent of the sheep are raised in the Extreme North province.

In the maize-producing areas of southern Cameroon, two crops can be produced when rainfall is sufficient: a main crop planted in March-April and harvested in July-August, and a second crop planted in August-September and harvested in December-January. In 1984, second-crop maize accounted for less than 5 percent of total production in the main maize-producing provinces of the West and Northwest provinces, but made up about one-third of total production in the other southern provinces (Southwest, Littoral, Center, South and East). No second-crop maize is produced in the drier north. A single crop of maize is usually planted in June and harvested in October. Little hybrid maize is used in Cameroon because the infrastructure required for the production and distribution of hybrids does not exist. However, several development projects, particularly in the West and Northwest provinces, are attempting to provide hybrid seeds along with extension advice on cultivation practices. Notwithstanding these regional efforts, production increases have primarily been achieved through an expansion in cultivated area, and only to a lesser extent through intensification of production on existing area through the use of fertilizer and improved seed.

Land preparation and harvesting of foodcrops on traditional farms are typically labor intensive. Generally speaking, men are responsible for land clearing and women for land preparation, planting, weeding, and harvesting. Disease problems limit the use of draught animals in all but the north of the country, and the hand hoe is commonly used to prepare and cultivate land. In 1984, 85 percent of the farms employed hand cultivation only, roughly 2 percent had tractors, and the rest used draught animals, primarily cattle. Harvesting is also accomplished with hand tools. Soil fertility is traditionally maintained through intercropping, rotation and fallowing, but approximately 45 percent of farmers now use some animal manures or chemical fertilizers on crops in general, and 23 percent of all farmers use fertilizers on foodcrops in particular. Most farmers purchase chemical fertilizer for use on cash crops, particularly export crops.

The traditional sector is changing because of a rising demand for food created by the growth in population and urbanization. The limits to increases in production through area expansion and productivity improvements under traditional farming methods are being reached, and the use of modern inputs, such as fertilizer and mechanization, is expanding. With the oil boom and the growth of urban areas, foodcrop production has become increasingly profitable, especially in areas with good transport links to the southern cities. As a result, some farms are adopting more specialized, market-oriented farming systems rather than the more traditional diversified subsistence production. Further, a favorable relative price ratio between foodcrops and export crops has encouraged the growth of foodcrop production at the expense of export crops. The same forces are contributing to a breakdown in the traditional sexual division of labor, as more men become involved in aspects of foodcrop production that were previously the domain of women.

The "modern" farm sector in Cameroon is composed of government and privately owned plantations, where products such as oil palm, rubber, sugar cane, and bananas are grown. These are either processed domestically or are exported. Also included are government cotton and irrigated rice projects. Rice production in the traditional sector, typically rainfed, accounted for roughly 11 percent of the total in 1984. Most of Cameroon's rice is grown in two large irrigation projects: Société d'expansion et de modernisation de la riziculture à Yagoua (The Company for the Expansion and Modernization of Rice Production in Yagoua—SEMRY) in the far north and the Upper Nun Valley Development Authority (UNVDA) in the northwest of the country. SEMRY is the larger of the two, producing over 90 percent of modern sector rice through double cropping. Both of these projects lease plots of irrigated land to farmers. Farmers are provided with a technical package that includes mechanized land preparation and harvesting, fertilizers, and extension advice. Farmers are required to sell paddy to the project, which mills and markets the processed rice. At the time of sale, farmers reimburse the project for production costs.

In northern and western Cameroon, some maize is produced on large-scale private farms linked to processors. One company, Maiserie du cameroun (The Maize Processors of Cameroon—MAISCAM), is using modern production and processing techniques to provide maize for human consumption, animal feed, and the brewing industry.

Because of the lack of accurate time series data, no statistical analyses of supply response for grains or other foodcrops in Cameroon are available. It appears that with a few exceptions, notably rice and wheat, domestic production has kept pace with the demand generated by the growth in population and incomes. The increasing monetization of the economy has forced women, who often have no other way to earn income, to produce more foodcrops to meet family cash expenses and nutritional needs. Food production has also been stimulated by changes in the relative price of foodcrops to export crops. For many years, agricultural exports have been taxed. During the period 1979-84, for example, producer taxes equivalent to 30-40 percent of the export price for cocoa and coffee were levied through the pricing policies of the Office national de la commercialisation des produits de base (National produce marketing board – ONCPB), which handles exports of cocoa, coffee, cotton, and groundnuts (USAID, 1988). This, coupled with the strong demand for food, caused foodcrop prices to rise relative to export prices. For example, it has been estimated that the price of maize relative to robusta coffee increased by roughly 40 percent between 1969-71 to 1980-82 (Prasad, 1987). Low levels of food imports indicate that producers have responded to these price signals by increasing the output of food products.

Consumption: The Pressures of Change

Staple food consumption is regionally diverse in Cameroon. In the higher rainfall zones near the coast and in the south, root crops, such as cassava, yams, and macabo/taro, and plantains, are the major staples. Maize, is an important staple foodcrop in the west and northwest, while millet and sorghum are eaten in the drier northern part of the country.

The importance of regional variation in foodcrop consumption is reflected in the data obtained from Cameroon's first national household expenditure

TABLE 6.3 Regional Food Expenditures

Regions	Mean per capita expenditures (CFA francs)			
	Nonfood	Food	Total	Cereals
Forest	67,121	81,462	148,583	8,179
High Plateau	53,608	82,385	135,993	14,077
Savanna and Steppes	33,578	136,716	170,294	58,287
Coast	70,212	81,911	152,123	7,578
Yaounde	329,056	125,408	454,464	16,052
Douala	240,278	123,102	363,380	17,571
All Cameroon	75,097	105,083	180,180	27,881

Regions	Share of mean per capita expenditures (percent)			
	Nonfood	Food	Total	Cereals
Forest	45	55	100	10
High Plateau	39	61	100	17
Savanna and Steppes	20	80	100	42
Coast	46	54	100	9
Yaounde	72	28	100	12
Douala	66	34	100	14
All Cameroon	42	58	100	27

Source: Unpublished data from the 1983/84 Household Expenditure Survey.
Note: The regions are based on agro-climatic characteristics. The Forest Region includes the provinces of Center, South, and East. The High Plateau includes the West and Northwest provinces and the Departments of Mungo and Mémé in the Littoral Province. The Savanna and steppes includes the Adamoua, North, and Extreme North Provinces, and the Coast Region includes the Southwest Province and the remainder of the Littoral.

survey conducted in 1983/84. Results of this survey can be used to examine regional food expenditure patterns (Table 6.3). The data are divided into the major urban centers of Yaoundé and Douala plus four agro-ecological zones: Forest, which includes the Center, East, and South provinces; High Plateau, the West and Northwest provinces and the departments (counties) of Moungo and Mémé in the Littoral province; Savanna and Steppes, which includes the Extreme North, North and Adamoua; and the Coast, which is comprised of the province of the Southwest and the remainder of the Littoral.

Mean per capita expenditures on food include monetary expenditures as well as the value of home production consumed by the household. Beverages and food consumed away from home are included in the nonfood category. Table 6.3 shows that, for the country as a whole, food accounts for 58 percent of the value of total expenditures, while expenditures on cereal and cereal products represent 27 percent of food expenditures. There are many regional differences; for example, in the higher-income urban areas of Yaoundé and Douala where all food is purchased in the market, the percentage of expenditures devoted to cereal (12 percent) and cereal products (14 percent), is lower than the national average. It is also lower in zones such as the Coast and Forest, (10 and 9 percent) where plantains, roots, and tubers take the place of cereals as the primary staples. Cereal expenditures in the north (Savanna and Steppes zone), accounting for 42 percent of food expenditures, are particularly high because grains, especially millet and sorghum, are the major staple in this drought-prone region. Cereal consumption, particularly maize, is also very important in the Hauts Plateaux, accounting for 17 percent of total food consumption.

Some of the regional differences in patterns of grain consumption are weakening as transportation improves, modern services spread (particularly electricity), and tastes and preferences shift with urbanization and growth in incomes. The most important change is the growing preference for bread made from imported wheat and wheat flour, and for rice obtained from domestic and foreign sources.

Bread consumption in Cameroon has risen dramatically during the last two decades, increasing by roughly 4 percent per year per capita. Once largely eaten by urban elites and expatriates, bread is rapidly becoming a major staple. In 1972, for example, Cameroon had a population of 6 million, and there were 68 bakeries (boulangeries) with an annual output of about 40 thousand tons of bread. By 1988, with a population of 11 million, the number of bakeries rose to over 400, and the output of bread had quadrupled. Average annual per capita consumption is now roughly 25 kilograms per person. There are three principal reasons for the rise in bread consumption. First, bread is a convenience food, particularly in urban areas where time for food preparation shrinks when more members of the household, especially women, join the full-time labor force. Many traditional staples, such as cassava, involve substantial

TABLE 6.4 All Food and Bread Price Indices, 1975/76=100

	All food	Bread
1975/76	100	100
1976/77	116	100
1977/78	136	100
1978/79	149	110
1979/80	156	120
1980/81	175	146
1981/82	205	159
1982/83	238	164
1983/84	269	164
1984/85	259	170

Source: GICAM, 1988.

preparation and cooking time. Second, bread is relatively inexpensive. Data in Table 6.4 indicate that since 1975/76, the price of bread rose at less than half the rate of food prices in general. Third, rural electrification made the establishment of bakeries outside of large cities practical and led to growth in demand outside urban areas.

Convenience in preparation, as well as changes in consumer preferences, are major factors in the increasing consumption of rice in Cameroon. Rice is not as consumer-ready as bread, but still takes much less time to prepare than many other traditional staple foods. Also, rice easily substitutes for the starchy staples (cassava, yams, macabo/taro) used in most traditional dishes. In addition, as in the case of bread, in real terms the price of rice trended downward during the last decade, while the price of traditional staples remained constant.

Reliable estimates of price and income elasticities of demand for rice and bread are not available, but it appears that per capita consumption of nontraditional grains is highly responsive to increasing incomes.

The brewing industry in Cameroon also uses significant quantities of domestic and imported grains. Beer is made primarily from malting barley, which is imported from Europe, and from maize, which is imported or purchased domestically. The rapid growth in consumer income in the early 1980s led to a substantial expansion in the brewing industry. However, when the rate of economic growth declined with the fall in petroleum earnings in the mid-1980s, beer consumption also fell. In 1986, the peak year, total consumption was 5.6 million hectoliters; equivalent to roughly 150 thousand tons of grain.[3] In 1989, the industry estimates that consumption will not exceed 4.8 million hectoliters (equivalent to less than 130 thousand tons of grain).

Grains are also used in increasing amounts in the livestock feeding industry. The growth in consumer income has resulted in more demand for meat and livestock products, particularly chicken and eggs. Feed mills that serve the poultry industry are centered in the coastal region of the South where the main urban consuming centers are located, the transportation network is well developed, there is easy access to ports for imported ingredients, and where the flour milling industry provides a ready source of byproducts for use in feedstuffs. Feed mills use domestically produced maize and bran from flour mills. In 1988, an estimated 39,600 tons of maize and 9,600 tons of bran were used in poultry feeds. Although brewers' grains are also available, the feed industry does not make use of them because there is no established system to collect, dry, and distribute this byproduct.

Trade: Growth of Rice and Wheat Imports

Cameroon is largely self-sufficient in most foodstuffs, importing mainly rice, wheat, and meat. Growing consumer preferences for staple foods that are quicker to prepare have resulted in increasing imports of wheat, wheat flour, and rice. In addition, changes in relative prices have encouraged consumers to substitute bread and rice for traditional staples. Even though rice is domestically produced, imports rose because consumers prefer imported rice, which is thought to be of higher quality than domestically grown rice and is substantially cheaper. Most domestically grown rice is produced in government irrigation projects in the extreme north, which makes transport costs to the relatively more urban south quite expensive. Price is an important factor in the competitive position of domestic rice production and is discussed in more detail below. During the oil boom, imports of grain-based products, particularly poultry meat, surged, but with the economic recession and government restrictions aimed at saving foreign exchange, these have since fallen.

The foreign trade statistics of Cameroon are subject to substantial inaccuracy because of the substantial underreporting of trade caused by inefficiencies in the bureaucracy, fraud, and smuggling. Sometimes there are strong incentives for importers to avoid paying import tariffs. This leads to the underreporting of imports. Another contributing factor is unreported trade between Cameroon and neighboring countries, particularly Nigeria and Chad. Important trade routes have existed between these countries for hundreds of years. National boundaries established by the colonial powers in the late 1800s arbitrarily divided tribes and villages into different nations but did not erase commercial linkages. Thus, drought, war, exchange rate fluctuations, and government policies in neighboring countries affect cross-border trade. Official trade statistics do not capture much of this trade, and goods that are officially imported for trans-shipment to other countries (e.g., Chad) may fail to be delivered. These factors mean that the official statistics on Cameroon's international trade in grain are often dubious. Informal conversations with grain traders lead the authors to conclude that the wheat and wheat flour

TABLE 6.5 Grain Imports and Exports (tons)

	Average 1977/78-80/81	Average 1981/82-84/85	1985/86	1986/87	1987/88
IMPORTS					
Wheat	65,171	60,739	13,720	29,650	48,650
Wheat flour	27,195	19,637	32,830	91,247	149,592
Rice	25,152	34,332	32,810	58,710	79,787
Other cereals	1,304	2,224	1,010	11,998	1,998
Other flour[a]	5,543	11,736	4,931	991	22,247
EXPORTS					
Wheat flour	608	2,055	428	1,052	4,306
Rice	2,156	8,544	2,893	1,692	2,169
Other cereals	79	5,677	415	263	263

Source: Ministry of Planning, Department of Statisitics and National Accounts.
[a]Primarily semolina.

import figures are relatively accurate, but that rice imports are substantially underreported because of the financial returns from evading import duties. With these caveats, some broad conclusions can be drawn from the official data.

The major imported grains in Cameroon are wheat, which is brought in either as grain or flour, and rice (Table 6.5). Small amounts of hard wheat are imported, but most imports are wheat flour or soft wheat used in the production of French-style bread. In recent years, the import mix has shifted from wheat grain to wheat flour, primarily because imported flour is cheaper than domestically milled flour. The reasons for this are discussed below. Some barley malt is imported for use by breweries. In the past, maize was imported for brewing because of difficulties in assembling sufficient quantities from smallholders and an unreliable domestic supply. Maize imports are currently prohibited in order to save foreign exchange and to promote the use of domestically produced maize in the brewing industry. The private company, MAISCAM, referred to earlier, is attempting to meet the industrial demand for maize and maize products by producing on its own properties in the North and by purchasing additional grain from smallholders. Other grains and grain products, such as millet and sorghum, and semolina are imported under food aid programs for re-export to other countries, particularly Chad, where civil war has reduced agricultural production. All grains are imported by private individuals or companies into Cameroon under a licensing system operated by the Ministère de Commerce et de l'Industrie (Ministry of Commerce—MINDIC). The principal source of wheat and wheat flour imports is France. Rice comes primarily from Thailand and Pakistan.

As mentioned above, there are substantial but unknown amounts of maize, rice, millet, and sorghum either exported or re-exported to neighboring countries. The amounts vary with the exchange rate and supply conditions in neighboring countries. Grain exports are also subject to licensing restrictions. Sometimes exports take place under licence. For example, the rice marketing agency, SEMRY, uses exports to dispose of rice that it cannot sell in the domestic market. Most of the time, exports occur without a license and are not reflected in trade statistics. As an indication of the possible magnitude involved, in 1987 MINDIC issued licenses for 40 thousand tons of rice imports. Conversations with private traders lead the authors to conclude that actual imports were as high as 240 thousand tons. Domestic consumption was probably about 99 to 130 thousand tons. The difference between these consumption and import estimates gives a rough idea of the volume of private and unlicensed trade in rice with the neighboring countries of Chad, Nigeria, and the Central African Republic.

Marketing: Traditional and Modern Systems

The marketing of foodstuffs in Cameroon is dominated by small-scale private traders. The most recent comprehensive survey of the food marketing system available (GFS-Midas, 1986) estimates that over 90 percent of foodcrop producers in Cameroon market at least part of their own produce. Producer/retailers are dominantly female and deal in small quantities of produce. Some only market their own produce, while others buy small quantities from other farmers. This group of sellers, known as "buyam sellams," typically does not own any means of transport. Farmer/retailers represent about 60 percent of the agents supplying foodstuffs to the capital city and are important in the marketing of domestically produced grains, particular sorghum and maize. They are less important in the marketing of domestically produced rice and imported rice and wheat.

There are no uniform grades or standards for food products in Cameroon, and only meat is inspected. There are regulations on weights and measures, but these are not used in most retail markets (Ayissi, et al., 1988).

The major source of maize and sorghum for urban and nonproducing areas is the marketed surplus of small farmers. Only a small percentage of total production enters the marketplace because the vast majority is consumed by producing households. Twenty-three percent of the country's maize production and 7 percent of its sorghum production is marketed (Table 6.1), although these percentages vary considerably by province. In general, a higher percentage of maize is grown for sale than sorghum. Approximately 37 percent of all farms harvesting maize sold some for cash (Table 6.1), compared to 13 percent of farms producing sorghum. Producers sell surplus grains either directly to consumers, or to intermediaries who assemble and transport the grain and sell it to other intermediaries, retailers, or consumers. There is virtually no commercial market for sorghum outside the north, and most of this is in the few urban centers. Most sales are made directly to consumers. Thus, the market for grains, especially sorghum, is thin.

According to survey data, sorghum is transported an average of only 30 kilometers from production to consumption centers (GFS-Midas, 1986). Wholesalers handle roughly 50 percent of the marketed maize crop, and the average transportation distance is 75 kilometers. Post-harvest losses, due to inadequate storage and handling, are significant for both crops. Estimates range as high as 80 percent for sorghum and from 25 to 50 percent for maize.

Food wholesalers, who are the next stage in the marketing chain, are legally supposed to have fixed business premises and are required to handle minimum quantities of produce. These provisions exist so that the government can regulate the activities of wholesalers and food prices, but the legal provisions are widely ignored (Ayissi, et al., 1988). About 60 percent of all wholesalers are unregistered (GFS-Midas, 1986) and, as in the case of foreign trade, this results in a substantial proportion of grain being handled outside of government regulations. Food wholesalers are most active in the urban areas of the south, handling both domestic and imported grains. However, they frequently handle other food and nonfood products and engage in other activities such as transportation. Wholesalers deal, on average, with two types of food products. For over 50 percent, the major food product handled is cereals; however, this is more likely to be rice or wheat flour than domestic maize (GFS-Midas, 1986). The proportion of the total volume of cereals marketed by each type of marketing agent is not known, but producers and small-scale marketing intermediaries are an extremely important component of the marketing system for sorghum and maize in Cameroon.

Imports and large-scale production make the rice and wheat marketing systems different from those for sorghum and maize. The rice marketing system is broadly similar in structure to that for wheat and is covered in the policy section below. As indicated above, there has been strong growth in the consumption of wheat, and the wheat marketing system has become an increasingly important factor in the food supply of the country. Imports of wheat grain are made by a parastatal organization, the Société camerounaise des minoteries (Cameroon Milling Company — SCM) (Figure 6.2), which mills the grain in its own mill in the port city of Douala. SCM's milling capacity, however, is not sufficient to satisfy domestic demand for wheat flour, and the balance of demand is imported. A second mill is currently being constructed by a private company and is scheduled to be in production during 1990. Unlike wheat grain, imports of wheat flour are made by private importers/wholesalers under licenses issued by the Ministry of Commerce. The process of allocating licenses is extremely politically sensitive. The Ministry of Commerce does not inform SCM or make public the amount allocated nor the names of the recipients of import licenses. The authors were told by individuals involved in

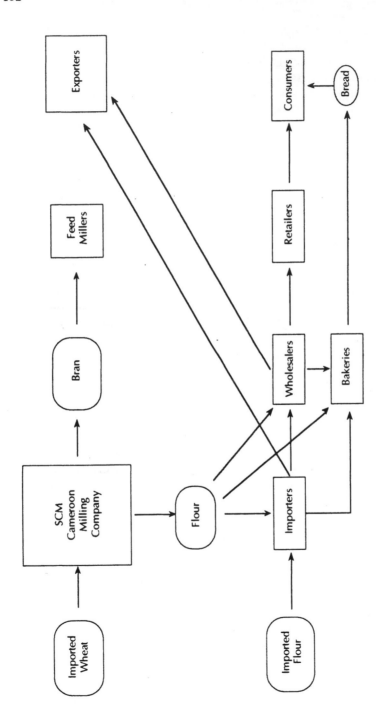

FIGURE 6.2 The Wheat Marketing System in Cameroon

the system that licenses are sometimes issued to individuals who may not import the grain. This is because the licenses have value to those who actually wish to import grain and can be sold by the license holders to these individuals or companies for a profit. The lack of communication between wheat and wheat flour importers and the SCM has resulted in irregular supplies. Sometimes too much wheat flour is available on the domestic market, and at other times there can be shortages.

Imported wheat flour is generally less expensive then domestically produced wheat flour, partly, it is alleged by the government, because imported flour is subsidized by suppliers such as the European Community. To assure that the SCM can sell its more expensive wheat flour, the government operates a system called "jumelage," which requires importers to purchase an amount of domestically milled flour equivalent to a percentage of the licensed import quantity. The purchase requirement is not always enforced, however, particularly if the government feels that supplies are likely to become scarce, resulting in ready markets for SCM flour without government intervention. In addition, importers sometimes ignore the requirement because sanctions are rarely enforced. As in other aspects of government regulation in Cameroon, the legal controls that exist have only a limited effect upon the behavior of the grain marketing system.

The number of importers varies with market conditions, but typically numbers no more than fifteen. Importers sell to wholesalers, who in turn sell to bakeries and retail outlets. Some importers also wholesale. Wholesalers, retailers, and other traders export unofficially to neighboring countries, such as Chad and Nigeria, depending on market conditions (see section on trade).

Importers prefer to purchase imported rather than domestic flour because it is considerably cheaper, and it believed to be of higher quality. For example, even with a 30 percent import tariff, in mid-1989, imported flour was about 15 percent cheaper than domestically milled flour. The government sets a maximum retail price for flour and bread and this is a further incentive to seek the lowest cost supplies.

The system for importing and distributing rice is very similar to that for wheat and wheat flour, except that rice is also domestically produced. SEMRY, a parastatal organization, manages production, storage, milling, and distribution of domestic rice. As indicated earlier, most of the domestically produced rice originates from government irrigation projects in the north. Rice importers are supposed to purchase an amount equivalent to roughly 30 percent of their import license from SEMRY. As in the case of wheat flour, however, imported milled rice is preferred by importers to domestic rice because of price and quality differences. SEMRY rice is not competitive with imported rice because of high costs of production and transport, and poor quality. In 1988, producers were paid 78 CFAF/kg for paddy, which translated into an ex-mill price of 130 CFAF/kg. It costs about 30-50 CFAF/kg to transport rice from the far north to

TABLE 6.6 Major Grain Policy Instruments in Cameroon

Instrument	Millet/sorghum	Maize	Rice	Wheat
PRODUCTION/CONSUMPTION				
Producer guaranteed price			X	
Deficiency payments				
Government purchases	X[a]		X	
Production quota				
Input subsidies				
— credit		X	X	
— fertilizer/pesticide		X	X	
— irrigation			X	
— machinery/fuel			X	
— seed		X		
Crop insurance				
Controlled consumer price	X[a]	X[a]	X[a]	X[a]
TRADE				
Imports				
— tariff		X	X[b]	X[b]
— quota			X[c]	X[c]
— subsidies				
— licensing		X	X	X
— state trading				
Exports				
— taxes				
— restrictions				
— subsidies				
— licensing		X	X	X
— state trading				
OTHER				
Marketing subsidies				
— storage	X[a]		X	
— transport	X[a]			
— processing			X	
State marketing	X[a]		X	

[a]Largely inoperative or with limited effect.
[b]Plus additional stabilization fund taxes.
[c]Parallel purchase requirement for domestic products.

the population centers in the south. After adding the government-allowed profit margin, SEMRY rice costs roughly 40 percent more in the south of the country than imported rice, which in 1989 sold at a retail price of 110 CFAF/kg. Because of the difficulty of finding a market, SEMRY rice is often stored in paddy form for substantial periods of time, resulting in a drier, less-glutinous rice, which consumers do not like. Old and inefficient milling equipment also causes a large proportion (frequently over 50 percent) of broken grains. Estimates obtained by the authors from traders suggest that less than 25 percent of domestic rice is of equivalent quality to imports. As in the case of flour, domestic retail prices are controlled; however, the controlled prices do not provide a profitable margin to importers. This is one of the major reasons why there are such substantial unrecorded imports of rice. Importers find it more profitable to smuggle rice through the port of Douala, rather than pay import duties and be forced to buy domestically produced rice.

Policies: Form Versus Substance

Cameroon, like most other countries, maintains a variety of institutions and laws that affect the food system (Table 6.6). Both production- and consumption-oriented policies are used to promote food self-sufficiency and to ensure "fair" producer and consumer prices. There are a myriad of laws and regulations that permit the state to intervene in most aspects of food production, marketing and transportation, but these laws are seldom enforced. The reasons for non-enforcement are complex, but include inefficiency and corruption, as well as impracticality. In many cases, the government lacks the institutional capacity to enforce existing laws and regulations. Government institutions and regulations primarily influence modern sector production and distribution of rice, bread, sugar, table oils, meat and imported food, but have only a limited effect on traditionally produced crops, because of the difficulties associated with enforcement. For most commodities, therefore, supply and demand determine prices within a private, competitive system dominated by a large number of small traders.

Consumer Price Stability

A devastating drought that affected most of West Africa in the early 1970s provided the impetus in Cameroon for the development of policies to control food prices. Laws were introduced that gave the state the power to intervene in price determination and the sale and transport of food and nonfood commodities. The motivation for these laws was threefold: first, to stabilize prices; second, to prevent alleged speculation and unfair trading practices by middlemen; and third, to prevent local shortages during drought. Legislation, enacted in October 1972 and modified in subsequent years, gives the Ministry of Commerce and Industry the power to regulate the marketing system and set prices for all goods in the economy, including all food products. These laws exist at the national, provincial, and departmental (county) levels.

At the provincial and departmental levels, local authorities are authorized to regulate prices of locally produced goods by establishing maximum and minimum prices (Table 6.6). Legal sanctions exist to punish wholesalers and retailers who fail to respect these prices, but these are rarely applied. Local authorities can theoretically manage the flow of products by issuing "permits de collection" (collection permits), which a marketer must possess to transport food products from one region to another. At the national level, MINDIC can set prices and regulate inter-provincial flows of food products by issuing food product export licenses (authorization d'exporter des vivres), although, as indicated earlier, substantial unlicensed export trade takes place. In addition, MINDIC sets the allowable cost structure and profit margin for a number of imported and locally produced goods.

Despite this array of rules and regulations, with the exception of rice, bread, beer, table oil, sugar and meat, effective government intervention and control of the food system is minimal. Continued growth in smallholder production has kept Cameroon relatively food self-sufficient, and has limited upward pressure on prices. Consequently, there has been little incentive to exercise greater control over the food system. Further, with the exception of some areas in the North, Cameroon has not experienced food shortages since 1984, lessening concern about food security. The large number of producers/ retailers in open air markets makes it almost impossible to implement price controls for most commodities. Most marketers find it easier, and presumably cheaper, to pay the bribes necessary to avoid obeying the rules. Long frontiers and limited government personnel further reduce the ability of the state to implement regulations. Only food commodities that are processed in large-scale plants, such as beer and table oil, are affected by price controls.

In the 1970s, two government food marketing agencies were created to help stabilize prices, especially in urban areas and the drought-prone north. The government established the Mission de développement des cultures vivriers, maraîchères et frutières (The Mission for the Development of Food, Vegetable, and Fruit Crops—MIDEVIV) in 1973 and Office céréalier (The Cereals Office—OC) in 1974. MIDEVIV has a broad mandate to improve food production and marketing. Its responsibilities include conducting food studies, extension, improving foodcrop market infrastructure, facilitating the development of food production areas ("Green Belts") around large urban areas, and marketing selected foodcrops in larger urban areas, particularly the capital. The Cereals Office was organized because of food supply problems in the northern provinces created by drought. The main objective of the OC is to stabilize prices of cereals produced in the north, primarily millet, sorghum, and rice. OC is charged with: (a) purchasing grains at harvest time and reselling later to stabilize consumer prices; (b) distributing certified varieties of seeds; and (c) maintaining stocks to cushion inter-seasonal variations in supply. The OC was not awarded a monopoly over grain purchases and sales in the North, but it was anticipated that once in operation it would control 10-15 percent of the region's cereal market.

Neither MIDEVIV or OC achieved their objectives. Their market share never reached 10 percent of the volume of cereals sold in the areas in which they operate. Both organizations were plagued by waste, poor grain quality because of inadequate storage and handling, and high operating costs. In the case of OC, the millet and sorghum market is very "thin," and only a small percentage of total production is marketed (Table 6.1) Poor management and a lack of financial control exacerbated the problems. Unofficial estimates suggest that the government provided 700 million CFAF per year in subsidies to these organizations. Further, OC and MIDEVIV were unable to compete with private traders on price, quality, or quantity. For example, in 1982/83, the price of a 100 kg bag of domestic rice sold by OC was 14,500 CFAF, while it sold for 13,678 CFAF in the open market (Essama, 1984). MIDEVIV and OC activities are currently being evaluated as part of a comprehensive government review of parastatal operation. It is likely that this evaluation will result in MIDEVIV's liquidation and substantial changes in OC's structure.

Increasing Domestic Production and Achieving Self-Sufficiency

After the drought in the early 1970s, the Cameroon government initiated a number of actions to promote food self-sufficiency. In an attempt to expand domestic production of imported grains, such as rice and wheat, large-scale, government-controlled production schemes were launched (such as the rice projects in the north). More recently, in certain provinces regional development authorities responsible for the general development of both food and export crop production were established. Both approaches incorporated the expansion of agricultural research activities, the use of input subsidies and production credits, and improvements in extension, seed multiplication, and marketing infrastructures.

Until the late 1970s, little research was done on foodcrops. Instead, emphasis was placed on tropical export crops such as cocoa, coffee, cotton, and rubber. In the late 1970s, the government launched several new projects to increase research on traditional foodcrops, as well as on rice and wheat. One of the largest of these is the National Cereals Research and Extension Project mandated to conduct agronomic research on all major cereals grown in Cameroon. While these new research efforts resulted in limited advances, no major breakthroughs in terms of improved varieties or production practices occurred for a number of reasons. Because Cameroon is so climatically diverse, research results cannot be uniformly applied throughout the country. Seed multiplication systems do not function reliably except in certain limited areas, forcing farmers to depend primarily on their own stocks of local seed varieties. The extension service, which is not linked to the research establishment (exceptions being those agents working with regional development authorities), lacks both a viable technical package to extend to farmers and trained personnel and equipment. Finally, poor marketing infrastructure, a lack of on-farm storage, and large post-harvest losses reduce the incentives for producers to adopt new production technologies, despite favorable grain prices. Research and extension alone cannot rectify these problems; they also require long term public and private investment.

Delivering chemical inputs has been a particular problem. In the 1970s, the government invested heavily in the Société camerounaise des engrais (Cameroon Fertilizer Company—SOCAME) to produce agricultural chemicals domestically. Plagued by technical and management problems, and heavily in debt, the factory closed in 1981. All chemical inputs are now imported. In 1984/85, total use of chemical fertilizers was 105 thousand tons, of which 64 thousand tons were distributed at subsidized prices. The government distribution system handles about 60 percent of the total fertilizer used, parastatals and cooperatives account for roughly 30 percent, and the private sector distributes the remaining 10 percent. Demand usually exceeds supply. Transportation and coordination difficulties sometimes result in supplies arriving late or not at all. Often, the wrong kind of fertilizer is delivered. Under a program funded by the U.S. Agency for International Development (USAID), the government is phasing out subsidies by 1991 and will leave the marketing of chemical inputs entirely to the private sector (Steedman, 1988).

Historically, the government provided input subsidies to smallholders for fertilizers, pesticides, and improved seeds primarily to produce cocoa, coffee and cotton, the major export crops. Only recently, under the auspices of the regional development projects, were limited input subsidies made directly available to foodcrop producers. Given the favorable price ratio between foodcrops and export crops, some fertilizer may have been diverted by farmers to foodcrops, including maize. However, little fertilizer or other modern inputs are used in the production of millet and sorghum.

Self-Sufficiency in Major Urban Staples

To ensure stable supplies and prices of important urban food staples, such as rice and wheat, and to save foreign exchange, the government intervenes in both supply and demand for these staples. In the 1970s, the government launched efforts to achieve self-sufficiency in rice and wheat production. In 1975, an attempt was made to introduce wheat production in northern Cameroon. The Société de développement du blé (The Wheat Production and Processing Company—SODEBLE) was formed to develop domestic production and processing capacity for wheat and, to a lesser extent, maize. SODEBLE was supposed to develop 18 thousand ha of land, but by 1987, only 500 ha of maize and 50 ha of wheat were under cultivation. As in the case of many other parastatal organizations in Cameroon, SODEBLE was plagued with technical, managerial, and financial problems that resulted in numerous reorganizations. Substantial government subsidies were required to keep the organization afloat. In 1988, based on the recommendations of the government's review of parastatals, SODEBLE was dissolved, and the wheat production effort abandoned.

To aid in the achievement of self-sufficiency in rice production, the government formed SEMRY, UNVDA, and the Société de développement de la riziculture dans la plaine de Mbo (Company for Rice Development in the Mbo Valley—SODERIM). Rice is grown under irrigation in these projects using the same organizational model. Each leases irrigated plots to local farmers, providing land preparation, extension assistance, and chemical fertilizer (ammonium sulfate and urea). Farmers are required to sell all output, minus their family requirements, to the project at harvest. From the proceeds, the farmer pays a charge to the project for land preparation, irrigation water, and other inputs. The project organizations mill the rice and sell it to wholesalers.

During the past two decades, the government made several attempts to protect domestic rice production from lower-cost imports. A "jumelage" system similar to the one for wheat and wheat flour was implemented in the late 1970s. This system required rice importers to buy a quantity of domestically produced rice equal to a fixed proportion of their import volume; usually 1 kg of domestic rice for every 2 kg of imported rice. As in the case of wheat, the purchase requirement was not always met by importers. Substantial problems began to emerge in the early 1980s when additional rice area was brought into production and double cropping was introduced, resulting in a five-fold increase in domestic rice production. At this point, SEMRY began to amass large stocks. There were few willing buyers because imported rice was more profitable, particularly if ways could be found to avoid the import duty and the requirement to buy domestic rice. During most of the 1980s, rice importers made significant profits because the price for imported rice was lower than the retail rice price.

In 1988, the jumelage system was replaced by a Price Equalization Fund (Caisse de péréquation) set up to impose a tax on rice imports equal to the difference between the price of imports (including tariff and port charges) and a fixed selling price for domestic rice. The tax receipts accrued from this system were to be placed in a fund used to subsidize domestic rice production. However, to date, this fund has not operated as intended. Ineffectual and corrupt implementation of the tax system, the government's decision to lower the tax from 44 CFAF per kg to 10 CFAF per kg, and the waiver or tolerated nonpayment of import taxes have rendered the fund inoperative (van de Walle, 1989).

In 1989, the Ministère de l'agriculture (Ministry of Agriculture—MINAGRI) announced that rice producers will no longer be obligated to sell their output to SEMRY, although they will continue to have access to inputs through the projects. Government planners hope that the private sector will be more successful in marketing rice than SEMRY.

Future success depends on the three projects overcoming their infrastructural and technical difficulties. At a minimum, production and transportation costs must be reduced for domestic production to compete with imports. At present, all three receive large government subsidies, while only SEMRY and, to a much lesser extent UNVDA, produce irrigated rice. The original intent was for farmers to pay the full cost of services and inputs provided, but in recent years the costs of technical assistance, land preparation, chemical inputs, and irrigation have been born increasingly by the government. In August 1987, about 80 thousand tons of paddy rice were in SEMRY storage facilities, an amount that would have required 13 months to mill at full capacity, as well as an additional one thousand tons of milled rice. Despite government subsidies approaching 8 billion CFAF, the organization's total debt was estimated at over 11 billion CFAF.

Using World Bank data, van de Walle (1989) estimates that the world price of rice will remain at about $US 250 per ton in the foreseeable future. At current exchange rates in Cameroon, this would be equivalent to a price of about 75 CFAF/kg of milled rice. If SEMRY rice is to compete, costs must be reduced by 40 percent. With the existing price relatives and given a double-cropping production maximum of 85 thousand tons, SEMRY would still generate a deficit of 5 billion CFAF. Even with substantial reforms, it will be almost impossible for SEMRY to reduce sufficiently the ex-mill price of domestic rice for it to be competitive with imported rice in the south. Also, it is not clear whether the Caisse de Péréquation will be able to perform its intended function.

Rice policy in Cameroon is at a critical stage. Importers are unhappy with the stabilization tax because it is an added cost that cannot all be passed on to consumers. In addition, a liquidity crisis in the banking sector created by the fall in the country's earnings from petroleum exports has reduced loan funds, forcing importers to self-finance or use high-cost, informal capital sources. As the situation now stands, importers cannot make a profit unless they smuggle rice into the country to avoid the border taxes. Fraud at the port, and reportedly in the MINDIC, contributes to the problem by reducing the incentive to implement meaningful reform. The current operation of the rice import system makes it difficult for honest importers to stay in business and does little to solve the problem of the lack of the country's competitiveness in rice production.

Conclusions

In many respects, Cameroon is one of the more economically successful countries in Africa. Blessed with abundant natural resources and relative social and political stability, the country's economy has grown rapidly. The exploitation of petroleum resources led to particularly rapid economic expansion during the early 1980s, which has now ended because of a fall in export earnings. Fiscally conservative economic policies and the monetary discipline of a fixed exchange rate also contributed to low inflation and economic stability, but the fixed exchange rate limits the government's ability to influence the country's competitive position internationally. Despite rapid population growth, domestic food production has kept pace with demand. Significant increases in the production of maize and more modest increases in cereal substitutes, such as plantain, manioc and yams, helped to maintain food price stability.

An attractive price ratio between export and foodcrops, particularly maize and plantain, also contributed to increases in food production. However, changes in consumer preferences and relative prices have resulted in a substantial increase in the consumption of wheat and rice, which cannot be produced competitively in Cameroon.

There are a large number of governmental institutions—ministries, parastatal organizations, and integrated development projects—involved in the food system in Cameroon. Many of these institutions have overlapping responsibilities and jurisdictions. This has created a complex, fragmented "official" system, which in many cases is costly and inefficient, and creates rent-seeking opportunities. The traditional foodcrop system, which provides most of Cameroon's food supplies, operates autonomously, largely outside of this framework. It consists of many participants moving a large volume of goods sold in open markets at prices determined by supply and demand. Attempts to control food prices have a limited effect because of the difficulties involved in influencing the traditional sector, and because the incentives to evade controls are too great. The food system in Cameroon has operated successfully without effective government controls, and government institutions have had little impact on food system performance.

Despite its successful record, the traditional system is under pressure because of the continued high growth in population and increasing urbanization. There seems to be no viable technological package on the horizon that will lead to substantial cost reductions and output increases in those grains, such as maize, in which Cameroon appears to have a comparative advantage. Continued rapid population growth, urbanization, and dietary changes are leading to growing demand for convenience foods, such as rice and wheat flour for bread, much of which is imported. Without reductions in production or transport costs or a devaluation of the CFA franc, domestically produced rice will not compete successfully with imported rice. The decline in overseas earnings from oil is leading to a renewed emphasis on agricultural exports to generate foreign exchange, which will increase competition for resources, particularly land and fertilizer, with domestic food production. This means that rice and wheat imports can only be made at the expense of other products. Steps could be taken to make domestic products more attractive to consumers by promoting the processing of maize and manioc into consumer-ready products; but the thinness of the markets for these commodities and the consequent supply uncertainty for processors, as well as the lack of suitable processing technologies, needs to be overcome. There is a continuing need to improve national infrastructure, particularly the transportation system, but this is difficult during the current period of financial stringency. The record suggests that the government will be more successful in overcoming pressures on its food system by helping to stimulate private sector activity rather than expanding the role of the public sector. This will require the government to actively implement meaningful reform, something that it has not been willing or able to do in the past. The challenge facing the country is to recognize the constraints that market forces impose upon food policy options, and to work with, rather than against, these forces.

Notes

1. Unless otherwise indicated, the split years used in this chapter are Cameroon July-June fiscal years.
2. Cameroon is part of the French franc zone (Communauté financière africaine). The CFA franc has been fixed in value against the French franc at the rate of 50 CFA francs = 1 French franc. In recent years, this has meant that 300-350 CFAF = 1 US dollar.
3. Based on estimates provided by the brewing industry.

References

Ayissi, M.J.P., S.J. Nkwain, and R.M. Njwe. *Food Marketing/Transportation System in Cameroon*. Report prepared for the U.S. Agency for International Development. Dschang, Cameroon, January 1988.

Essama Nssah, B. *Impact of Pricing and Related Policies on Agricultural Production in Cameroon*. Report prepared for the U.S. Agency for International Development. Yaoundé, Cameroon, 1984.

G.F.S.-Midas. *Etude de la Commercialisation des Produits Vivriers: Synthèse et Proposition. Rapport principal*. Yaoundé, Cameroon, Ministry of Commerce, 1986.

Groupement Interprofessionnel pour l'Etude et la Coordination des Intérêts Economiques au Cameroun (GICAM). *L'économie Camerounaise, Exercice 1986-87*. Bulletin GICAM No. 199, Douala/Yaoundé, Cameroon, 1988.

Ministère de l'Agriculture (MINAGRI). *1984 Agricultural Census*. Unpublished document. Yaoundé, Cameroun.

Prasad, R. *Performance and Potential of the Agricultural Sector*. Unpublished paper prepared for a World Bank review of the Cameroon Agricultural Sector. Washington D.C., April 1987.

Steedman, C. *Rural Development Planning and Budgeting in Cameroon*. Report prepared for the U.S. Agency for International Development and the World Bank. Development Alternatives Inc. Ann Arbor, Michigan. April 1988.

U.S. Agency for International Development. *Agricultural Sector Briefing Paper*. Yaoundé, Cameroon, January 1988.

van de Walle, N. *Rice Politics in Cameroon: State Commitment, Capability and Urban Bias*. Princeton University, unpublished manuscript. 1989.

World Bank. *World Development Report 1988*. New York: Oxford University Press, 1988.

7

Colombia

Alvaro Silva Carreño and Rodolfo Alvarado

The Republic of Colombia, a country of 29 million people located in the northwest corner of South America, is the oldest democracy in Latin America. The population has risen substantially from 12 million in the early 1950s, although the growth rate has begun to decline. Over the same period, the economy moved from a predominantly rural foundation with over 60 percent of the population residing in the countryside, to an urban base with more than 65 percent of the population now living in urban areas. Recent projections suggest that the urban population will rise to 70 percent of the total by the year 2000 (SAC, 1986a).

In spite of social conflict, since the mid-1950s the Colombian economy has expanded continuously in terms of output and employment (Thomas, 1985). A diversified agricultural sector met most of the rapidly growing population's demand for food and raw materials with a decreasing share of the labor force. In 1990, agriculture accounted for roughly 22 percent of gross domestic product (GDP), compared to 25 percent in 1970. Despite this relative decline, the agricultural sector still employs 28 percent of the labor force.

Coffee has dominated the Colombian agricultural economy since the beginning of the century. In 1990, it still accounted for 17.4 percent of the total value of agricultural output and 28 percent of the country's exports. The livestock (dairy, poultry, and pork) industry makes up 36 percent of agricultural GDP and is growing rapidly. Grains represent 17.3 percent of the value of agricultural production, but Colombia is not self-sufficient in grain products. In 1990, grain imports totalled almost US$ 102 million; 31 percent of agricultural imports and 1.4 percent of total imports (Ministry of Agriculture, 1988).

Grain Production in Colombia

Land Availability and Use

Colombia lies just north of the equator. The Andes mountains, extending along the Pacific Coast of South America, branch into three ranges while

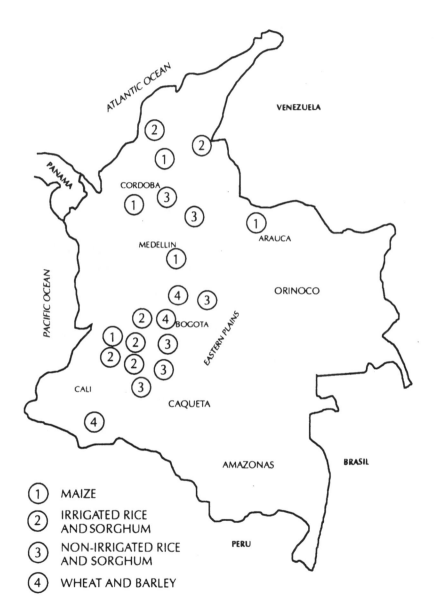

FIGURE 7.1 Grain production regions in Columbia

crossing the country, generating climatic diversity – from tropical conditions in the low valleys, to the mild "coffee zone belt" at middle altitudes, to cold mountain tops. Soils in the Andes are volcanic in origin and naturally fertile, and two rainy and two dry seasons allow year-round production (Figure 7.1). Only 14 million hectares of the country's land area of 112 million hectares are suitable for semiannual or plantation crops and another 19 million for grazing (Instituto Colombiano Agropecuario-ICA, 1984). Most of the remaining 79 million hectares are located in the hot and humid tropics of the Amazonas, Orinoco and the Pacific. This area is thickly forested and is home to a small and scattered population of Indian tribes and minority groups.

The Andes mountains are the center of national political and economic power, and most of the country's industrial and agricultural production takes place there. The mountains and their valleys cover only 26 percent of Colombia's surface area but generate 70 percent of agricultural GDP. Out of the 6.6 million hectares suitable for semiannual crop cultivation (including grains), 6.2 million are located in the tropical valleys and lowlands of the Andes (ICA, 1984). The hilly lands of the coffee zone belt include 7.4 million hectares appropriate for plantation crops (coffee is planted on one million hectares), but only 174 thousand hectares suitable for semiannual crops. The colder and higher altitude areas in the Andes mountains contain roughly 650 thousand hectares appropriate for agricultural production, but just 220 thousand support semiannual crops (Table 7.1). About 2 million hectares in total are planted to semiannual crops each year, less than 3 percent of the nation's land area and less than 30 percent of the land judged to be suitable for semiannual crops (Table 7.2).[1]

In 1987, grains were planted on roughly 1.3 million hectares. Over 90 percent (1.24 million hectares) of the grain crop was grown on tropical lands, with 50 percent in maize, 29 percent in rice, and 21 percent in sorghum. In the cold mountain climates, 41 thousand hectares were planted to wheat and 47 thousand hectares to barley. Grains occupy only a small percentage of the land judged suitable for their cultivation in tropical climates. There is considerable scope for the expansion of grain production in the tropical zone, with attendant opportunities for economic growth and employment. However, poor transportation, absence of irrigation and drainage, and lack of marketing and processing facilities are major bottlenecks to the growth of grain area and production in tropical regions. At present, increases in the planted area of one crop frequently lead to reductions in others. Rice, sorghum, sugarcane, soybean, and cotton crops compete for land and machinery in the most developed tropical regions of Cauca Valley, Tolima, Huila, and Eastern Plains, in spite of the vast surrounding frontier of undeveloped land. Wheat and barley producers face a land shortage in cold areas because of the expansion of high-value dairy, potato, vegetable, and cut-flower operations and because of rapid urban development in the Andes plateaus. Maize is widely grown in the

TABLE 7.1 Agricultural Land Capacity in Colombia ('000' ha)

	Tropical land	Coffee zone	Andes[a]	Total
Cropped area	6,205	7,414	656	14,274
Irrigated	3,172	76	251	3,499
Rainfed	3,033	7,337	405	10,775
Pasture	18,645	51	699	19,395
Total	24,850	7,465	1,355	33,669

Source: ICA, 1984.
[a]Highlands in the Andes Mountains.

TABLE 7.2 Area Planted in Semiannual Crops, 1985-91 ('000' ha)

	1985	1986	1987	1988	1989	1990	1991[a]
Grains							
Rice	326	303	353	363	475	487	429
Maize	541	552	623	664	759	837	921
Sorghum	192	227	259	266	239	273	270
Wheat	45	46	41	38	46	57	54
Barley	31	38	47	53	50	54	55
Total grains	1,134	1,166	1,323	1,385	1,569	1,706	1,729
Other							
Beans	132	138	121	128	132	165	152
Cotton	196	190	173	229	189	207	259
Peanuts	4	4	4	5	5	3	6
Potatoes	139	156	162	170	173	161	150
Sesame	27	29	19	12	14	13	14
Soybeans	54	78	65	61	93	116	118
Tobacco	7	8	10	8	9	7	6
Total other	559	603	553	614	615	672	705
Grand total	1,694	1,769	1,876	1,999	2,184	2,379	2,434

Source: Instituto de Mercadeo Agropecuario (IDEMA) and Ministry of Agriculture, 1991.
[a]Preliminary.

tropical and remote frontier lands of Cordoba, Arauca, Caqueta, Guaviare, and Meta. Maize not only faces competition from lucrative but illegal cocaine plantations, but fields are frequently damaged by guerrilla warfare.

Growth in Grain Production

There are three major farm types in Colombia. Large corporate enterprises grow and process sugarcane and palm oil. Family-owned commercial farms produce rice, sorghum, cotton, and soybeans. Peasant farmers cultivate small plots of land, remaining competitive with the larger commercial farms by producing labor-intensive crops such as coffee, tobacco, maize, beans, plantains, potatoes, and cassava.

Rice, maize, and sorghum are the major grains produced in Colombia. In 1990, rice accounted for 7.3 percent of agricultural GDP, maize accounted for 5.8 percent, and sorghum 2.9 percent. The remaining grains accounted for less than 1 percent of agricultural GDP. Rice and sorghum production occupy the most developed zones in the tropical lands and valleys: Cauca Valley, Tolima, Huila, Eastern Plains, and the North Coast States. Two rainy seasons in the tropical valleys and a few irrigation schemes make it possible to grow two crops per year on the same land. The growing seasons run from June to October and January to March (Montes, Candelo, and Muñoz, 1985). The bulk of the rice produced (95 percent) is long, thin-grained varieties; The remainder is long, broad-grained varieties. Rice types are further classified according to the percentage of broken kernels. Sorghum was introduced into Colombia in the early 1960s. Recently, American and Australian hybrid seeds have become dominant because of their precocity and high yields. Sorghum is classified according to color as red or white, and is also graded depending on the proportion of damaged and broken kernels.

Rice and sorghum are typically planted on mechanized farms either by farmers or tenants. The size of farms in Cauca Valley, Tolima, and Huila varies between 40 and 60 hectares; while in the Eastern Plains and the Atlantic Coast savannas and valleys, the most common farm size is between 60 and 80 hectares. In some regions of the Atlantic and Pacific Coast, traditional rice cultivation systems still predominate. Farmers grow small plots of rice on 3 to 5 hectares, and most of the production is processed and consumed by the farm family.

Rice production increased by just under 5.5 percent per year between 1970 and 1990 (Table 7.3). In 1970, production amounted to 700 thousand tons, rising to 2,060 thousand tons in 1990. Under the influence of modern technology and more productive varieties, yields almost doubled between 1970 and 1987 to over 5 tons per hectare. Sorghum production grew at an annual rate of 9.8 percent during the same period, rising from about 200 thousand tons in the early 1970s to 600 thousand tons in the mid 1980s. Most of this increase is attributable to expansion in area, which jumped from 54 thousand hectares in 1970 to almost 270 thousand hectares in 1990. Little increase in yield occurred, remaining at roughly 2.5 tons per hectare.

Maize is traditionally one of the most important Colombian agricultural products. It occupies a large portion of the total area planted to grains and is an essential part of rural and urban diets. Maize is commonly classified into six groups according to kernel characteristics: dent, flint, flour, sweet, pop, and waxy (Hiddink and Joosten, 1986). The most common maize in Colombia, the flint type, accounts for about 90 percent of domestic production. Two-thirds is white maize and one-third is yellow maize. Three to 4 percent of maize

TABLE 7.3 Colombia: Grain Balances, 1970-91

Year	Area ('000 ha)	Production ('000 tons)	Yield (tons/ha)	Imports ('000 tons)	Exports ('000 tons)	Stocks ('000 tons)	Apparent consumption Total ('000 tons)	Per capita kg
Rice								
1970	257	702	2.7	300	8	183	787	38
1975	373	1,614	4.3	0	116	306	1,536	64
1980	321	1,655	5.2	0	60	283	1,658	62
1985	326	1,708	5.2	0	57	248	1,667	57
1990	485	2,060	4.2	0	85	469	2,000	60
Maize								
1970	661	877	1.3	28	0	36	879	41
1975	573	723	1.3	0	15	129	834	35
1980	614	854	1.4	29	0	73	884	33
1985	541	763	1.4	60	0	75	803	28
1990	837	1,213	1.4	14	0	54	1,219	36
Sorghum								
1970	54	118	2.2	14	0	4	133	6
1975	134	335	2.5	0	0	33	357	15
1980	206	431	2.1	206	0	183	542	20
1985	192	499	2.6	136	0	124	545	18
1990	273	777	2.8	0	0	36	854	26
Wheat								
1970	46	54	1.2	317	0	80	363	17
1975	31	39	1.3	326	0	55	368	15
1980	38	46	1.2	540	0	128	550	21
1985	45	76	1.7	659	0	55	738	25
1990	57	105	1.8	655	0	51	797	24
Barley								
1970	51	87	1.7	59	0	43	104	5
1975	76	122	1.6	13	0	55	122	5
1980	63	110	1.7	39	0	21	201	8
1985	31	60	1.9	95	0	9	150	6
1990	54	100.	1.9	94	0	1	199	6

Source: Instituto de Mercadeo Agropecuario (IDEMA) January 1991.

production is classified as the flour type, and the remaining 7 percent as special sweet types, which are consumed in an immature state (*chocolo*). Maize is mainly produced in areas with tropical climates, where two crops per year can be harvested. Maize farms, which are generally smaller than 10 hectares, use abundant family labor to grow not only maize, but also beans, cassava, and plantain. Under this system, maize production has been stagnant since the early 1970s, with little increase in average yields.

Wheat and barley are also grown in Colombia. These crops are cultivated by small farmers on the cold plateaus of Boyaca, Cundinamarca, and Nariño. Because of climatic conditions, domestically produced wheat is generally "soft" wheat with an average protein content of 10 percent, although processors prefer hard wheats. The heterogeneous character of Colombian wheat production increases assembly and industrial processing costs. Colombian researchers have not developed varieties of wheat suitable for cold climates, and only in recent years were efforts made to produce varieties adapted to tropical climates. Wheat is graded on the basis of the proportion of damaged and broken kernels and on the *puntaje* test, which measures density by weighing a specified volume of grain. The brewing industry, the most important consumer of barley, classifies the product into two grades according to percentage of broken kernels. In addition, after purchase, a fermentation proof determines suitability for beer production. Rejected grain is sold to the animal feed industry.

Technological Change

The 1960s were characterized by heavy government investment in agricultural research, extension, and irrigation and drainage facilities (Thomas, 1985). Yields of the major grains, particularly rice, grew at a rapid rate until the mid-1970s. Since then, growth has slowed as government investment was shifted to mining, electricity, roads, and services required by the growing urban population (Montes, 1986). Agricultural research expenditures fell from 0.46 percent of agricultural GDP at the beginning of the 1970s, to 0.25 percent in the early 1980s. Research facilities and scientific personnel were cut back because of a reduction in government funding. However, various private organizations also fund research programs in Colombia, and their support is increasing. These include the Centro de Investigaciones para la Agricultura Tropical (The Tropical Agriculture Research Center—CIAT), which researches rice, cassava, and livestock.

Investment in irrigation and drainage, which are important for rice and sorghum production, also fell drastically since the mid-1970s. Public expenditures on land improvement dropped from 2 percent of agricultural GDP in 1970, to less than one-half percent in 1985 (OPSA, 1986). With the support of World Bank loans, the Instituto de Hidrologia, Meteorologia y Adecuacion de Tierras (The Institute for Hydrology, Meteorology and Land Improvement-

HIMAT) is rehabilitating irrigation districts to maintain and improve existing facilities. These districts are mainly located in the valleys of Tolima, Huila, and Cauca and in the North Coast States. Rice, sorghum, sugarcane, and cotton are the main beneficiaries of improvements in irrigation and drainage facilities.

Rice yields increased dramatically when new varieties and their associated "technological package" of fertilizers, pesticides, irrigation schemes, and mechanization were introduced. This "Green Revolution" was supported by government investment in irrigation, research, and development activities (Montes, Candelo, and Muñoz, 1985). Between 1970 and 1982, rice production grew at an annual rate of 9 percent and yields at 5 percent. However, domestic consumption increased more slowly, constraining further production growth. Although Colombia's rice yields are above international averages, high domestic costs and a drop in international prices restrained exports during the early 1980s. Inventories increased, the real price and profitability of rice dropped, and cultivated area decreased from 386,000 hectares in 1982 to 350,000 hectares in 1987. A peak of 485,000 was reached in 1990.

As indicated earlier, maize production and yields are stagnant. Only a small portion of the crop is produced using modern technology in the Cauca Valley and a few regions in North Coast. Maize is cultivated mainly in the remote lowlands by more than 100 thousand settler families on small plots using labor-intensive technology. Although new varieties and cultivation practices are available, the distribution network for modern inputs and the extension service are weak and costly, and hence, the rate of adoption of new varieties and technologies is slow.

Wheat production has been a story of failures in Colombia. In spite of growth in wheat consumption, domestic production has consistently fallen short of the growth in demand. Output increased at an annual rate of just under 3.4 percent in the period 1970-90, but Colombia is increasingly dependent on imports to satisfy domestic requirements (Table 7.3). Up to the mid-1950s, policy-makers were committed to achieving self-sufficiency. In the early 1960s, the government changed its policy because it appeared that Colombia did not have a comparative advantage in wheat production. United States food aid shipments under Public Law 480 of 1954, which subsidize American wheat exports, also contributed to the shift in wheat policy. A conscious decision was made to import rather than to produce wheat domestically (Candelo, 1986). Peasants in the Andes Mountains continue to grow a small amount of wheat, equivalent to less than 10 percent of domestic consumption in 1987.

Sorghum production expanded rapidly to meet the demands of the animal feed industry. Almost all sorghum is produced on mechanized farms, and hybrid seeds are widely used. Colombia is self-sufficient in sorghum, contrasting with barley production, which is insufficient to meet the demands of the brewing industry. Local production suffered yellow rust attacks in the mid-1970s, and area cultivated and yields dropped drastically. Colombia had to increase imports of barley and malt to satisfy domestic demand, and although barley output has begun to increase, the country is a long way from self-sufficiency (Table 7.3).

Cereal Consumption

Population increases and per capita income growth have contributed to an expanding demand for cereals. Rice, maize and wheat, the primary foodgrains, make up about one-fourth of all calories and proteins consumed by Colombians (Table 7.4). Sorghum is the main source of calories for the production of animal feed concentrates.

Over the last two decades, rice has become a staple in urban and rural areas, with a doubling of per capita consumption. Rice provides about 14 percent and 12 percent, respectively, of the total calories and protein in the average diet (Table 7.4). However, the rate of increase in demand for rice has stagnated because of slackening population growth and substitution of wheat products for rice, partly due to changes in preferences and partly to consumer response to relative prices. Rice is mainly consumed as a cooked staple for lunch and supper. A small percentage of rice and rice byproducts are used by the animal feed and brewing industries.

For centuries maize has been a staple foodgrain, especially for the poor, and per capita consumption in rural areas is still larger than in the cities (Table 7.4). Only a small percentage of maize and maize byproducts enter the animal feed industries. *Arepas* (maize bread), cakes, pudding (*natilla*), cereals, many flour/maize soups, and porridges are traditional and typical dishes all over the country. Since 1970, total maize consumption has fluctuated around 900,000 tons per year. Small milling companies buy approximately 600,000 tons, which they make into maize flour, broken maize, and other byproducts. The maize processing industry uses the balance for the production of starches, precooked flour, and byproducts, including maize oil, animal concentrates, and glucose syrup.

Per capita consumption of maize declined during the 1970-90 period (Table 7.3) as consumers began to substitute wheat for maize. As a result of international increases in wheat prices, relative prices favored maize consumption from 1974-79, while declines in the price of wheat favored wheat consumption from 1980-90. However, maize consumption may expand in the future. The growing number of women with jobs outside the home has led to great changes in consumption patterns, resulting in increases in the use of convenience foods in urban areas. Maize competes favorably with wheat in this regard because maize bread, for example, is commonly eaten with roasted and fried chicken, which is one of the most popular "fast foods" in the cities. The higher international prices of wheat in the late 1980s may also encourage local production and consumption of maize. In addition, efforts are being made to

TABLE 7.4 Calories and Protein Supplied by Various Foods in Colombia, 1981 (percent)

	Total Protein			Total Calories		
	Total	Urban	Rural	Total	Urban	Rural
Meat	26	28	22	5	6	4
Milk	14	14	14	6	7	5
Rice	12	12	13	14	15	14
Maize	6	5	9	7	5	8
Potatoes	3	3	3	4	4	4
Beans	6	6	6	2	2	2
Bread	4	5	2	4	5	2
Noodles	3	3	3	2	2	2
Sugar	1	1	1	17	17	17
Plantain	2	2	3	6	6	7
Oil	0	0	0	14	16	11
Cassava	1	a	1	3	2	5
Other	22	20	23	14	12	19
Total	100	100	100	100	100	100

Source: Pardo, 1981.
a Less than 0.5
Detail may not add because of rounding.

substitute maize for wheat in bread baking. Yellow maize is frequently used as the base in the manufacture of poultry feed concentrates, and it is likely that the poultry industry will continue to expand rapidly.

Wheat is now a staple foodgrain, especially in urban areas, with consumption growing at over 4 percent annually. Wheat is the main raw material for the bakery and pasta-goods industries, although substitutes are encouraged through the Mixed Flour Program (SAC, 1986b). The main objectives of this program are to save foreign exchange and to encourage domestic maize and rice production. A small proportion of imported wheat and wheat byproducts are used for animal feed.

Sorghum consumption has grown quickly because of rapid development of the livestock industries, mainly chicken, eggs, and pork. Sorghum, used exclusively as a raw material in the animal feed concentrate industry, has displaced maize because of relative prices. Sugar and palm oil also join maize as significant calorie sources in feed concentrates. The brewing industry buys most of the domestic and imported barley. Only a small percentage of local production is used by millers to manufacture products for humans and for animal feed. Per capita consumption of barley increased from 5 kg in 1970 to 7 kg in the late 1980s.

In sum, most available rice, maize, and wheat supplies are consumed directly by humans. Sorghum provides most of the calories used by the animal feed industries, complementing protein provided by soybeans, cottonseed, and fish meal. Processed animal feed is chiefly used (87 percent) by poultry and pork enterprises. Although the rate of population growth will probably continue to decline, it will still provide the major impetus behind growth in grain consumption. Rice consumption will likely increase at about the same rate as population growth, unless a new variety is developed that can be used in animal feeds. Sorghum consumption is presently growing very rapidly because of the fast annual growth in chicken (4 percent), egg (2 percent), and pork (4 percent) consumption. The recent adoption of a set of economic adjustment measures in Venezuela may reinforce this trend by reducing the flow of imports of these products into Colombia (see section on trade policy below). Maize consumption will increase at a rate of 2-3 percent per year. Wheat consumption will likely grow at 2.5-3 percent for bread and 2 percent per annum for pasta.

The Grain Marketing System

Colombian grain marketing channels are complex because of many social, geographical, storage, and processing factors. The main participants in the marketplace are consumers, retailers, wholesalers, processors, and farmers. Most grain transactions are made using direct face-to-face negotiations between market participants. A commodity exchange system was recently introduced for spot transactions, but there is no auction system. Prices are determined by supply and demand in a marketplace affected by government policies and in an environment characterized by poor market information unevenly distributed among participants (Silva, Ramirez, and Bustamante, 1986).

The Instituto de Mercadeo Agropecuario (Government Agricultural Marketing Institute—IDEMA), the government participant in grain marketing, regulates the market by buying, selling, and distributing grains and oilseeds. It is also responsible for maintaining stocks; importing and exporting; and establishing support prices for rice, sorghum, maize, wheat, barley, beans, sesame, and soybeans. Support prices are set so that farmers can recover production costs. These prices are announced at the beginning of the planting season. The government installed in 1990 decided to minimize the role of IDEMA.

The support price is uniform throughout the country, and IDEMA participates in the market like an ordinary buyer. Farmers are under no obligation to sell grain to IDEMA. If the market price rises above the support price, farmers will generally seek alternative outlets; however, when the support price is greater than the market price, the government's share of marketed grains rises. In general, IDEMA buys less than 10 percent of domestic grain production because, in most cases, the free-market price exceeds the support price (Figure 7.2). The major exception is in remote, newly colonized areas far from consumption centers. Since late 1990, real support prices are being reduced.

IDEMA is charged with maintaining adequate stocks to regulate supplies between harvest seasons. The organization owns silos and warehouses and sometimes rents additional space from private elevators. These facilities will be sold to the private sector. Most storage facilities are located in cities and ports, with little situated near production areas. Industrial processors, IDEMA, producer's cooperatives, and twelve elevator firms called "*Almacenes Generales de Deposito*" (AGD) (subsidiaries of private and official banks) own the bulk of storage capacity; most farmers, on the other hand, own no grain drying or storage facilities. The AGD issues deposit certificates specifying the quality and quantity of stored grain to owners. These can be bought and sold in the marketplace and may be returned to the bank as loan collateral under the "*Bonos de Prenda*" system. This system allows banks to make loans at a subsidized interest rate using the deposit certificate and rediscounting part of the loan through the Central Bank. Processors and IDEMA are the main users of this system (Thomas, 1985).

Located in the main rice-producing areas of Tolima, Eastern Plains, Huila and the Atlantic Coast States, 360 rice mills are the principal paddy buyers (Figure 7.3). The mills maintain their own drying and storage facilities using capital from equity, commercial loans, and subsidized storage credit obtained under the Bonos de Prenda system mentioned above. Once rice is milled, it is sold to wholesalers at terminal markets, who in turn sell to retailers, including

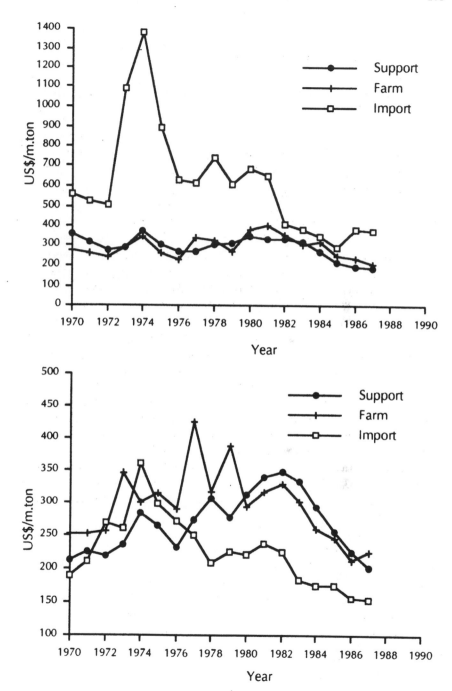

FIGURE 7.2 Prices of Major Grains in Colombia, 1970-90 *Source:* Based upon data from Instituto de Mercadeo Agropecuario (IDEMA), 1988.

supermarkets, independent neighborhood stores, consumer-owned cooperatives, and other retailers (Figure 7.3). Millers also sell white rice directly to supermarkets and cooperatives. Because IDEMA owns no significant milling capacity, processing must be contracted with mills. IDEMA markets milled rice to consumers through its own retail outlets and to cooperatives and supermarkets in larger cities. Byproducts from the milling process are sold to brewing, baking, and animal feed industries (Figure 7.3).

A grain commodity exchange began operation in 1980, but it handles only a small percentage of all grain transactions. IDEMA is a strong supporter of the commodity exchange, accounting for 80 percent of its transactions in 1987. The organization uses the exchange system to sell maize, rice, sorghum, wheat, and grain byproducts. The exchange is a stock company, and its partners include growers' groups, processing industry and trader associations, IDEMA, and brokers. Growers, processors, and traders hire brokers to buy or sell through the commodity exchange. Recently an arbitration mechanism was established to deal with grain transaction disputes. Local futures markets still do not exist.

Because maize is grown on thousands of small farms, marketing is costly and complex. Three buyers predominate: large-scale processors, numerous small milling companies, and IDEMA. Processing and milling industries commonly do not purchase maize directly from farmers, relying instead on rural assemblers and urban wholesalers. The processing industries produce starches, flours, arepas, feedstuffs, instant soups, and maize oil, which move into wholesale and retail distribution channels and to the feed industries. Maize is marketed via wholesalers and retailers to low- and medium-income households, where it is eaten in baked products, arepas, and soups. IDEMA sells maize directly to maize-based industries or through the commodity exchange. Milled maize is also offered at IDEMA's special retail outlets.

In the case of sorghum, farmers market directly to feed industries or to assemblers, who sell the crop to feed processors or to the vertically-integrated poultry industry. The commodity exchange is occasionally used to market a small percentage of sorghum production. In some regions where the marketing infrastructure is poorly developed, farmers sell directly to IDEMA. Approximately 90 percent of the wheat consumed in Colombia is imported by IDEMA or the mills. The agency also purchases about 40 percent of locally grown wheat at the support price. It generally sells purchased wheat through the commodity exchange or directly at prices that cover costs and tariffs. Wheat marketing will be privatized in the near future. Mills produce flour and semolina for the bakery and pasta industries, while byproducts flow to the animal feed industries.

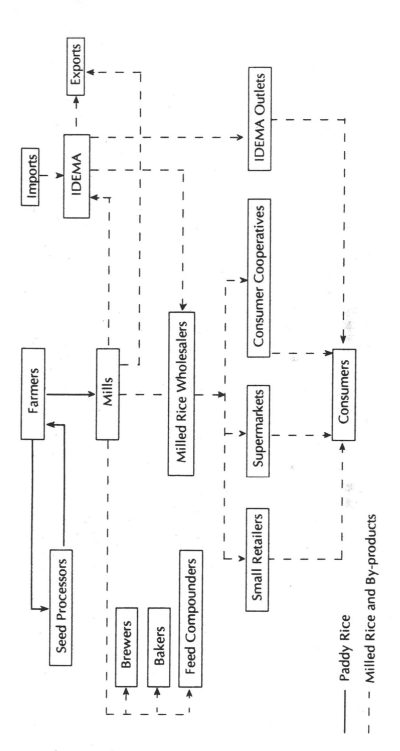

FIGURE 7.3 Marketing Channels for Rice and Rice By-products in Colombia

International Trade

As indicated earlier, Colombia's agricultural exports are dominated by coffee, although a variety of other products, including bananas and cut flowers, are also important. Colombian grain exports are insignificant; however wheat imports are important on both a dollar and volume basis. Rice exports grew rapidly until 1977, but lower international prices and an overvalued peso eroded profitability.[2] Reduced government investment in irrigation and research and extension also resulted in higher domestic production costs compared to competitors.

Grains account for a significat share of total food imports. In 1987, imports of food products amounted to $US438 million, while total cereal imports cost $US 97 million (22 percent), of which $US75 million were wheat imports (Table 7.5). Imported hard and soft wheat account for about 90 percent of domestic consumption, compared to 25 percent in 1954. Traditionally, Colombia's main wheat suppliers are the United States and Canada between May and November, and Argentina in the remaining months. In recent years, a number of other countries, including France and Saudi Arabia, shipped subsidized wheat to Colombia, encouraging the United States to offer counter subsidies through the Export Enhancement Program (EEP) to protect its market share. Feed quality wheat is generally imported from Canada, and it is occasionally substituted for sorghum by the animal feed industry. Barley imports, mainly from Canada, are made directly by the brewing industry, and these increased during the 1980s.

IDEMA and private processors imported grains from major international traders on a FOB or CIF basis. However, since mid-1991, grain imports were liberalized and the private sector will realize this activity and so IDEMA's role will be minimized.

Public Policy

Up to the 1930s, the Colombian economy was predominantly directed by free market forces, with little state intervention. The economic instability generated by the Great Depression in the 1930s caused major social upheavals, which resulted in the ascendance of the Liberal (Social Democratic) Party and a constitutional reform that legalized direct state intervention in the economy. Since then, the Colombian economy has generally been characterized as mixed, with both state and private participation. The state not only establishes objectives and priorities through economic planning, but works in tandem with the private sector in production and trade activities. Industrialization, a major goal, was stimulated through import substitution policies (SAC, 1986b).

After the constitutional reform, several new policy instruments and institutions were created. These included foreign exchange controls and import and export quotas, tax reforms to make income distribution more equitable and to

TABLE 7.5 Grain Imports, 1986-87 (Million $US FOB)

	1986		1987		1988		1989		1990	
	Value	Percent	Value	Percent	Value	Percent	Value	Percent	Value	Percent
Wheat	68.0	76.9	75.3	77.6	69.6	72.0	88.5	73.8	62.3	
Barley	8.7	8.9	7.5	7.7	5.0	5.2	9.6	8.0	17.9	17.6
Malt	7.6	8.6	10.0	10.3	17.3	17.8	18.6	15.5	17.9	17.6
Maize	0.0	0.0	0.2	0.2	4.7	4.9	0.0	0.0	0.0	0.0
Oats	4.2	4.7	4.1	4.2	3.0	3.1	3.3	2.8	3.5	3.4
Total grain imports	88.5	100.0	97.1	100.0	96.6	100.0	120.0	100.0	101.6	100.0
Total food product imports	383.0		438.1		380.9		237.5		325.1	

Source: Instituto de Comercio Exterior (INCOMEX).

increase public revenues, and special incentives to promote exports. Many institutions were created to implement agricultural policy: the Ministry of Agriculture in 1932; the Agricultural Bank (Caja Agraria) in 1931; IDEMA in 1944; the Agrarian Reform Laws and the institute charged with their implementation in 1936 and 1962; the Colombian Research Institute (ICA) in 1963; the Fondo Financiero Agropecuario in 1973; and the Hydrology, Meteorology and Land Improvement Institute (HIMAT) in 1974.

As state intervention in the economy grew, associations to protect the interests of the private sector sprang up. In the case of grains, farmer and processor associations include: the National Rice Growers' Federation (Federacion de arroceros) (FEDEARROZ) created in 1947; the National Association of Rice Millers (Asociacion de Molineros de arroz) (MOLIARROZ) created in 1959; the National Association of Rice Manufacturers (Asociacion de Industriales del arroz) (INDUARROZ) created in 1969; the National Federation of Grain Producers (Federacion Nacional de cerealistas) (FENALCE) created in 1962; the National Association of Grain Manufacturers (Asociacion de Industriales Manufacturerosole Cereales) (ADIMCE) created in 1970; the Federation of Wheat Millers (Federacion de Molineros de trigo) (FEDEMOL) created in 1950; and the Federation of Feed Manufacturers (Federacion de Fabricontes de Alimentos concentrados para animales) (FEDERAL) created in 1969 (Bejarano, 1985).

Major political and economic reforms are underway in the early nineties. A new political constitution was democratically launched in July 1991 in which the principles of government intervention were kept. In contrast, the government policy shifted toward an open economy characterized by the prevalence of free-market forces and less government intervention.

Macroeconomic Policy

Macroeconomic policy has traditionally had a major impact on Colombian agriculture. Between 1950 and 1967, as industrialization gathered pace, Colombia entered a period of intense urbanization. The policy instruments used to promote industrialization were restrictions on imports, high tariffs, and an overvalued exchange rate. These measures discriminated against agricultural exports, but the agricultural sector benefited from the availability of cheap imported machinery, credit subsidies, and public investment in research and land improvement.

From 1967-75, a more balanced development strategy gave priority to industrial and agricultural exports, while protecting selective domestic industries from competing imports. In 1967, export promotion was added to the list of government goals. Before 1967, multiple exchange rates were fixed for long periods, even during times of inflation. This produced severe balance of payments deficits, leading to sporadic and massive devaluations of the peso.

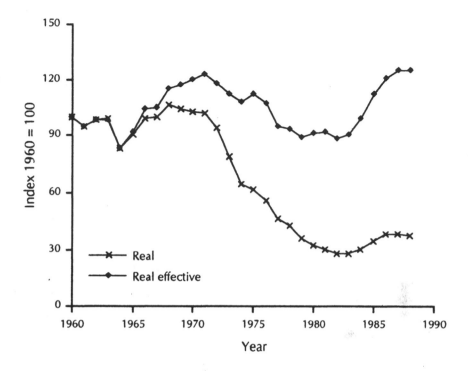

FIGURE 7.4 Colombian Exchange Rates. *Source:* Based on data from IMF, 1989. (Note: The real rate is the nominal rate divided by the Colombian wholesale price index; the real effective rate is the real rate multiplied by the U.S. producer price index. A decline in these indices denotes an appreciation of the currency, a rise denotes depreciation.)

In 1967, a "crawling peg" system was introduced to adjust the exchange rate to provide regular and less drastic adjustments. As a result, uncertainty associated with the exchange rate was substantially reduced, and the value of the currency depreciated in both nominal and real terms (Figure 7.4). Measures to promote exports were introduced, including a tax rebate certificate for minor exports; subsidized credit from the Export Promotion Fund (PROEXPO) financed by a tax on imports; and the "Plan Vallejo" system, essentially a customs rebate on the import content of industrial exports (García and Montes, 1987). Between 1968 and 1975, the depreciation in the real exchange rate increased the profitability and volume of exports. After 1975, however, the export promotion rebates and cheap credit failed to compensate for exchange rate appreciation induced by growth in coffee export receipts and the government's fiscal deficit.

In the late 1970s, rice and even maize exports were profitable. However, during the 1970s the increasing overvaluation of the Colombian peso, the coffee bonanza, the fiscal deficit, and a dramatic reduction in government investments in agriculture (referred to earlier) lowered Colombia's competitiveness in grain production. This situation was further aggravated by declining prices in the world market caused by excess supply. Since 1984, several important economic adjustments were implemented, including currency devaluation, cuts in government expenditures, and tax reform. As a result of these measures and rising international grain prices, Colombia became more competitive in grain production (Silva and Alvarado, 1988). However, reduced public investment in research, extension, and irrigation, which occurred in the 1970s, may have resulted in more costly grain production and higher consumer food prices.

Many studies indicate high rates of return to investment in agricultural research and extension activities (White and Havlicek Jr., 1982). In Colombia, the return is also estimated to be high: rice, 58 percent; soybeans, 79 percent; wheat, 11 percent; potatoes, 68 percent; barley, 53 percent; and palm oil, 30 percent. At present, these crops do not receive a sufficient share of research funds. Their internal rate of return for research can be five times the opportunity cost of public funds, which has been estimated at 10 percent (Romano, 1987). The internal rate of return for irrigation projects reaches rates of return as high as 25 percent.

Trade Policies

As a whole, the Colombian agricultural sector is strongly trade-oriented, historically due to the importance of coffee in the economy. Other products such as flowers, cotton, bananas, tobacco, sugar, beef, shrimp, fish, cocoa, plantain, fruits, vegetables, rice, wood products, and processed items also contribute to agricultural exports. Because most Colombian agricultural products are tradeables, macroeconomic policies, such as exchange rate management and export or import incentives, directly influence the performance of the agricultural sector. Imports of agricultural products account for about 7 percent of total imports.

Because of the economic significance of coffee and Colombia's important position in the world market, macroeconomic and agricultural policies are closely linked to the coffee industry. In contrast to coffee, grain exports are negligible, and wheat is virtually the only grain imported. Consequently, domestic concerns drive trade policy for grains. As a price-taker in the world wheat market, the country tries to take advantage of high inventories in the international marketplace, buying only what is needed to fill domestic shortages between crops. Domestic inventories are relatively small, approximately equivalent to two weeks of domestic consumption at the beginning of crop harvests and up to three months of consumption during the final harvest period. The widespread practice of growing two crops each year, feasible because of ample rainfall, makes it possible to maintain low inventories of staple grains (Silva, Hernandez, and Candelo, 1986).

The same basic objectives and policy instruments pertaining to grain trade have been in use since the mid-1940s. The basic strategy is to cushion domestic grain producers and consumers from major price swings on both the international and local markets, to provide economic stability, and to attain and maintain self-sufficiency in grains, with the exception of wheat. The main trade policy instruments are tariffs on grain imports to protect domestic producers (33 percent for private importers and 10 percent for IDEMA); quotas on imports of grains and oilseeds established annually and revised whenever necessary to fill the gap between local consumption and production; state purchases and sales on international markets through IDEMA; and quotas on grain exports to control domestic prices and inflationary pressures (Table 7.6).

In 1968, Colombia signed the Andean Pact, which created a common market with four other countries: Bolivia, Ecuador, Peru, and Venezuela. Colombia's membership in the Andean Common market has had a marked effect on grain trade. The maintenance of unrealistic multiple exchange rates in Venezuela (7.5 bolivares per US dollar, 14.5 bolivares per US dollar, and the market rate of 40 bolivares per US dollar at the end of 1988) and Colombia's protection of domestic producers generated large price differentials among the member countries. Consequently, recorded and unrecorded inflows of products, such as wheat flour, eggs, chicken, feed concentrates, rice, and precooked maize flour, from member countries increased during the 1980s. The Venezuelan government enacted drastic economic adjustment measures in 1989. Multiple exchange rates were eliminated and subsidies to the agricultural sector reduced substantially. Food prices rose sharply, and the flow of Venezuelan products to Colombia was suddenly terminated. Volatile macroeconomic policies in neighboring countries and the use of grain export subsidies in the EEC, Saudi Arabia, and the US encourage Colombian policy-makers to continue intervening in the country's trade in grains to avoid sharp domestic price fluctuations and to protect domestic agriculture. In 1991, the intervention in grains dropped and the price bonds system was set up as the basic stabilization instrument. Price bonds are calculated from international prices of the last 60 months and actualized twice a year.

Price and Income Supports

An agricultural stabilization program comprising grains and oilseeds was initiated in 1944. The program included support prices, tariffs and quotas on imports and exports, and government participation through IDEMA in the grain market (as a storage agent, buyer, seller, and an investor in marketing facilities).

TABLE 7.6   Major instruments of Grain Policy in Colombia

Instrument	Maize	Rice	Sorghum	Wheat
		Commodity		
PRODUCTION/CONSUMPTION				
Producer guaranteed price	X	X	X	X
Deficiency payments				
Government purchases	X	X	X	X
Production quota				
Input subsidies				
— Credit	X	X	X	X
— Fertilizer/pesticides				
— Irrigation		X		
— Machinery/fuel				
— Seed				
Crop insurance				
Controlled consumer price				
TRADE				
Imports				
— Tariff	X	X	X	X
— Quota	X	X	X	X
— Subsidies		X		
— Licensing				
— State trading	X	X	X	X
Exports				
— Taxes				
— Restrictions		X		
— Subsidies		X[a]		
— Licensing				
— State trading	X	X	X	X
OTHER				
Marketing subsidies				
— Storage	X	X	X	X
— Transport				
— Processing				
State marketing				

[a]Small tax rebate and export credit.

Regional and temporal price differences caused by transportation and storage costs are common in the Colombian markets. Price stabilization programs are crafted so as to help prevent extreme price fluctuations because of bumper crops. However, the programs are not designed to smooth out inequalities resulting from regional and temporal price variations.

A grower-funded development and stabilization program for rice, modeled after the long-running Coffee Fund, was established in 1963. The Coffee Fund is financed with funds collected from coffee growers when international prices are high. Money is returned to growers when prices are low. The Fund is a public entity, but the government regularly establishes contracts for its administration with the Coffee Federation, a grower organization. The Rice Fund was financed by a tax of one Colombian cent per kilo up to 1983, and 0.5 percent of the farm price thereafter. With the enactment of Public Law 68 in 1983, other cereals were also included. As in the case of rice, monies are obtained from a 0.5 percent tax on grower receipts and are administered by grower associations (FEDEARROZ and FENALCE) through government contracts. However, the new grain funds have yet to build up sufficient resources to support price stabilization programs. Currently, the funds chiefly finance private extension programs. Plans are in place to begin grain marketing activities in the early 1990s that will help to curb extreme price fluctuations. IDEMA's price stabilization role will gradually diminish and eventually disappear.

Consumer price controls were used temporarily in special situations as policy instruments in Colombia. Some basic food items (milk, wheat flour, and cooking oil) were occasionally subject to price controls, as were some agricultural inputs, such as pesticides and fertilizers. Presently, neither grains nor any foods are under price control.

The state intervenes in the grain economy to achieve both short- and long-run objectives. The primary short-term objective is to reduce fluctuations in domestic prices and supply, and to shelter the economy from frequent and large price changes in international markets. In the long run, the goal is to improve productivity and marketing opportunities. Support prices are fixed by the government, taking into account production costs. Price are set such that farmers still earn a minimum level of profitability, even in bumper harvest years. International prices (evaluated as C.I.F. plus unloading, plus administrative and financial costs) are taken into account when setting price supports. However, support prices have not followed international prices very closely, chiefly because either the exchange rate was overvalued for long periods or international prices fluctuated drastically (Figure 7.2). Support prices have often been used to isolate domestic prices from changes in international prices and from shifts in the exchange rate (Schuh, 1985). Monetary policies, inflation, the level of international prices, and exchange rate policy all influence the level of support prices in real terms.

Conclusions

Grain production, prices, imports, exports, and policy in Colombia are extremely sensitive to changes in international prices and to variations in the exchange rate. Because of its large contribution to GDP, employment and the balance of payments, the agricultural sector both affects and is affected by macroeconomic variables. Macroeconomic policies in Colombia have had a major impact on the profitability of grain production in Colombia and international competitiveness. Between 1967-75, export promotion was emphasized and a "crawling peg" exchange rate system was adopted. However, because of the coffee bonanza, after 1975 currency overvaluation occurred, and the agricultural sector lost its competitive advantage. In the late 1980s, as a result of a major devaluation, increases in international grain prices, and renewed government investment in research and irrigation, domestic grain production gradually recovered its international competitiveness. This is being tested in a more open economy in the early nineties.

The Colombian urban and rural sectors are ill-equipped to handle large and sudden changes in prices and incomes because futures markets do not exist, crop insurance has only recently become available, grain drying and storage facilities are insufficient, and only the small portion of cultivated land that is irrigated and drained can be switched easily between competing crops in response to changes in relative prices. Because of these problems, the government still maintains some programs that promote stability. After almost fifty years, several stabilization instruments are still in use, including support prices, tariffs on imports and exports and government participation (IDEMA) in grain storage, purchase and sale. However, major changes toward an open economy and freer market forces are underway and new stabilization instruments such as price bonds were adopted.

The driving force behind price stabilization policies in Colombia is the perception that a free market would result in unacceptable domestic market conditions in terms of farm incomes and price stability. Fluctuations in international prices, unpredictable subsidy policies in supplying countries, unstable trade and foreign exchange policies in neighboring countries, and droughts and floods expose domestic low-income consumers and producers to unacceptable levels of price risk. There is no futures market in Colombia that would allow farmers and processors to hedge against this risk. These are major factors that underlie the stabilization programs, which represent a "second best policy" to deal with difficult conditions that are beyond the direct control of the Colombian government. The instability that appears inherent in grain markets is viewed in Colombia as symptomatic of serious market failure. Faced with this prospect, the Colombian government chooses to protect the welfare of both farmers and consumers by some intervention to stabilize domestic grain supply and prices even in the midst of instrumenting an open economy model.

Notes

Since this chapter was written, there have been significant policy reforms in Colombia's grain economy.

1. If an adjustment is made to account for multiple cropping (which occurs on 30

percent of the cropped land), only 20 percent of available land is in use.

2. In 1970, there were roughly 19 pesos to the US dollar. In 1988, the corresponding figure was roughly 300 pesos.

3. See the United States chapter in this volume for a discussion of the Export Enhancement Program.

4. In practice, tariffs act like a cushion because when international prices are low, tariffs are strictly enforced by IDEMA to protect Colombian farmers, and when international prices are high, tariffs are often relaxed to protect consumers.

5. See Chapter 1 for a discussion of changes in international market conditions and prices for grains.

References

Bejarano, J. "Economía y Poder." Bogotá: Editorial Presencia, 1985.

Candelo, R. "Análisis y Perspectivas Económicas del Cultivo del Trigo." *Revista Nacional de Agricultura* 877 (1986): 124-54.

García, J. and G. Montes. "Coffee Boom, Government Expenditures and Relative Prices in Agriculture. The Colombian Experience." Washington, DC: IFPRI, 1987.

Hiddink, G.B. and F.J. Joosten. "Economics of Maize in Colombia." Wageningen: Department of Economic Development, Agricultural University Wageningen, 1986, (Mimeo).

Instituto Colombiano Agropecuario (ICA). "Plan de Adopción y Transferencia de Tecnología." Documento PLANTRA No. 1. Bogotá, 1984.

Instituto de Mercadeo Agropecuario (IDEMA), Ministerio de Agricultura. *Anuario Estadísticas del Sector Agropecuario.* Bogotá: OPSA, 1988.

International Monetary Fund (IMF). *International Financial Statistics Yearbook.* Washington, D.C.: 1989.

Ministerio de Agricultura (Ministry of Agriculture). *Anuario Estadísticas del Sector Agropecuario.* Bogotá: Oficina de Planeamiento, 1988.

Montes, G. "La Política Monetaria, el Crédito de Fomento y el Sector Agropecuario." *Revista Nacional de Agricultura* 876 (1986): 101-8.

Montes, G., R. Candelo, and A. M. Muñoz. "La Economía del Arroz en Colombia." *Revista Nacional de Agricultura* 871 (1985): 147-208.

Oficina de Planeamiento del Sector Agropecuario (OPSA). "Incidencia del Precio de los Insumos en los Costos de Producción Agropecuario." Bogotá, 1986, (Mimeo).

Pardo, F. *La Situación Nutricional de la Población Colombiana.* E4-33. Bogotá: DRI-PAN, 1981, (Mimeo).

Romano, L. "Economic Evaluation of the Colombian Agricultural Research Systems." Stillwater, Oklahoma: Oklahoma State University, 1987, (Mimeo).

Schuh, G. E. "Strategic Issues in International Agriculture." Washington, D.C.: World Bank, 1985, (Mimeo).

Silva, A. and R. Alvarado. "Proceso de Ajuste y Políticas Agropecuaria y Alimentaria: Algunas Reflexiones sobre la Experiencia Colombiana." *Revista Nacional de Agricultura* 885 (1988): 103-24.

Silva, A., D. Hernandez, and R. Candelo. "El Manejo de la Estacionalidad en el Sistema Alimentario como Instrumento de Estabilización y Crecimiento." *Revista Nacional de Agricultura* 866 (1986): 109-33.

Silva, A., J. Ramirez, and D. Bustamante. "Notas Sobre la Problemática del Sistema de Mercadeo de Alimentos y Alternativas de Mejoramiento." *Revista Nacional de Agricultura* 877 (1986): 21-32.

Sociedad de Agricultores de Colombia (SAC)(a). "Perspectivas de Desarrollo Agropecuario: Agricultura Horizonte 2000." *Revista Nacional de Agricultura* 874 (1986): 119-208.

————(b). "Programa de Harinas Compuestas para la Incorporación de Cereales de Producción Nacional." *Revista Nacional de Agricultura* 877 (1986): 82-115.

Thomas, V. *Linking Macroeconomic and Agricultural Policies for Adjustment with Growth: The Colombian Experience.* Baltimore, MD: The Johns Hopkins University Press for the World Bank, 1985.

White, F.C., and J. Havlicek, Jr. "Optimal Expenditures for Agricultural Research and Extension: Implications of Underfunding." *American Journal of Agricultural Economics* 60/1 (1982): 47-55.

8

China

Colin A. Carter and Simei Wen

China is the world's largest producer and consumer of grain. The production and availability of grain is given the highest priority by government planners. The Chinese economy as a whole is still dependent on grain and grain products.[1] For example, grain production occupies about 80 percent of the total cropped area and provides an equal proportion of calories in the Chinese diet. Moreover, more than 50 percent of the labor force is engaged in grain-growing activities, and about 40 percent of the gross value of agricultural output is attributable to the grain economy.

The agricultural sector in China has long been confronted with the pressures created by a large and growing population, a fixed land area, and a relatively small amount of cultivable land per capita. It is remarkable that a country with 20 percent of the world's population has managed largely to achieve self-sufficiency in food production with only 7 percent of global arable land. The Chinese economy continues to depend on the agricultural sector to generate a major portion of the gross value of economic output. The sector remains traditional in the sense that roughly 80 percent of the population lives in rural areas and 65 percent of the labor force works in agricultural enterprises. The remaining rural work force is employed in light industry. However, as in many other developing countries, agriculture is gradually becoming less important in the overall economy. Agriculture's contribution to the gross value of output of industry and agriculture fell from 70 percent in 1949 to about 21 percent in 1987. Crop farming (including all grains and cash crops) is the most important agricultural activity, contributing approximately 60 percent of the value of gross agricultural output, while animal enterprises contribute 23 percent.

China's agricultural and food sectors have undergone a number of important changes since the end of World War II. Immediately after the Communists gained power in 1949, a distributive land reform was carried out throughout the country (excluding Tibet). The land reform was based on the premise of providing "land to the tiller," and by the end of 1952 when the land reform program was complete, close to one-half of the country's arable land was redistributed, and over 60 percent of the rural population benefited (Wong, 1973). The rural poor also benefited from redistribution of other capital assets, buildings, and draft animals. Although this process was a little bloody, it helped the ruling Communists consolidate power, gain economic strength by eliminating large inequalities in the ownership of capital assets, and provide economic incentives to the poor rural masses. During this period, the economy was basically free market-oriented, and the state played a minor role in resource allocation in the rural economy. The subsequent cooperativization (1953-57) and the following communization (1958-62) movements did not lead to the economic gains that Chinese leaders had hoped for, but instead to a near collapse of the rural economy. The "Readjustment Period" began in 1963, and by the end of 1965, the rural economy was near normal. However, the following ten-year "Cultural Revolution" almost ruined the Chinese economy in general, and the food and agricultural sector in particular. After the death of Mao Zedong in 1976, major policy changes became possible and politically desirable. Beginning in 1978, the commune system was gradually replaced by the production responsibility system (Lin, 1988), which began a process of decentralization of economic decision making in China's rural economy.[2]

All sectors of the Chinese economy grew rapidly following the introduction of a package of economic reforms affiliated with the production responsibility system in late 1978. Over the 1978/79 to 1986/87 period, national income grew at an annual rate of 9 percent per year, while the industrial sector expanded at 10 percent per year. During the same nine-year period, agricultural output grew at a slower rate of 6 percent, and grain output grew at 4 percent per annum, compared to 2 percent between 1957-78.

China is an agriculturally and climatically diverse country. It is useful to group the provinces, which are the major administrative unit, into regions. The regions in Figure 8.1 were delineated by Barker, et al. (1982), and the dividing lines are based primarily on differences in climatic and growing conditions. The two most important grains produced in China are rice and wheat. In general, most of the rice is produced in the southeast and most of the wheat in the northeast.

Growth in Grain Production

Total grain output in China increased from 161 million metric tons in 1952 to 405 million metric tons in 1987 (Table 8.1), an increase of 152 percent over thirty-five years. Rice, wheat, and maize are the principal grains. Rice accounted for 43 percent and wheat for 22 percent of total grain production in 1987. The

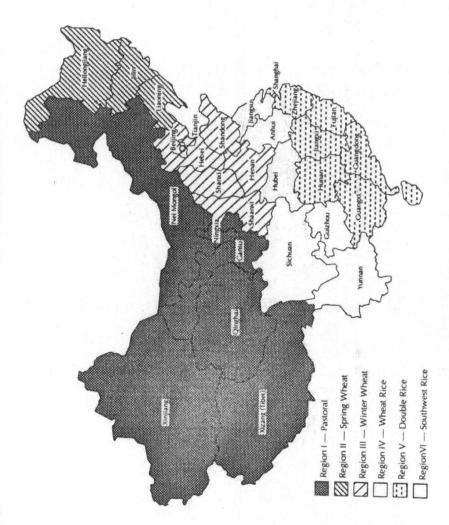

FIGURE 8.1 Agricultural Regions of China

Region I — Pastoral
Region II — Spring Wheat
Region III — Winter Wheat
Region IV — Wheat Rice
Region V — Double Rice
Region VI — Southwest Rice

TABLE 8.1 Grain Production, 1952-87 (million metric tons)

	1952	1957	1965	1970	1978	1979	1980	1981	1982	1983	1984	1985	1986	1987
Rice	68.4	86.8	87.7	110.0	136.9	143.8	139.9	144.0	161.6	168.9	178.3	168.6	172.2	174.4
Wheat	18.1	23.6	25.2	29.2	53.8	62.7	55.2	59.6	68.5	81.4	87.8	85.8	90.3	87.8
Maize	16.9	21.4	23.7	33.0	55.9	60.0	62.6	59.2	60.6	68.2	73.4	63.8	71.2	79.8
Soybeans	9.5	10.0	6.1	8.7	7.6	7.5	7.9	9.3	9.0	9.8	9.7	10.5	11.6	12.2
Potatoes	13.1	17.5	19.9	26.7	31.7	28.5	28.7	26.0	27.1	29.2	28.5	26.0	25.2	28.2
Millet	11.6	8.6	6.2	8.8	6.6	6.1	5.4	5.8	6.6	7.5	7.0	6.0	—	—
Sorghum	11.1	7.7	7.1	8.2	8.1	7.6	6.8	6.7	9.5	8.4	7.7	5.6	—	—
Total grain[a]	160.6	190.7	194.5	240.0	304.8	332.1	320.6	325.0	354.5	387.3	407.3	379.1	391.1	404.7

Sources: SSB, 1987:55; 1987:28; SSB, 1988:26; USDA, 1984:12.
[a]Total grain and potatoes for 1952 and 1957 are adjusted to the 5:1 potatoes to grain conversion rate used in official statistics for 1964 and later. Rice is unmilled, dried paddy; maize is computed on a shelled basis; and beans are added based on their weight after removal from pods and drying.

increases in total grain output between 1952 and 1987 are mainly attributable to improvements in rice, wheat, and maize production.

Between 1949 and 1986, grain output grew at an average annual rate of 2.7 percent, compared to a population growth rate of about 1.8 percent. After the founding of the People's Republic of China in 1949, grain production grew rapidly (at a rate of 3.5 percent per annum) as China recovered from the devastation wrought by World War II. The new government reduced the tax burden on farmers, encouraging increased grain production. The growth in production subsequently slowed in the late 1950s and 1960s. In the 1978-86 period, growth picked up again and reached 3.2 percent annually, largely because of the effect of incentives created by economic reforms introduced in late 1978. These reforms had two major policy components for agriculture: higher prices and greater farmer freedom to make production decisions.[3] The price of government-procured grain, on average, was raised only 17.5 percent during the thirteen years prior to the reforms, but there was a 20 percent increase in prices in 1979 alone. In addition to raising prices, the government also introduced the **production responsibility system**, which gave farmers added freedom to respond to higher prices (see the policy section below for a more extensive discussion of the economic reforms).[4] During the 1978-86 period, all of the production growth came from improvements in yields rather than from the addition of more land. Grain yields increased at a rate of 4.2 percent per year, which is above the long-run trend (of about 3.4 percent per annum).

Looking at the mix of grains in China, the annual growth rates of two of the major grain crops, rice and maize, have been similar to increases for total grains. However, the output growth rates of both soybeans and wheat during the reform period of 1978-86 were greater than during 1952-57. This may reflect government efforts to stimulate production of crops that have an important role in China's grain trade (soybeans are exported and wheat is imported).

Total cropped area declined by about 6 million hectares or 4 percent over the 1978-86 period (Table 8.2).[5] About 0.5 million hectares a year were lost to nonagricultural uses such as the expansion of small towns and cities; increases in the number of rural homes because of improvements in rural living standards; expanded construction of infrastructure such as rural roads and high- ways; and the addition of more fish ponds as cropped land was shifted to relatively more profitable fish farming. Between 1978 and 1986, the area devoted to grain crops decreased by about 9.6 million hectares (8 percent), while industrial crops increased by about 5.9 million hectares (41 percent) (Table 8.2). Farmers chose to grow industrial crops because they yielded greater returns than grain crops. Since the implementation of the 1978 reforms, farmers now have more choice in deciding which crops to grow, and they readily respond to differences in profitability by shifting land from one crop to another.

TABLE 8.2 Area Sown to Various Crops, 1978-86 ('000' ha)

	1978	1980	1982	1984	1985	1986	1978-86 (percent change)
Grain crops	120,587.0	117,234.0	113,462.7	112,884.0	108,845.3	110,951.5	-8
Rice	34,420.9	33,878.7	33,071.3	33,178.7	32,070.0	32,266.1	-6
Wheat	29,182.6	29,228.0	27,955.3	29,576.7	29,218.0	29,630.9	+2
Maize	19,961.1	20,352.7	18,543.3	18,536.7	17,694.0	19,123.7	-4
Soybeans	7,143.7	7,226.7	8,418.7	7,286.0	7,718.0	8,303.1	+16
Potatoes	11,796.3	10,153.3	9,370.0	8,988.0	8,572.0	8,696.6	-26
Industrial crops	14,440.1	15,921.3	18,794.0	19,288.0	22,379.3	20,323.9	+41
Cotton	4,866.4	4,920.0	5,828.7	6,923.3	5,140.0	4,353.8	-11
Oil-bearing crops	6,222.3	7,928.7	9,343.3	8,677.3	11,800.0	11,482.5	+85
Sugar	879.5	922.0	1,115.3	1,230.0	1,485.3	1,479.9	+68
Tobacco	783.9	512.0	1,124.0	897.3	1,312.7	1,158.5	+48
Other crops (vegetables, etc.)	15,076.9	13,224.0	12,498.0	12,049.3	12,392.7	12,879.3	-15
Total cropped area	150,104.0	146,379.3	144,754.7	144,221.3	143,617.3	144,154.5	-4

Sources: SSB, 1986a:71; SSB, 1987:27.

Within the grains subsector, the area sown to soybeans rose by 16 percent, while that of wheat increased by only 2 percent during the 1978-86 period. This also reflects a change in the relative profitability of these crops. The decline in area sown to other grains and related crops varies to some degree, with potato area decreasing the most at 26 percent. This indicates that returns to the farmer were unfavorable, likely caused by a decline in potato consumption.

Grain yields increased from 1 metric ton/ha on average to over 3.5 metric tons/ha between 1949 and 1986 (Figure 8.2). In 1986, milled rice yields reached 5.3 metric tons/ha, wheat 3.0 metric tons/ha, and maize 3.7 metric tons/ha. China's rice and wheat yields rank among the highest in the world. Between 1952 and 1986, grain yields grew at an average annual rate of 2.9 percent. As mentioned above, the yield growth rate increased to 4.2 percent during the 1978-86 reform period, while the area sown to grain continually declined. Thus, the growth in production was solely attributable to improved yields. Technological change has also made an important contribution to yield increases in China since 1949.

Technological Advances

Chinese farmers have constantly faced the pressures created by a large, densely settled population and a limited amount of cultivable land. As a result, farmers place a great deal of emphasis on raising per hectare yields. Yields are maximized by using high levels of inputs and by increasing the number of crops harvested per unit of cultivated land. Sophisticated seed breeding, water control, land preparation, and fertilizer-using technologies have a long history of development in the country. Evidence indicates that, as early as the seventeenth century, rice yields in China reached 2.3 metric tons/ha, a peak that many South and Southeast Asian countries did not surpass until 1979 (Hsu, 1982). In the nineteenth and twentieth centuries, there were no significant advances made in grain production technology until the early 1950s.[6] In fact, grain yields in the first half of the twentieth century were likely no higher than in the 1700s.

When modern inputs such as chemical fertilizers, pump irrigation and tractors were introduced to China's grain production system in 1949, they supplemented rather than supplanted traditional inputs such as organic fertilizers, canal irrigation, and draft animals. The founders of the People's Republic of China committed themselves to strengthening the traditional but already highly developed agricultural infrastructure, which had suffered from a century or more of neglect and substantial war damage. Work included reconstructing the water conservancy system, selecting and improving pure strains of the best existing seed varieties, and reorganizing and enlarging the "Four-tier Agricultural Research and Extension Network." New technology was disseminated from the central government, working down through

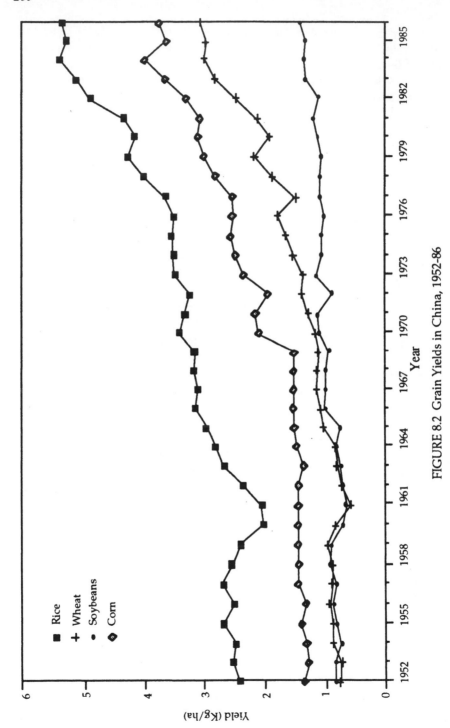

FIGURE 8.2 Grain Yields in China, 1952-86

provincial, county, and finally township (commune) levels. In addition, information was also fed back up from the grass roots to higher administrative levels.

To improve farming methods, the famous Eight-Character Charter was promulgated by Mao Zedong and issued to all levels of government in late 1958. This included: (a) soil improvement and deep plowing; (b) additional fertilization; (c) increased irrigation; (d) popularization of suitable strains; (e) dense planting; (f) pest control; (g) improved field management; and (h) innovation in the area of farm tools. (These principles are known in Chinese as tu, fei, shui, zhong, mi, bao, guan, and gong.)

As a result of this program and related efforts, substantial technological advances were made in grain production, and modern inputs were rapidly adopted by farmers. For example, per hectare chemical fertilizer use amounted to over 135 kg in nutrient terms in 1987 compared to less than 1 kg in 1952. Approximately 44 percent of the total area sown was irrigated in 1987 compared to only 18 percent in 1952. However, between 1978 and 1987, the rate of growth of modern input usage slowed down (compared with the earlier 1965-78 period), while irrigated area as a percent of total cropped area remained unchanged and power machine-tilled area as a proportion of the total declined from about 41 to 35 percent. These changes may well reflect more efficient agronomic practices such as more intensive farming techniques resulting from the household production responsibility system currently practiced in China. More recent data also suggest that higher prices for agricultural products (introduced as part of the 1978 reforms) have also played a large part in explaining improvements in yield. In the future, there may be room for improved technology to contribute to additional yield increases because high-yielding varieties of rice, wheat, and maize have not yet covered all the areas sown to these crops, even though major grain yields are already as high, on average, in China as among other leading producers.

The most important technological breakthroughs occurred in plant breeding in the 1960s with the development and distribution of dwarf, high-yielding rice strains and hybrid maize. In the early 1970s, another breakthrough was made when domestically developed F1 hybrid rice was introduced to Chinese farmers. Since 1978, its use has been actively promoted, and it is now widely used throughout China. Improved strains of wheat and maize were also developed and released following the success with rice. High-yielding varieties (HYV) of foodgrains, coupled with intensive farming techniques, have relaxed, to a great degree, the constraints imposed by the limited land base.

Chinese grain production has increased continuously since 1949 and particularly since 1978. However, sustaining growth in the future will be difficult because improvements in grain output will depend solely upon increases in land productivity. Considering the relatively small size of farming units and the large amount of labor available in China's grain sector, biological/chemical

TABLE 8.3 Per Capita Consumption of Selected Foods, 1978-85 (kg)

Year	Urban consumer			Rural consumer		
	Grains	Oils	Meat	Grains	Oils	Meat
1978	—	—	13.7	248.0	1.3	5.8
1979	—	—	17.4	258.0	—	6.5
1980	—	—	19.0	257.0	1.4	7.6
1981	145.4	4.8	18.6	256.0	1.9	8.7
1982	144.6	5.8	18.7	260.0	2.1	9.0
1983	144.5	6.5	19.9	260.0	2.2	10.0
1984	142.1	7.1	19.9	267.0	2.5	10.6
1985	131.2	6.4	20.2	257.0	2.6	11.0

Sources: USDA, 1987; SSB, 1987.

(labor-augmenting) technologies rather than mechanical (labor-displacing) technologies are the preferable way to bring about increased grain output. However, policy decisions, such as those affecting the availability of inputs and input and output prices, will also have an important impact on the future level of output.

Importance of Foodgrains in the Chinese Diet

Foodgrains provide most of the energy, protein, and other nutrients in the average Chinese diet. Over 80 percent of total calories come directly from grains. According to the World Bank (1985), daily per capita energy availability (from all foods) in China is 2,580 kcal (1980-82). This not only exceeds the average for other low-income countries, but also that for middle-income countries (2,560 kcal). Daily per capita protein availability in China is 68-70 g, and this also surpasses the average for developing countries. However, only about 10 percent of protein intake is derived from animal sources, as compared to 35 percent for the world and 21 percent for developing countries (World Bank, 1985).

Grain consumption is much higher in the countryside than in the cities (Table 8.3). In 1985, per capita consumption of grain in the cities was 131 kg per annum, compared to 257 kg in the countryside. Furthermore, grain consumption patterns have changed dramatically since the institution of the 1978 economic reforms. The reforms resulted in a rapid rise in consumer incomes, which led to growth in per capita consumption of preferred cereals (wheat and rice), meat, and edible oils. For example, in rural areas, per capita consumption of wheat and rice increased by 70 percent over the 1978-85 period (USDA, 1987). Meat consumption almost doubled in rural areas and increased by about 50 percent in urban areas (Table 8.3).

FIGURE 8.3 Major Grain Policy Initiatives, 1949-85

Time period	Policy initiative
1949-53	Basically free market economy. Prices for both consumers and producers determined by the market. The state food agency (Grain Bureau - Liangshi ju) does some trading.
1953-57	State dominates marketing, but a limited private sector remains. Government intervenes through price policy and buffer stocks. State policy with regard to purchase and sale of grains uniformly applied throughout China without regard to region.
1958-62	State assumes a monopoly position on trade; no private sector involvement. Price and area are both determined by state organs.
1963-65	Private sector trade allowed, but only to a limited extent.
1966-78	Private sector involvement totally eliminated. State area planning strictly enforced.
1979-85	Uniform government purchase and sale of grains continues but quotas are reduced. Private sector is allowed to trade grains. Grain prices determined by the government for state purchases only. Area planning abandoned.
1985-	Uniform government purchase and sale replaced by contract sales. Quotas further reduced, and private sector's role increased.

Note: Consumer prices in urban areas have been set by the government since 1953.

Carter and Zhong (1988) estimated income elasticities for grain and meat consumption in China using 1952-86 data.[7] For urban areas, they estimated the income elasticity for foodgrains to be 0.15 and 1.16 for meat. These results are consistent with meat as a "luxury" item in China. For rural areas, they estimated the income elasticity of demand to be 0.37 for foodgrains and 0.97 for meat. These estimates suggest that for a given increase in per capita income, foodgrain consumption in rural areas will rise more than twice as fast as in urban areas. On the other hand, rural meat consumption will grow at about 80 percent the rate of consumption in urban areas. As incomes continue to increase, growth in meat consumption in percentage terms will outpace the growth in foodgrain consumption. In several areas of China, most notably the south, intensive livestock production based on the use of feedgrains has begun to develop in recent years. It is likely in the future that the Chinese will gradually consume more grain indirectly through animal products.

Marketing System and Institutions

Government Involvement in Grain Marketing and Price Formation

Since 1949, the Chinese government has been involved in grain marketing and grain pricing (Figure 8.3). State procurement of grains (mainly wheat, rice, and maize) was initiated in 1953, and this set the stage for eventual total government control over foodgrains. State procurement policies were based on the principle that because foodgrains are scarce staple commodities, their production and distribution must be directly controlled by the state to ensure food security.

Grain producers had to turn over a certain amount of their produce to the state food agency (Grain Bureau – *Liang Shi ju* – operated by the Ministry of Commerce) according to a procurement quota at pre-established prices. Grain procurement quotas delivered to the state were based on the planned acreage cropped to grains and the average productivity of the land calculated according to previous records. Under the commune system, grain procurement quotas were assigned to the commune and then reallocated to the production team. Because peasants were simply laborers in the team and shared the revenues earned by the team, they did not individually sell grains to the Grain Bureau. Quota prices were also uniformly set by the state across the country (with some minor differences across provinces).

In turn, the state food agency distributed rationed quantities of subsidized foodgrains to urban consumers at prices lower than what farmers were paid by the state. But not all "urban" consumers are eligible for food rationing. The Chinese government has maintained very strict residence regulations to curb the flow of peasants migrating into the cities. However, for a variety of reasons, there are some people who reside in cities or towns but lack a legal residence certificate and hence are not eligible for rationed food. For those with legal residence certificates, ration cards are issued by the local Grain Bureau specifying the amounts of certain food items that can be bought at state-set low prices (including rice, wheat flour, and edible oil). The cards even list specific food shops where the rations should be bought.

The Central Government (i.e., Ministry of Commerce) is responsible for all grain procurement through regional offices. The Ministry then redistributes the procured grain to provincial governments (to the provincial Grain Bureaus), which in turn control the urban ration shops. The Ministry sets the criteria for rationing, but the provincial governments administer the program. In some cases, the provincial governments must purchase grain from sources other than the central government. Provincial governments buy from other provinces, local free markets, and foreign sources. Prior to 1978, no private grain sales were permitted in free markets. Therefore, the state monopolized all aspects of marketing, including transportation, storage, and processing, and prices were set by the planning office of the Ministry of Commerce rather than by the marketplace.[8]

When controls over the private sector were relaxed and free markets permitted to reopen in 1978, farmers began to sell surplus grain on the free market, but only after they fulfilled their procurement quotas. At the beginning of the 1985 crop year, the state government introduced a new "contracting" arrangement for state grain purchases in place of the procurement quota system. In principle, the new contract system gives farmers more freedom to decide how much grain to sell to the state. In practice, the state continues to maintain a certain degree of control in the sense that the central government still has targets for grain production and procurement. It tries to use the local bureaucracy to enforce these targets. At the beginning of each crop year, the local state food agencies (operated by the Ministry of Commerce) reach a contract agreement with each producer that specifies a certain volume of grain to be delivered at the state price. The food agency will also purchase over-contract deliveries at a higher price, which is also set by the state. Farmers' decisions on how much of their over-contract surpluses to sell to the state are influenced by the relationship between the state over-contract price and the free market price. Under this dual marketing arrangement, the state food agency is the primary market outlet for Chinese foodgrains, with the rural and urban free markets handling the residual.

In general, producers prefer to sell surplus grain (i.e., surplus after contract deliveries and after home consumption) at the free markets, where prices are about 50 percent higher than the average state prices. Usually (since the early 1980s) the state purchases about 28-30 percent of the grain harvest, and 60 percent is consumed by the farm family. The remaining 10-12 percent becomes "surplus," which can be sold on the free market. Although the current contract system resembles the former procurement quota system in terms of farmers' relative freedom to market grains, quantities delivered to the state declined from over 90 percent of surpluses in the early 1980s, to around 50-60 percent in 1985/86. (Before the reform, "surplus" grains could not be legally sold in free markets. Control was gradually relaxed.) This implies that farmers now have more power to make decisions about the disposition of surplus grain. However, the contracting system still has some compulsory elements in the sense that the contract is reached not on a completely voluntary basis, but with some degree of bureaucratic enforcement.

Free Markets for Grain

After the 1978 reforms, many local free grain markets were established in the same locations as traditional farmers' markets, including local town fairs and urban, streetside agricultural produce markets. In addition, some major grain trading centers were re-established, including markets in Changsha in Hunan,

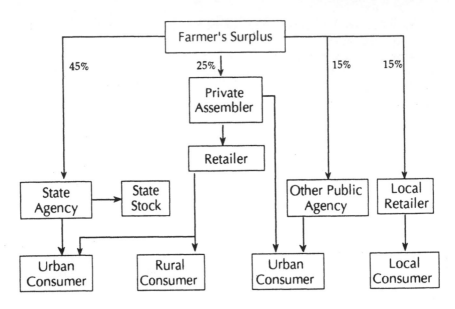

FIGURE 8.4 Grain Marketing Chain, China. *Note:* Percentages of total marketed grain are rough estimates based on personal observations made at local Changsha markets in 1987.

Wuhu in Anhui, and Wuxi in Jiangsu, all located in the east and southeast of China where grain production is relatively specialized. These markets are relatively well-equipped facilities, encompassing trading yards, storage houses, and grading supervisors. They trade milled and unmilled rice, wheat, flour, and soybeans. Grain traders at these markets include local grain producers, part-time farmers who bring grains from surplus to deficit areas for resale, those buying grain for their own family consumption, and urban consumers who work in the cities but do not have a food ration card because their legal residence is in a rural area. Participants at free grain markets also include private "specialized" marketing agents, individual rural people or urban "unemployed" who ship goods from relatively abundant to relatively scarce areas for profits. These individuals only recently began operation after the relaxation of government controls over the private sector. Moreover, some local government-administered food agencies also participate in interregional grain trade. Direct regional reallocation of grains by the central government is declining, so local governments in grain-deficit areas must purchase some of their foodgrain requirements from other regions. Sometimes, this is accomplished via the private markets. Although complete data are not available, recorded interprovincial grain shipments (outside the central planning system) increased from approximately 2 million metric tons in 1980 to over 5 million metric tons in 1985.

Although data are not available to quantify precisely the percentage flows of grain through various sales outlets, the main actors and their relative positions can be determined. As an example, we show the flow of grains through marketing outlets in the Changsha area in 1987 (Figure 8.4). Farmers sold approximately 45 percent of their surplus to the state food agency and 40 percent to local free markets. Twenty-five percent was sold to private assemblers, 15 percent to other public agencies, such as neighboring provincial government food agencies, and 15 percent directly to local consumers, both rural and urban.

Transportation, Storage, and Processing

Although transportation, storage, and processing of grains have remained government monopolies since 1979, this has gradually been changing. The government monopoly over handling grains means that the state is a major player in this regard although farmers do store, and to a lesser degree, process grains. There are still no available data that indicate the relative role of the private sector in these activities, but it is generally believed to be growing, particularly with regard to grain storage.

Grain grading is relatively unimportant in China. The Ministry of Commerce set up a formal grading system, which is only used by the ministry when it sells grain. Grain in the free markets is seldom graded, and what is done is based on visual inspection to determine differences in quality.

Beyond the local level, the free market in grains is not interregionally well integrated in the sense that marketing participants and the products traded are mainly from local or nearby areas. Although government policy now permits the private sector to merchandise grains between provinces (since the introduction of economic reforms in 1978), the backward marketing infrastructure, such as poor roads, lack of bulk transportation means and information inefficiencies, makes it difficult for the private sector to move grain to distant locations. Thus, few private traders are engaged in interregional and interprovincial grain transactions.

In China, interprovincial flows of commodities are chiefly transported by the state-owned railway system. Its facilities are unevenly distributed among provinces and often cannot meet bulk transportation requirements. Storage structures for agricultural products are also owned and managed by the state. Because of the sheer size of the system, it is difficult for the government to run the rail network efficiently. For example, when there is a bumper harvest, the local Grain Bureau is often unable to purchase and store all the grain that farmers want to sell.

The lack of an efficient marketing infrastructure is one of the major con-

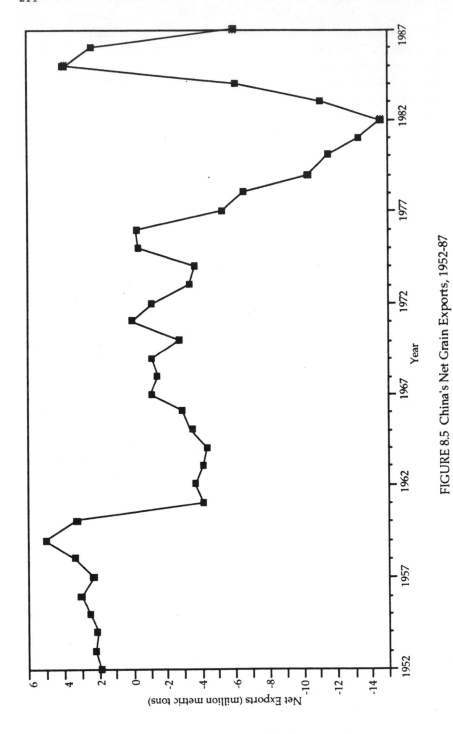

FIGURE 8.5 China's Net Grain Exports, 1952-87

straints to the development of China's grain economy. For example, when grain quota requirements were relaxed, many of Guangdong Province's farmers switched to the production of high-value export crops such as subtropical vegetables and fruits, resulting in regional grain deficits. Instead of purchasing grain supplies domestically from surplus provinces, the provincial government of Guangdong imports grain supplies because of high domestic transport costs and frequent bottlenecks. Government actions to improve the marketing infrastructure would have a desirable impact on grain production and rural incomes.

China's Position in International Grain Trade

Because of its size, population, and the relative importance of grain in its economy, China has had a considerable impact on world grain trade. It was a net exporter of grains in the 1950s (Figure 8.5), while in the 1960s, it began importing wheat to make up for the shortfalls experienced during the "Great Leap Forward".[9] During the 1970s and early 1980s, China was one of the world's largest grain importers (Figure 8.5). From 1978 through 1983, China consistently accounted for over 10 percent of world wheat imports. With bumper domestic harvests in 1985 and 1986, China once again became a net exporter of grain. However, 1987 and 1988 harvests fell below the levels attained in the mid-1980s, and China was importing grain again.

Carter and Zhong (1988) argue that grain exports during the 1950s can be explained by increased grain production during the recovery period after the 1949 revolution and China's requirement for foreign exchange to fuel the development of an industrial complex. There was a severe food crisis in the late 1950s and early 1960s that virtually eliminated all grain stocks. As a consequence, imports were required during the 1960s to rebuild stocks and to meet growing domestic demand. Imports were primarily resold in urban areas.

Grain imports were quite stable throughout the 1960s and into the mid-1970s (Figure 8.5), ranging from 3.5 to 4.3 million metric tons each year. This quantity represented only about 3 percent of total production, but accounted for 10-15 percent of state retail grain sales. In 1979, imports jumped to 12.4 million metric tons and continued to rise in succeeding years. In 1982, imports reached 16 million metric tons, with net imports of 14.6 million metric tons. Interestingly, near record grain harvests were produced during the same period that imports reached new levels. Large grain imports were necessary for a number of reasons including (a) the resale of foodgrains to rural areas to allow farmers to adjust to the economic reforms; (b) a rising demand for grains to produce meat and liquor because of the increased consumer demand for these products; (c) transportation bottlenecks in China; and (d) the government's desire to rebuild stocks to a level approximately equal to an eight-month supply. Stocks were considered adequately rebuilt by the end of the 1980s.

TABLE 8.4 Grain Trade and China's Trade Balance, 1970-86

	1970	1973	1975	1976	1977	1978	1979	1984	1985	1986
				- - - Exports (million $US)[a] - - -						
Grain	110	445	720	450	395	350	335	720	1,362	1,312
Total, agr.	980	2,175	2,855	2,670	2,735	3,255	3,920	5,907	6,696	7,593
Total, all exports	2,155	5,100	7,130	7,260	8,110	10,120	13,750	26,139	27,350	30,942
				- - - Imports (million $US)[b] - - -						
Grain	280	1,015	885	560	1,105	1,460	1,945	1,712	997	1,082
Total, agr.	600	1,750	1,355	950	2,110	2,650	3,750	5,069	5,117	5,145
Total, all imports	2,210	5,025	7,395	6,025	7,120	11,185	15,560	27,410	42,252	42,916
Exchange rate (Yuan/$US)	2.46	1.99	1.86	1.94	1.86	1.68	1.55	2.74	3.51	3.72

Sources: Surls, 1982:188, 193; SSB, 1986:569-572; SSB, 1987:89-91. Some minor adjustments were made to the data presented in the latter source.
[a]Export values, F.O.B. basis. All figures are rounded to the nearest $US 5 million.
[b]Import values, C.I.F. basis. All figures are rounded to the nearest $US 5 million.

In the mid-1980s, China was running an average annual foreign trade deficit of $US 12.1 billion, although agricultural trade maintained a surplus, that in 1986 amounted to $US 2.4 billion (Table 8.4). Wheat accounts for 85 percent of all grains imported into China. Since 1961, on average, China imported about 9 percent of all wheat sold on the world market. China imports wheat from several different countries and regions of the world, including Argentina, Australia, Canada, the EC, and the United States. Australia and Canada were the largest suppliers during the 1980s. China sharply reduced wheat imports from the United States, partly in retaliation for the U.S.-imposed import quotas on cotton and textiles. Textiles are China's leading export item in value terms.

On the export side, high-value rice and soybeans provide foreign exchange, while maize is also an important export crop in some years. Between 1970 and 1985, grain exports averaged 2.72 million metric tons per year, of which 1.3 million metric tons were rice exports and .4 million metric tons were soybean exports. Coarse grain exports, which are more volatile, peaked at 6.3 million metric tons in 1985. Chinese rice exports are destined primarily for Asian and East European nations and Cuba. Major markets for coarse grains are Japan, South Korea, and the USSR. China also exports soybeans and soybean meal to Indonesia, Japan, Malaysia, the USSR, and Thailand. In 1986, soybean exports totaled 1.4 million metric tons, and soymeal exports amounted to 1.1 million metric tons.

Carter and Zhong (1988) analyze future import needs in light of the rising demand for meat and other grain-fed animal products. They predict that China will import about 30 million metric tons of grain (annually) throughout the 1990s, but feedgrain imports will gradually replace wheat. The USDA (1988) predicted Chinese grain production to the year 2000 and then compared their results with Chinese State Council (CSC, i.e., the central government) projections made in 1985. The USDA does not use an explicit model; instead it makes assumptions about the planted area and yield for each major crop. USDA analysts predict that about 500 million metric tons of grains will be produced in 2000. This estimate is similar to that generated by the CSC. We believe these production targets are attainable. However, even with a level of production of 500 million metric tons, imports would still be required to meet demand at current levels of consumer prices.

Public Policy

For a brief overview, see Figure 8.3 which shows the historical evolution of grain policy in China (see also the section on the marketing system and institutions.) Also, additional policy information is provided in Figure 8.6. This summary reflects policies in place in the late 1980s. As discussed throughout this chapter, the government continues to be highly involved in China's grain markets. Major policy instruments include government purchases and sales,

FIGURE 8.6   Major Grain Policy Instruments in China

Instrument	Commodity		
	Maize	Rice	Wheat
PRODUCTION/CONSUMPTION			
Producer guaranteed price	X	X	X
Deficiency payments			
Government purchases	X	X	X
Production quota			
Input subsidies			
- credit	X	X	X
- fertilizer/pesticide	X	X	X
- irrigation	X	X	X
- machinery/fuel	X	X	X
- seed		X	
Crop insurance			
Controlled consumer price		X	X
TRADE			
Imports - tariff			
- quota			
- subsidies			X
- licensing			
- state trading	X		X
Exports - taxes			
- restrictions			
- subsidies			
- licensing			
- state trading	X	X	
OTHER			
Marketing subsidies			
- storage	X	X	X
- transport	X	X	X
- processing		X	X
State marketing		X(a)	X(a)

Notes: (a) food rations in urban areas.

subsidized consumer (urban) prices, input subsidies, and state-controlled foreign trade.

When the Communists rose to power in 1949, the industrial sector was near collapse and growth in China's agricultural economy was at a standstill. As part of its development strategy, the government planned that income generated by agricultural surpluses would be used to fund industrial expansion.

Primary objectives that served as a basis for the crafting of government policies for the grain economy were (a) maintenance of a stable supply of foodgrains for urban consumers at low prices and equitable income distribution among urban consumers; (b) growth of rural incomes; (c) improvement in agricultural efficiency and a rise in agriculture's contribution to the national economy; and (d) the maintenance of social and political stability.

Between 1949 and 1953, state influence over grain production was indirect and depended chiefly on farmer responsiveness to changes in relative prices. The state met the growing urban demand for grain by purchasing in rural markets. However, the terms of trade favored farmers, and the government found that it was spending large sums of money to obtain needed supplies (Lardy, 1984). As a result, in late 1953, the state introduced a system of compulsory quota deliveries at fixed prices. Within a year of introducing the quota procurement system, the government, under the pressure of escalating urban grain requirements, initiated a system of coupon rationing in the cities to distribute state-procured cereals equitably to the urban population. However, coupon holders still paid less for grain than the state gave producers. Because the quota price was very low relative to free market prices, farmers had little economic incentive to complete grain deliveries.

Collectivization, launched in the mid-1950s, provided the institutional and administrative means to meet policy objectives. However, collectivization also meant fewer incentives for farmers because of the egalitarian distribution of economic returns. Massive rural construction activities were carried out using local labor, which was not paid. Further, when communal income was distributed, shares were based on the number of hours worked without adjustment for differences in productivity. Rural unpaid labor was used to construct needed infrastructure, including irrigation projects, land terracing, and rural roads. The government supplied modern inputs, such as chemical fertilizer and pesticides, to farmers through local supply and marketing cooperatives at comparatively low prices, although the supply of some products was rationed.

The effect of government policies on the grain economy prior to 1978/79 are summarized by Barker, et al. (1982: 166-67).

...China maintained low food grain prices and rationed basic necessities so as to insure a more equitable distribution of supply. Although the government subsidized grain prices to consumers, producer prices remained well below international levels and low relative to prices of non-grain crops. An adequate supply of grain was insured through production planning based on area targets.... Utilization of collective labor for land and water development..., coupled with significant technological advances for the major grain crops, allowed the government to maintain low prices.

However, these policies proved expensive because the government maintained low prices for rationed grain in the urban areas and incurred large and regular losses. The policies also had adverse effects on other crops and led to a distorted allocation of resources. For example, compulsory grain acreage planning by the government, coupled with the "taking grain as a key link" policy, squeezed other crops that could have had a comparative advantage in some areas. Similarly, the administered price system and government area planning quotas influenced farmers' decisions on what and how much of various crops to produce. Market signals, which normally lead to comparatively efficient production, had no place in this system. In late 1978, after a series of debates among policy-makers and academics on the economic and social consequences of policies, the Chinese government adopted radically new measures that gave a higher priority to the agricultural sector and increased its allocation of resources. Policy changes were made in four major areas: (a) price incentives to farmers were improved; (b) a larger share of the government budget was allocated to agriculture to finance large infrastructural projects and poverty relief expenses in rural areas, and an increased supply of industrial goods were earmarked for agricultural enterprises; (c) quota deliveries to the state were reduced, while free markets were reopened; and (d) rural institutions (including physical infrastructure and social organization) were reconstructed, such as the abolition of the commune system and the establishment of the production responsibility system in production to allow greater resource mobility and flexibility in production decision making.

In 1979, the first year of the new policies, quota prices for grain were raised 20 percent, and the above-quota prices were set at 50 percent above the quota price. By 1985, the state procurement price for grain was, on average, 67 percent higher than in 1978. The reopening of free markets and reduced quota deliveries raised grain producers' incomes. Moreover, abandonment of the commune system and implementation of the production responsibility system provided grain producers with badly needed freedom to make decisions based on local economic conditions. Prices rose, not only for grain crops, but also for non-grain crops. However between 1978 and 1987, prices for major grain crops relative to non-grain crop prices remained almost unchanged (Table 8.5). Because the prices of inputs such as chemical fertilizer are still controlled by the government, and only a small share of their fertilizer disposition comes from the government supplies, the grain-fertilizer price ratio does not tell us much about farmer responsiveness to changes in relative prices (Table 8.5). The 1987 ratio was lower than the 1978 one, but data show that ratios in 1980-84 were much higher than in 1978.

The long-range effect of these policy instruments on agriculture as a whole and on grain production in particular is difficult to predict. However, the provision of price incentives, the reopening of free markets, and the creation of new and revitalization of older institutional arrangements in rural China made an obvious contribution to the acceleration of growth in grain yields that occurred after 1978.

On the consumption side, government policies continue almost unchanged

TABLE 8.5 Relative Price Ratios of Major Commodities, 1952-87

Commodity		1952	1957	1965	1978	1980	1985	1987
	Fertilizer[a]	0.37	0.51	0.96	1.14	1.52	1.13	1.03
Grain[b]	Other chemicals[a]	0.06	0.08	0.14	0.19	0.25	0.11	0.09
	Raw coal[a]	6.99	8.44	8.07	8.78	11.48	9.93	8.17
	Rice	1.63	1.55	—	1.20	—	1.24	1.17
	Maize	1.84	1.59	—	1.44	—	1.47	1.30
Wheat[c]	Soybeans	1.11	1.02	—	0.67	—	0.66	0.58
	Peanuts	0.85	0.67	—	0.54	—	0.44	0.43
	Rapeseed	1.09	0.67	—	0.50	—	0.46	0.48
	Cured tobacco	0.20	0.19	—	0.20	—	0.28	0.24

Source: SSB, 1986: 637-640, 1988: 791-93.
Note: The same unit of measure is used to compare commodities (kg, m. ton, etc.).

[a]Prices used to calculate the ratios are average state-set retail prices. Fertilizer and chemical ratios are based on standard weights; for example, ammonium sulfate with 20 percent nitrogen.

[b]Prices for grain as a category refer to the weighted average of procurement prices.

[c]The ratios are all computed on the same basis using the relationship between salt and the individual commodity. For example, in 1952, 1 kg of wheat had the same value as 70 kg of salt, while the same unit of rice exchanged for 43 kg of salt, so the wheat/rice price ratio was 1.63. Salt is chosen as a converter because the price of salt is more stable than almost any other consumer good in China.

since the mid-1950s. The government is committed to supplying rationed quantities of staple foods to urban consumers at low subsidized prices. Since the mid-1950s, the rationed retail prices of foodgrains remained virtually unchanged, while state budget allocations for food price subsidies soared from 1.9 billion yuan in 1961 to over 20 billion (or $US 5.8 billion) in 1985 (Table 8.6). By 1985, urban consumer grain subsidies accounted for over 10 percent of the national budget (Table 8.6). In addition, agricultural subsidies ranged from 26 to 40 percent of the total budget. If data were available that allowed the added costs of subsidized imports to be included, consumer subsidies might exceed 20 percent of the budget. The subsidy to urban grain consumers is very costly, and the Chinese government recognizes that steps must be taken to reduce the budgetary burden.

TABLE 8.6   Agricultural and Grain Subsidies in China, 1950-85

Year	State budget expenses (billion yuan)	Agricultural subsidies[a] (billion yuan)	% of budget	Consumer grain subsidies[b] (billion yuan)	% of budget
1950	6.52	0.03	0.00	0.00	0.00
1951	13.31	0.00	0.00	0.00	0.00
1952	18.37	0.00	0.00	0.00	0.00
1953	22.29	0.22	0.00	0.00	0.00
1954	26.24	0.05	0.19	0.00	0.00
1955	27.20	0.06	0.20	0.00	0.00
1956	28.74	0.07	0.24	0.00	0.00
1957	31.02	0.07	0.22	0.00	0.00
1958	38.76	0.08	0.21	0.00	0.00
1959	48.71	0.94	1.93	0.00	0.00
1960	57.23	0.10	0.17	0.00	0.00
1961	35.61	10.35	29.06	1.91	5.36
1962	31.36	8.08	25.76	2.89	9.21
1963	34.23	5.18	15.13	2.38	6.96
1964	39.95	4.58	11.46	2.40	6.01
1965	47.33	4.23	8.94	2.05	4.33
1966	55.87	4.18	7.48	2.10	3.75
1967	41.94	5.50	13.11	2.38	5.68
1968	36.13	2.91	8.05	2.49	6.89
1969	52.68	3.17	6.02	2.72	5.16
1970	66.29	7.02	0.59	3.04	4.59
1971	74.47	7.94	10.67	2.44	3.28
1972	76.66	9.53	12.43	2.97	3.87
1973	80.97	10.82	13.37	2.95	3.64
1974	78.31	14.01	17.89	3.30	4.21
1975	81.56	15.19	18.62	4.18	5.13
1976	77.66	20.12	25.90	4.99	6.43
1977	87.45	17.91	20.48	4.94	5.65
1978	112.11	16.00	14.27	3.63	3.24
1979	110.33	27.64	25.05	7.33	6.64
1980	108.52	33.84	31.19	10.80	9.95
1981	108.95	44.12	40.50	13.48	12.37
1982	112.40	41.65	37.06	14.72	13.10
1983	124.90	49.99	40.03	18.28	14.64
1984	150.19	48.48	32.28	20.50	13.65
1985	186.64	49.98	26.51	20.40	10.93

Source: Carter and Zhong, 1988: 44.
[a]Agricultural subsidies include consumer subsidies on imported grain.
[b]Edible oil subsidies are also included in this column. Consumer grain subsidies include consumer subsidies on domestic grain.

The Thirteenth Congress met in late 1987 and reaffirmed the government's commitment to the production responsibility system. The government planned no major changes in its grain pricing policy. In the latter part of 1988 (because of three successive years of stagnant grain production), the government

TABLE 8.7 Grain Prices and Price Ratios

Year	Actual procured price (1) (Yuan/100 kg)	Free market price (2) (Yuan/100 kg)	Ration retail price (3) (Yuan/100 kg)	Procured: Free market (1)/(2) (percent)	Procured: Retail (1)/(3) (percent)	Free market: Retail (2)/(3) (percent)
1980	31.56	55.40	28.34	57	111	195
1981	33.02	54.60	28.10	61	118	194
1982	33.92	56.00	28.34	61	120	198
1983	35.44	53.80	28.88	66	123	187
1984	36.24	48.60	29.04	75	125	169
1985	37.28	47.90	28.18	78	132	170
1987	50.90	85.30	44.20	58	117	193

Sources: Ministry of Commerce for free market prices; SSB, 1988: 789, 94.
Note: Grain prices for all the categories in the table are weighted averages for all grains.

announced an 18 percent increase beginning in 1989 in the price of grains marketed to the state. At the same time, no changes were made in consumer prices. In order to increase grain production further, the government planned to increase the availability of agricultural credit and to guarantee a supply of agricultural inputs such as fertilizers and other chemicals (Li, 1989). They also planned the introduction of more market mechanisms to provide producers with clearer economic signals, consisting of a package of measures including more use of the free market for grains and the diffusion of new varieties. As shown in Table 8.7, however, since 1980 both the procurement and rationed (retail) prices of grain did not significantly move toward the free market price. In the short run, grain prices for both producers and consumers will not reflect the true (relative) scarcity of the commodity. The private sector, however, will play an increasing role in setting prices for foodgrains. State budget allocations for agriculture and grain imports will probably not exceed current levels. If even part of the growing demand for grain is to be met domestically, Chinese grain producers must receive higher prices. Urban grain prices must rise, but this will be difficult because increased consumer prices will require simultaneous hikes in urban wages if a reduction in the urban standard of living is to be avoided. Success will also depend crucially on the degree and speed of technological advance and infrastructural improvement in rural China.

Political events in the spring and summer of 1989 make China's future even more problematic (Joint Economic Committee, 1989). In addition, inflationary pressures and a growing trade deficit will affect economic performance, as well as the food balance sheet. Progress on the badly needed price reforms may be stalled for at least a few years. Immediate concern over the trade balance and a foreign exchange shortage may put a lid on grain imports. In view of its recent political unrest, China's policy toward freer international trade may change, and this will affect imports. The current government has announced a desire to move back toward "self-reliance" as part of its development goal.

Notes

1. In official Chinese statistical sources, "grain" is understood to include rice, wheat, maize, soybeans, sorghum, millet, potatoes, and other coarse grains. We use this definition because available Chinese data are collected and reported in this manner.

2. For more detail on this and the following discussion, see B.F. Johnston (1989), Perkins and Yusuf (1985), and Wen (1989).

3. McMillan, et al. (1989) and Lin (1987) both estimated the separate contributions made to higher agricultural output by the simultaneous effects of higher prices and the production responsibility system. McMillan, et al., estimate that 41 percent of the growth in output (1979-84) was attributable to the production responsibility system. Lin's figure was 60 percent for 1980-83.

4. Lin (1988) provides a good discussion of the production responsibility system.

5. There are no reliable and continuous data available on specific cropped areas prior to 1978. Therefore, Table 8.2 does not cover as long a period as is reported in Table 8.1.

6. The basic reason for stagnation was skewed distribution of land ownership and rural wealth. Before 1949, rich landlords, who accounted for less than 10 percent of the population, owned over 70 percent of China's cultivated land, and nearly all means of agricultural production. Land tenants and smallholders were extremely vulnerable to rises in taxes, which took as much as 70 percent of output.

7. The original estimates published in the Carter and Zhong book were later updated in a paper published in *Economic Development and Cultural Change*. The updated figures are reported here.

8. See Sicular (1987) for a detailed discussion of pricing policy in China.

9. The "Great Leap Forward" was an ill-founded attempt to rapidly modernize China, which caused severe economic hardship, and, when coupled with climatic disasters, resulted in actual starvation.

References

Barker, R., R. Sinha, and B. Rose (eds.). *The Chinese Agricultural Economy*. Boulder, CO: Westview Press, 1982.

Barker, R., D.G. Sisler, and B. Rose. "Prospects for Growth in Grain Production." In R. Barker, R. Sinha, and B. Rose (eds.), *The Chinese Agricultural Economy*. Boulder, CO: Westview Press, 1982.

Carter, C. and F. Zhong. *China's Grain Production and Trade: An Economic Analysis*. Boulder, CO: Westview Press, 1988.

Carter, C. and F. Zhong. "China's Past and Future Role in the Grain Trade." *Economic Development and Cultural Change* 39(July 1991): 791-814.

Hsu, R.C. *Food for One Billion: China's Agriculture Since 1949*. Boulder, CO: Westview Press, 1982.

Johnston, B.F. "The Political Economy of Agricultural Development in the Soviet

Union and China." *Food Research Institute Studies* 21/2 (1989): 97-137.

Joint Economic Committee, U.S. Congress. *Hearings on Agricultural Reforms in the Soviet Union and China.* Washington, DC: U.S. Congress, September 1989.

Lardy, N.R. "Prices, Markets, and The Chinese Peasants." In C.K. Eicher and J.M. Staatz (eds.), *Agricultural Development in the Third World.* Baltimore, MD: Johns Hopkins University Press, 1984.

Li, P. "A Report on the Government Work." *People's Daily.* April 6, 1989.

Lin, J. Y. "Household Farm, Cooperative Farm, and Efficiency: Evidence from Rural Decollectivization in China." Discussion Paper No. 533, Economic Growth Center, Yale University, New Haven, CT. March 1987.

_____. "The Household Responsibility System in China's Agricultural Reform: A Theoretical and Empirical Study." *Economic Development and Cultural Change* 36 (April 1988): S199-S224.

McMillan, J., J. Whalley, and L.J. Zhu. "The Impact of China's Economic Reforms on Agricultural Productivity Growth." *Journal of Political Economy,* 97/4 (1989): 781-807.

Perkins, D. and S. Yusuf. *Rural Development in China.* Baltimore: John Hopkins University Press, 1984.

Sicular, J. "Food Price Policy in China." Paper presented at the conference on *Comparative Food Price Policy in Asia,* Los Banos, Philippines, January 26-27, 1987.

The State Statistical Bureau of China (SSB). *The Statistical Yearbook 1986 (Zhongguo Tongji Nianjian 1986).* Beijing: China Statistical Publishing House, 1986.

_____a. *The Statistical Yearbook of Rural China 1986 (Zhongguo Nongcun Tongji Nianjian 1986).* Beijing: China Statistical Publishing House, 1986.

_____. *A Statistical Survey of China 1987 (Zhongguo Tongji Zhaiyao 1987).* Beijing: China Statistical Publishing House, 1987.

_____. *A Statistical Survey of China 1988 (Zhongguo Tongji Zhaiyao 1988).* Beijing: China Statistical Publishing House, 1988.

Surls, F.M. "Foreign Trade and China's Agriculture." In R. Barker, R. Sinha, and B. Rose (eds.), *The Chinese Agricultural Economy.* Boulder, CO: Westview Press, 1982.

United States Department of Agriculture, Economic Research Service (USDA, ERS). *Agricultural Statistics of the People's Republic of China, 1949-82.* Statistical Bulletin Number 714. Washington, D.C.: October 1984.

_____. *China: Situation and Outlook Report.* RS-87-8, Washington, D.C.: July 1987.

_____. *China: Agriculture and Trade Report.* RS-88-4, Washington, D.C.: June 1988.

Wen, S. "Development of the Cooperative Economy in Rural China." In John W. Longworth (ed.), *China's Rural Development Miracle: With International Comparisons.* Queensland: University of Queensland Press, 1989.

Wong, J. *Land Reform in the People's Republic of China.* New York: Praeger Press, 1973.

World Bank. *China: Long Term Issues and Options.* Washington, D.C.: 1985.

9

The Dominican Republic

Norberto A. Quezada

The Dominican Republic is a Caribbean country roughly 49 thousand km² in size occupying the eastern part of the island of Hispaniola, which it shares with Haiti. In 1989, the population stood at roughly 7 million with an average per capita income of $US 670. Despite a decline in agriculture's share of GDP from 23 percent in 1970 to less than 15 percent in 1987, the agricultural sector still employs about 40 percent of the total labor force (Quezada, 1988: 33). The agricultural sector accounted for more than 60 percent of merchandise exports until 1985, falling to 50 percent in 1987, primarily because of a decline in the value of sugar exports. The agricultural trade balance is, however, still positive, and agricultural trade continues to remain an important component of total trade.

The agricultural sector is driven by exports. In 1986, traditional export crops—sugarcane, coffee, cocoa, and tobacco—accounted for 34 percent of the value of agricultural production and 50 percent of crop production. In contrast, the cereal grains—rice, maize and sorghum—accounted for less than 9 percent of the value of agricultural production (13 percent of the value of crop production), significantly less than livestock products at 27 percent (Central Bank, unpublished data).

Since 1970, food imports in total have grown at about the same rate as nonfood imports, although grain imports increased rapidly. The rate of growth of grain production surpassed population growth, but per capita grain consumption consistently outpaced production gains. Increased use of feedgrains in the poultry industry helped account for some of the rise in consumption, while discrimination against maize and sorghum production through domestic price and trade policies also contributed to growing grain imports.

Rice is the most important staple food in the Dominican Republic, accounting for 31 percent of average caloric intake and 25 percent of all protein consumed.

Pasta and bread, which are manufactured from imported wheat, together contribute roughly 7 percent of food calories and 10 percent of protein intake, while maize and sorghum, which are used by the poultry industry, make up 4 percent of the calories and 12 percent of the protein consumed. Grains eaten in all forms account for 42 percent of caloric and 46 percent of protein intake, and represent 27 percent of household food expenses. Despite their minor role in the agricultural economy, grains are a very important part of the average person's diet. The Dominican Republic meets most of its rice requirements with domestic production, but depends on imports to supply the bulk of feedgrains; 85 percent of the maize consumed, for example, is imported. All wheat consumed in the Dominican Republic is imported.

Grain Production

The Dominican Republic is a tropical country situated between 18 and 20 degrees north of the equator. Warm temperatures throughout the year result in a twelve-month growing season when sufficient moisture is available. Although there is little seasonal variation in temperature, rainfall does change with the season. There are also sharp differences in rainfall between the windward and the leeward sides of four mountain ranges that cross the island. For most of the country, a rainy season lasts from May to October, while the rest of the year is dry. The northern part of the country, however, receives winter rainfall, permitting production of maize, beans, and other short-season crops. About 1.4 million ha of land are suitable for crop production; about 90 percent is currently cropped. An additional 1.2 million ha are in pasture and support livestock grazing.

Rice

Long-grain and nonglutinous rice are produced in the Dominican Republic. Rice, nearly all irrigated, occupies about 7 percent of agricultural land (100 thousand ha) and nearly half of irrigated area (FAO, 1985). After sugarcane, the second largest consumer of fertilizer is rice; about 90 percent of rice area is fertilized each year (SEA, 1984). Annual milled rice production totals about 320 thousand metric tons from two crops (Table 9.1). The larger crop (60 percent of the total) is harvested in the spring season, extending from May to August, with a smaller harvest is produced during the "winter season," extending from October to December. According to the Secretariat of Agriculture (SEA, "Plan Operativo," 1986), about 85 percent of total rice production is harvested in the higher rainfall Cibao Region in the northern part of the country (Figure 9.1).

Since 1970, rice production has expanded at an average annual rate of about 5 percent. Declines in production in 1985 and 1986 (Table 9.1) were primarily the result of a delay in producer payments made by the state agricultural

TABLE 9.1 Supply and Disposition of Rice and Maize, 1970-87

| Year | Area harvested ('000' ha) | Yield (m.tons/ha) | Production | | Imports[a] (Kt) | Beginning stocks (Kt) | Ending stocks (Kt) | Apparent consumption | |
			Rough (Kt)	Milled (Kt)				Total (Kt)	Per capita (kg)
Rice									
1970	83.0	2.6	213.4	136.5	0.0	38.1	41.8	132.8	33
1975	72.3	3.7	268.8	152.4	49.5	43.8	34.8	210.9	45
1981	111.3	3.7	406.8	258.9	60.7	32.0	82.1	269.5	46
1982	103.3	4.4	454.1	261.5	0.0	82.1	82.1	261.5	44
1983	119.3	4.3	508.6	322.8	0.0	82.1	85.2	319.6	52
1984	118.1	4.5	528.8	324.4	0.0	85.2	36.7	373.0	59
1985	110.3	4.7	518.7	320.5	22.0	36.7	51.4	327.8	51
1986	98.0	4.7	456.8	296.9	118.9	51.4	53.5	413.6	63
1987	101.6	4.8	486.2	316.1	0.0	53.5	45.4	324.2	48
Growth rates (percent per year)									
1970-87	1.2	3.7	5.0	5.1	16.7			5.4	2.2
1980-87	-1.3	4.1	2.7	2.9[b]	-100.0			1.2	-1.3

(continues)

Table 9.1 (*continued*)

| Year | Area harvested ('000 ha) | Yield (m.tons/ha) | Production | | Imports[a] (Kt) | Beginning stocks (Kt) | Ending stocks (Kt) | Apparent consumption | |
			Rough (Kt)	Milled (Kt)				Total (Kt)	Per capita (kg)
Maize									
1970	43.0	1.0	44.1		3.2	4.0	4.0	47.3	12
1975	73.6	0.6	42.7		37.1	37.9	6.0	111.7	24
1980	34.8	1.2	43.2		154.3	0.4	22.7	175.2	31
1981	29.1	1.4	39.9		151.0	22.7	36.5	177.0	30
1982	21.3	1.3	27.2		157.6	36.5	13.2	208.1	35
1983	27.2	1.0	27.8		212.2	13.2	51.2	212.1	34
1984	57.9	1.6	92.4		181.4	51.2	19.2	305.8	48
1985	40.5	1.6	66.1		261.9	19.2	0.2	347.0	54
1986	35.0	1.3	46.5		271.9	0.2	1.6	317.1	48
1987	33.8	1.4	46.3		272.2	1.6	1.0	319.1	48
Growth rates (percent per year)									
1970-87	-1.4	1.7	0.3		29.8			11.9	8.5
1980-87	-0.4	1.4	1.0		8.4			8.9	6.2

Sources: SEA, Rice Department, unpublished data; INESPRE, "Boletin Estadistico," 1985-88.
All figures are for milled grains, except rice area, yield, and rough production.
[a]The Dominican Republic exports no rice or maize.
[b]Rough and milled rice growth rates are different because of statistical error.

marketing agency, then the sole rice purchaser. Harvested area increased by just over 1 percent per annum during the same period, indicating that yield improvements were primarily responsible for production growth. Yields have risen by roughly 4 percent per year, and the rate of increase has accelerated since 1982. Rice yields in the Dominican Republic are above average yields obtained in other parts of Latin America (CIAT, 1985).

Rice accounts for approximately 11 percent of the value of crop production. About three-fourths of rice output is produced by 28 thousand irrigated farms, each less than 5 hectares in size. Roughly 40 percent of planted area is owned by the state but is farmed by settlers assisted by the agrarian reform agency. However, production from state land accounts for only 34 percent of rice output because of lower yields compared to privately owned farms. Because rice is so important in the average diet and because of the state's involvement in production, rice receives preferential treatment in terms of producer prices and access to new technology.

Maize and Sorghum

Maize and sorghum are primarily used as feedgrains in the Dominican Republic, with more than 95 percent of the crop consumed by livestock. Maize production is essentially stagnant, averaging about 45 thousand metric tons per year. Since 1970, average yields have increased at somewhat less than 2 percent per year, but declining harvested area resulted in virtually unchanged production (Table 9.1). Farmers find it more profitable to grow other crops, such as beans and cassava, or to use cultivable land for pasture.

To boost feedgrain output, the Dominican government encourages the production of sorghum, which is more drought tolerant than maize and can be grown successfully under rainfed conditions. Sorghum production has increased by about 8 percent per year since 1970 to over 50 thousand metric tons, largely because of an expansion in area. However, the use of more marginal land has resulted in a slight decline in average yields (Table 9.2).

Despite the increase in sorghum production, a declining share of domestic demand for feedgrains is met locally. Feed manufacturers and livestock producers would like to use more sorghum because it has 90 percent of the feeding value of maize (95 percent for ruminants) and is less expensive. However, because imported maize supplies are reliable and easy to obtain, they are used in place of sorghum. Maize imports have grown eleven-fold since 1970, to more than 250 thousand metric tons.

Most domestically produced maize is grown under rainfed conditions on poor land. Fertilizer is used on only 5 percent of the harvested area (SEA, 1984), and the domestic, open-pollinated varieties that predominate are low yielding. With irrigation, hybrids and fertilizer, yields could be much higher, but farmers use irrigated land and fertilizers to produce more profitable crops, such as tomatoes, melons, and other vegetables.

TABLE 9.2 Supply and Disposition of Sorghum and Wheat, 1970-87

Year	Area harvested ('000 ha)	Yield (m.tons/ha)	Production (Kt)	Imports (Kt)	Beginning stocks (Kt))	Ending stocks (Kt)	Apparent consumption Total (Kt)	Apparent consumption Per capita (kg)
Sorghum								
1970	4.6	3.1	14.0	0.0	0.1	0.1	14.0	4
1975	5.4	3.1	16.3	6.5	0.0	0.0	22.8	5
1980	9.3	2.7	25.1	0.0	0.3	6.9	18.5	3
1981	10.7	3.2	34.2	0.0	6.9	4.8	36.3	6
1982	13.3	2.6	34.6	0.0	4.8	6.2	33.2	6
1983	14.5	2.7	38.5	0.0	0.5	0.2	38.7	6
1984	15.9	2.7	42.9	2.0	0.2	1.1	44.0	7
1985	17.2	2.8	47.2	0.0	1.1	8.8	39.5	6
1986	16.4	2.8	45.6	0.0	8.8	8.4	46.0	7
1987	20.6	2.4	50.1	0.0	8.4	4.5	54.0	8
Growth rates (percent)								
1970-87	9.2	-1.3	7.8				8.3	5.0
1980-87	12.1	-1.5	10.4				16.6	14.0
				(Kt)				kg
Wheat								
1970				59.6	—	—	59.6	15
1975				91.7	—	—	91.7	20
1980				161.2	16.8	28.1	149.9	27
1981				165.9	28.1	47.2	146.7	25
1982				158.6	47.2	27.7	178.1	30
1983				174.8	27.7	30.6	171.9	28
1984				167.8	30.6	19.7	178.7	28
1985				243.0	19.7	26.7	236.0	37
1986				223.7	26.7	27.2	223.2	34
1987				249.5	27.2	27.2	249.5	37
Growth rates (percent)								
1970-87				8.8	7.3		8.8	6.0
1980-87				6.4			7.5	5.0

Sources: SEA, "Plan Operativo," 1987, 1988; INESPRE, "Boletin Estadistico", 1985, 1986; ONE, 1983, 1985; Molinos Dominicanos, unpublished data.

About equal volumes of sorghum and maize are produced, but sorghum is planted on only slightly more than half of the area planted to maize. In contrast to maize, sorghum is usually planted on larger, sometimes irrigated, farms, with the exception of the Oviedo region in the south, which has fertile soils and ample summer rains. Sorghum must be harvested during dry periods, encouraging the use of combines. Sorghum area continues to grow because the crop withstands drought; however, it is more difficult for smaller farmers to produce sorghum because mechanical harvesting equipment and timely harvesting services are not readily available.

Wheat

No wheat is produced in the Dominican Republic. Although several attempts have been made to introduce the crop, land suitable for wheat is already planted to high-value vegetable crops. Further, because of high temperatures, low yields can be a problem. However, in experiment station trials, yields as high as 4 metric tons/ha were achieved in the northwest. Although the government actively promotes white sorghum production for human consumption to reduce cereal imports (white sorghum requires special processing, but yields a flour that can be mixed with wheat flour and used in bread making), efforts have been unsuccessful because production subsidies were introduced simultaneously for rice and wheat, making the crop uncompetitive.

The principal wheat supplier is the United States. There is a single, state-owned wheat milling company with facilities in the capital city of Santo Domingo. A second mill site is in Puerto Plata. Since 1970, the volume of wheat imports has increased at an annual rate of roughly 9 percent (Table 9.2).

Grain Consumption and Self-Sufficiency

Although the total volume of grains consumed in the Dominican Republic is small in comparison to many other countries, grains are a very important part of the diet. In 1986, apparent consumption reached the 1 million metric ton mark for the first time. Of this, rice accounted for 41 percent, maize 32 percent, wheat 22 percent, and sorghum 5 percent.

In 1987, the proportion of total grain consumption produced domestically (the self-sufficiency ratio) was 43 percent (Figure 9.3). The ratio varies by crop—98 percent for rice, 15 percent for maize, 93 for sorghum, and 0 for wheat. Over the period 1970-87, the degree of grain self-sufficiency declined because the rate of increase in consumption (roughly 9 percent) was substantially above growth in production (less than 5 percent). The combined self-sufficiency ratio for domestically produced grains—rice, maize, and sorghum—fell from 100 percent in 1970 to 59 percent in 1987 (Figure 9.3).

The annual rate of growth in consumption for all grains between 1979 and

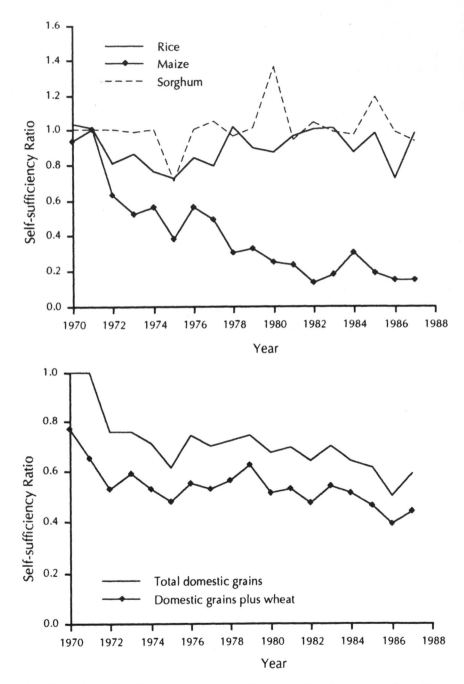

FIGURE 9.3   Self-sufficiency Ratios for Grains in the Dominican Republic.
Source: Tables 9.1 and 9.2.

1987 was close to 8 percent per year, while population growth was less than 3 percent. Demand for maize and wheat grew most rapidly at 12 and 9 percent, respectively. The rapid increase in maize use is attributable to sharply higher demand for livestock feed, particularly for poultry. However, as previously mentioned, maize production remained stagnant, forcing feed manufactures to resort to imports. Sorghum production increased rapidly, but not fast enough to meet the demands of the feed industry. Production of both maize and sorghum has been depressed by government policies, particularly substantial overvaluation of the domestic currency (see the discussion of grain marketing and government regulation below). Growth in wheat consumption resulted from changes in dietary preferences. As incomes have risen, consumers have switched from traditional staples, such as plantain, sweet potatoes and cassava, to bread and pasta products.

Between 5 and 8 percent of maize supplies are consumed directly by humans, mainly in the form of maize flour porridge and arepas (a "pancake" similar to a thick tortilla). Along with wheat bran, rice bran, and soybean meal, maize and sorghum are the main ingredients used in animal feeds. Increasing demand for chicken and eggs, which require substantial amounts of grain, has accounted for the bulk of growth in feedgrain use since 1970.

On a per capita basis, apparent annual rice consumption (51 kg in 1985) in the Dominican Republic is high relative to other Caribbean countries; in contrast, apparent per capita consumption of wheat (36 kg in 1985) is quite low (FAO, 1984). However, an unknown quantity of both grains is smuggled into Haiti because consumer prices are higher there. The Dominican government subsidizes the consumer price of wheat flour and, to a lesser extent, that of rice. The subsidy is enforced with price controls, and supply is maintained with imports paid with overvalued currency and low-interest loans, such as those granted under the US PL-480 food aid program (see the chapter on the United States). Without the subsidy, average per capita consumption would be lower. Annual per capita apparent wheat consumption averaged 36 kg in 1984-87, and apparent rice consumption averaged 55 kg (Tables 9.1 and 9.2).

A recent study of food consumption in the Dominican Republic (Rogers and Swindale, 1988) showed that wheat flour and rice are consumption substitutes. The cross-price elasticity between rice and pasta is 0.85. Plantain is less frequently consumed in place of rice, with a cross-price elasticity between rice and plantain of 0.12. Plantain and pasta were found not to be substitutes. Over time, the consumption of rice and wheat products is rising, while that of plantains and cassava is falling. This is due to a relative increase in the price of plantain and changes in dietary preferences in favor of rice and bread. Own-price demand elasticity was found to be -0.42 for rice and -0.19 for pasta. The demand for poultry products was found to be highly price and income elastic (Rogers and Swindale, 1988: 59-60).

Grain Marketing and Government Regulation

The Dominican government intervenes extensively in the grain economy. Major policy instruments include a producer guaranteed price, direct purchases of rice and sorghum, input and irrigation subsidies, government-controlled consumer prices, and state-controlled production on agrarian reform projects (farms owned by the state but farmed by settlers). A number of instruments are used to control trade, including tariffs, import permits and quotas, and an agricultural marketing agency with an import monopoly. Other indirect instruments are foreign exchange controls with currency overvaluation, credit subsidies, and government regulation of profits earned by marketing agents—private or governmental (Table 9.3).

Most aspects of grain marketing are regulated by the government. The method of control varies according to the crop, ranging from complete regulation of importing, milling, and distribution of wheat and wheat flour, to partial control over maize and sorghum imports. In 1987, rice marketing was liberalized after fourteen years of monopolistic control by a state marketing board. Import permits for maize and sorghum are awarded based on proof of parallel domestic procurement by the importer of these grains, although procured amounts need not be equivalent to the imports authorized.

The importance of grain in the diet is frequently used as the rationale for extensive regulation in the marketplace. A particularly popular argument is the need to protect the poor, who depend on grain even more than the average consumer, from high grain prices. However, government regulation also generates profits. Public interest arguments justifying continued government intervention may be valid, but personal gain may be equally important in maintaining the system.

Marketing Institutions

The main government agencies active in grain marketing are the Price Stabilization Institute (Instituto de Estabilización de Precios—INESPRE); the Agricultural Bank (Banco Agrícola—BAGRICOLA); the Dominican Wheat Mill (Molinos Dominicanos); and the Directorate General of Price Control (Dirección General de Control de Precios—DGCP). These agencies are headquartered in the capital city of Santo Domingo.

Instituto de Estabilización de Precios (INESPRE)

INESPRE, created in 1969, is the most important government agency active in managing food prices. It uses local procurement, direct imports, import control, and storage. The agency is organized as an autonomous body, but the Secretary of Agriculture heads the board of directors, and the President appoints the executive officer. INESPRE's statutory function is to smooth fluctuations in food prices between harvests to protect producers and consumers. However, in attempting to balance these often competing objectives, it sometimes creates price distortions in the marketplace.

INESPRE is involved in price management for many other foods in addition to grains. In 1986, its operations covered some 35 commodities, and it was the the only authorized buyer and seller of sugar, rice, and wheat bran. The organization owned a milk plant, several rice mills, a fish processing plant, five factories (producing condiments and packaging orange juice, taro, and legumes), two farms (one producing pork, the other aquaculture products), and even marketed generic medicines.

Despite its range of activities, the purchase and sale of rice, maize, and sorghum is INESPRE's main business. The organization is very active in all facets of domestic grain marketing (Table 9.4), even though its importance declined during the late 1980s when it was divested of its monopolistic powers. Rice procurement fell from 91 percent of total production in 1980 to 8 percent in 1987. INESPRE maize procurement as a percentage of total production declined from 17 to 3 percent of the total in the same period, while sorghum decreased from 20 to 10 percent (Table 9.4). INESPRE pays the support price, which is the same for all parts of the country, for procured grains. Maize's poor record is partly explained by the agency's inactivity in domestic procurement. Had it acquired more domestic production, fewer import permits would have been issued, thus encouraging feed mills to purchase domestic supplies. INESPRE's limited role in the maize market reflects the fact that the enforcement of the rice monopoly occupied most of the agency's resources.

INESPRE stores purchased grain in its own warehouses; some warehouse space is also rented from private entities. It typically sells to wholesalers and does not clean, dry, or store grain for farmers.

An import permit is required to bring foodstuffs into the Dominican Republic. INESPRE has a statutory monopoly over imports of all grains with the exception of wheat, which is handled by the Molinos Dominicanos. The organization authorizes grain imports by individuals or private companies, who obtain a special import permit from the Secretariat of Agriculture. The agency is also legally mandated to manage (imports, storage, and distribution) foreign food assistance (donations and concessional sales). The United States is an important supplier of maize and rice, purchased with short-term credits granted by the US Commodity Credit Corporation (CCC) and with long-term credits, which are part of the US PL-480 food aid program (see the chapter on the United States).

Total INESPRE purchases in 1985 amounted to $DR 609 million ($US 214 million at the official exchange rate), of which maize, rice, and sorghum accounted for 60 percent. INESPRE subsidized consumers by selling rice at a lower price than it paid producers.[1] The government transferred funds to the agency to cover part, but not all, of the difference. By the end of 1985, the organization's liabilities totalled $DR 309 million, and it had a negative net

TABLE 9.3 Summary of Grain Policy Instruments in the Dominican Republic, 1987

Instruments	Rice	Maize	Sorghum	Wheat[a]
PRODUCTION/CONSUMPTION				
Producer guaranteed price	X	X	X	
Deficiency payments				
Direct purchases	X	X	X	
Production quota				
Input subsidies				
Fertilizer/pesticide				
Machinery/fuel	X			
Seed				
Irrigation	X			
Government-controlled con-sumer price	X	b	b	X
TRADE				
Import — Tariff		X	X	
— Quota		X[c]	X[c]	
— Subsidy				
Export subsidy				
State trading (INESPRE)	X	X	X	
Concessionary imports				X
OTHER				
Marketing subsidies				
— Storage				
— Transport				
— Processing				
State agency (INESPRE)	X			X
Credit — Guaranteed loan				
— Int. rate subsidy	X	X	X	
Exchange rate policy	- - - Overvaluation[d] - -			
Government control over retail markup	X	X	X	X

[a]All wheat needs are imported.
[b]Maize/sorghum are feedgrains. Consumer prices for poultry and eggs are controlled.
[c]Import permits given in direct proportion to local purchases.
[d]Overvaluation is evident from the parallel (free) market rate. The official and parallel (free) markets were unified from 1985 to 1986. In 1987-1989, the official rate was again lower than the parallel rate.

worth of $DR 60 million. INESPRE failed to make timely payments to creditors and lost its access to US credit, including that provided by the CCC. Responsibility for PL-480 food imports was transferred by decree to the Technical Secretariat of the Presidency, a cabinet-level agency. However, the law granting import monopoly to INESPRE has not been changed.

In August 1986 under a new administration, INESPRE's mode of operation was radically changed. Responsibility for rice marketing was transferred to the Agricultural Bank (BAGRICOLA), sugar marketing to the sugar mills, and wheat bran marketing to the Dominican Wheat Mill and the Secretariat of Agriculture. Fees levied by INESPRE on private maize imports to raise revenue were abolished. However, the agency maintained its other operations and has become more active in various social programs, especially the "Popular Sales Program," involving distribution of food at subsidized prices to the poor, and the "Producer Markets," involving construction of farmers' market facilities. Rice is important in the first program but not in the second. As required by law, INESPRE maintains a monopoly over imports of edible oils.

The gross margins (markups) applied by INESPRE vary substantially by commodity (Table 9.4). The organization uses profits earned from the sale of some commodities to subsidize others, particularly rice. Small or negative profits in domestic rice marketing are largely offset by returns realized from the sale of vegetable oils and maize imported at the official exchange rate. Consequently, strong incentives exist to maintain high levels of vegetable oil and maize imports. However, because of cash flow problems and the loss of PL-480, INESPRE did not import maize in 1987. It continues to import edible oils and beans, adding large markups at sale. Even though the agency no longer monopolizes rice marketing, it still buys locally and distributes about 10 percent of total domestic consumption through the "Popular Basket" or "Popular Sales" programs of subsidized sales in poor areas.

The Agricultural Bank (Banco Agrícola — BAGRICOLA)

Agricultural credit was traditionally provided by the state bank, BAGRICOLA, but in the 1980s, private sector commercial and development banks became more active. However, BAGRICOLA is still the main government agency lending to the agricultural sector and is the sole source of agricultural credit for farmers on state-owned land. BAGRICOLA's share of agricultural credit decreased from 60 to 31 percent between 1980 and 1987 (Quezada, 1988: 30). Credit issued to grain producers accounts for more than half of the agency's loan portfolio. For example, in 1984, rice loans represented 55 percent and sorghum 3 percent of all its outstanding loans (BAGRICOLA, 1984).

Although BAGRICOLA's main role is to provide credit, it also participates in other aspects of grain marketing. In 1986, the agency inherited monopoly control over milled rice marketing from INESPRE, but high operating expenses, a costly bureaucracy, and the organization's inability to repay INESPRE's rice milling debts forced the government to further liberalize rice marketing in 1988. As a result, millers are no longer required to sell all of their milled rice to BAGRICOLA. Merchants can purchase rice directly from the mills, but still

TABLE 9.4   INESPRE's Participation in Grain Marketing, 1980-87 (percent)

Description	1980	1981	1982	1983	1984	1985	1986	1987
INESPRE procurement as a percentage of								
Rice								
Production	91	86	84	76	73	75	65	8
Imports	100	100	0	0	0	100	55	0
Consumption	90	88	88	79	79	82	61	8
Maize								
Production	17	16	6	10	14	9	6	3
Imports	100	99	92	97	73	14	1	0
Consumption	74	84	82	67	66	20	3	1
Sorghum								
Production	20	33	40	40	24	27	25	10
Imports	0	0	0	0	100	0	0	0
Consumption	14	45	50	39	15	13	33	14
Markups on								
Imports of:								
Rice	—	3	0	0	0	68	5	
Maize	—	13	31	46	42	4	54	
Procurement of:								
Rice	—	-18	5	6	-11	-26	32	-50
Maize	—	6	6	-4	-14	9	a	12
Sorghum	—	6	6	6	6	-7	3	14
Wheat bran	—	18	17	16	63	183	160	29

Source: INESPRE, Plan Operativo, 1985, 1986.
Negative margins indicate a loss to INESPRE.
aLess than 0.5 percent.

must obtain a permit to import. The agency continues to manage state rice stocks, to make decisions regarding rice imports, and to administer the price support system. State rice reserves are replenished with purchases made by BAGRICOLA from farmers at the support price. Some of this rice is used to supply INESPRE public distribution programs.

For maize and sorghum, BAGRICOLA issues certificates˙ verifying that importers have complied with the law requiring them to make parallel purchases of domestic grain. These certificates qualify the marketer to apply for import permits issued by the Secretariat of Agriculture. Before issuing the certificates, BAGRICOLA requires letters of credit from importers. These are used as prepayment for domestic grain delivered to the importer by farmers who owe money to the bank. This practice necessitates active involvement by bank officers in arranging harvesting, bagging, and transportation of domestic grain, distracting from their primary mission of credit management.

The Dominican Wheat Mill (Molinos Dominicanos)

Molinos Dominicanos (MD) is a 67 percent government-owned corporation that processes all imported wheat, and is the sole domestic supplier of wheat flour. The remaining 33 percent of MD's share capital is owned by private individuals, who have little say in corporate decisions. MD imports wheat duty free and is exempted from obtaining import permits. Flour is sold to bread and pasta makers at prices established by the government, as is wheat bran. Because it is the exclusive supplier, MD controls the amount of wheat available on the domestic market; thus allowing it to maintain state-mandated prices.

Up to 1984, MD sold wheat flour at a profit. When the Dominican peso was devalued in 1985, MD threatened to raise flour prices. At the government's order, it agreed to postpone the increase in exchange for cash transfers from the treasury. Low international wheat prices in 1986 and 1987 allowed the government to end the cash transfers. In 1988, MD was forced to double flour prices because of its precarious financial situation, which resulted from increases in the world wheat price, a further devaluation of the currency, and a lack of government financial assistance.

Directorate General of Price Control
(Dirección General de Control de Precios)

The Dirección General de Control de Precios (DGCP) controls the prices of major foodstuffs, including grain. The DGCP's objective, defined by law 13 of April 1963, is to protect the needy from "undue" price increases. DGCP sets maximum prices for twelve primary agricultural products, including rice, wheat flour, pasta, bread, and maize. Prices are based on costs of production (including manufacture) and are computed based on information supplied by the Secretariat of Agriculture (SEA), INESPRE, industry, and DGCP's staff. DGCP typically sets different prices for the various levels of the marketing chain. The normal markup for wholesalers is 4 percent, while 15 percent is allowed for retailers. DGCP inspectors and supervisors monitor wholesale and retail prices to ensure that marketers do not overcharge. Violators can be given citations and forced to sell at controlled prices in the presence of inspectors. Convicted violators can face fines or prison terms.

Other Institutions

Several other institutions are active in grain marketing and policy in the Dominican Republic. The Centro de Promoción de Exportaciones (Export Promotion Center—CEDOPEX) is in charge of promoting Dominican exports. It has the power to prohibit exports of most exportable commodities, including rice. The Central Bank fixes the exchange rate, while the Customs Authority collects import taxes. Grain is usually imported tax free, except for maize and sorghum, which are charged a 10 percent ad valorem tax on the f.o.b. value.

The Secretariat of Agriculture (SEA) is responsible for agricultural research, extension, training, planning, data collection, and analysis. It also rents machinery to farmers and distributes seeds and agro-chemicals; rice producers typically use these services more frequently than maize producers. The SEA also grants grain import permits (with the exception of wheat) with the concurrence of INESPRE. SEA manages a number of retail outlets in the capital city, where fresh produce and rice are sold. Prices are typically set below cost, with the losses borne by SEA.

Marketing Channels

Production from the more than 28 thousand rice farmers in the Dominican Republic usually moves through private millers (Figure 9.4). At one time, INESPRE operated a few mills, but procurement always remained below 10 percent of the paddy crop; BAGRICOLA now operates these mills, although legally they remain separate entities. During 1973-86, all privately milled rice was sold to INESPRE at fixed prices. In September 1986, BAGRICOLA assumed responsibility for rice marketing and trade, and in 1987, millers were allowed to market directly to wholesalers at prices determined by the government. BAGRICOLA continues to purchase paddy from small farmers and to distribute milled rice to wholesalers and retailers at controlled prices. Low-income consumers can obtain rice at below retail prices directly from BAGRICOLA and through other government programs, such as INESPRE's "Popular Sales Program," and at SEA's retail outlets (but not for rice). Rice imports are authorized by BAGRICOLA, even though SEA issues the permit; however, private imports have not yet occurred.

Maize is privately imported by feed manufacturers and livestock farmers who make up their own feed mixes, or by INESPRE. Private imports require clearance from INESPRE and permits from SEA, which are issued in proportion to importers' local purchases. Importers are required to submit letters of credit to INESPRE or BAGRICOLA to guarantee that they will purchase a fixed amount of domestic grain. Farm households sell to middlemen or to INESPRE, but the maize is eventually purchased by the same individuals that import. The agency usually buys only 5 to 10 percent of domestic production. Final buyers for imported and domestic maize include feed manufacturers and poultry and egg producers. Fewer levels are involved in sorghum marketing than in maize's case, because sorghum farmers produce a greater volume and usually deal directly with final buyers. INESPRE purchases less than 20 percent of total sorghum production at prevailing support prices and resells it after applying a fixed markup.

The State Wheat Mill directly imports wheat without INESPRE's assistance. Wheat flour is channelled to pasta manufacturers and bakeries using a quota system at prices set by DGCP. Flour is illegally re-exported to Haiti because grain prices there are not subsidized to the extent that they are in the Dominican

Republic. This increases the costs to the Dominican government associated with maintaining low wheat prices.

Grain Import Duties

Rice and wheat imports are subject to import tariffs and duties, but these are rarely levied because tax exempt agencies, such as INESPRE, BAGRICOLA, and Molinos Dominicanos, typically make the imports. INESPRE realizes a margin on sales of tax exempt maize and sorghum. This markup is referred to as the "differential" because grain is sold locally at a fixed price, but the acquisition price varies with time and point of origin. The "differential" is, ex-post, similar to the variable levy used by the European Community (see the EC chapter), except that it is not collected by customs. In some years (1984-1986), the "differential" was a fixed charge assessed against private importers who received import permits. After 1986, all imports of maize and sorghum were made by private individuals. Private imports were subject to a customs duty of 10 percent on the FOB value. In 1988, this charge was abolished, and import prices are now essentially the same as world market prices.

Grain Prices

From 1972-87, farm-gate prices were higher than support prices for rice and maize and lower than support prices for sorghum (Figure 9.5). In general, if INESPRE or BAGRICOLA purchase grain, they honor the support price. However, if farmers sell to private buyers, they may not actually receive the support price, even though private buyers are legally required to pay it. Government procurement is the major means used to enforce price supports. The support and farm prices of rice have generally been similar because INESPRE monopolized procurement until 1986, and BAGRICOLA continues to buy substantial quantities.

During 1972-80, the nominal producer price for rice increased by almost 7 percent per year; maize prices grew by nearly 11 percent per annum, while sorghum rose at an annual rate of 5 percent. The producer prices for all three grains grew at even faster rates since 1981 because of inflationary pressures and currency devaluation.

In general, grain pricing policy in the Dominican Republic favors consumers over producers by depressing prices through direct intervention or through exchange rate overvaluation. The degree of difference between domestic grain prices and border prices is influenced more by exchange rate policy than by direct control measures. Greene and Roe (1989) estimated an "equilbrium" exchange rate at which the supply and demand for foreign currency are equal. Their work indicates that the Dominican peso was substantially overvalued until the mid-1980s. Overvaluation can have a significant effect on agricultural production and consumption. The relationship on the producer side is

244

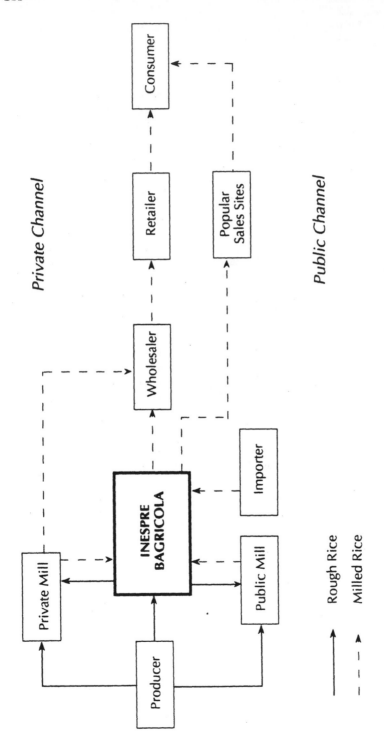

Private Channel

Public Channel

Rough Rice

Milled Rice

FIGURE 9.4 The Rice Marketing System in the Dominican Republic

illustrated using maize in Figure 9.5, but the same pattern applies to the other grains.

Using the nominal exchange rate, the import price of maize was above the average producer price until 1981, indicating a small implicit tax on domestic producers. Since 1981, producer prices were above the import price, suggesting no implicit producer tax. The situation is reversed if import prices are converted to domestic currency at the equilibrium exchange rate; then producer prices were consistently below import prices. During 1983-85, producer prices were 20-35 percent below import prices at the equilibrium exchange rate. The overvaluation of the peso means that domestic prices of commodities imported at world prices are kept lower than would otherwise be the case. This harms producers and reduces domestic production, although low prices benefit consumers. Various consumer price relationships are illustrated in Figure 9.6.

Consumer price subsidies emanate from two sources: small or negative markups on domestically produced and imported grains, and through exchange rate overvaluation. For rice, INESPRE's sales prices have either been close to or below the purchase price (Figure 9.6). In recent years, public sales prices tended to be below public purchase prices for rice, maize, and sorghum. For rice and maize, INESPRE sale prices have almost always been below border prices when exchange rate overvaluation is taken into account. Since 1981, the sales price of maize was above the import price at the official exchange rate, but substantially below the price of imports at the equilibrium exchange rate (Figure 9.6). The degree of implicit subsidy due to overvaluation averaged 20 percent over this period.[2]

Up to 1985, INESPRE sale prices for rice were higher than domestic purchase prices, but the marketing margin was still insufficient to cover costs. From 1985 to 1986, the explicit subsidy was at least $DR 247 per metric ton or 15 percent of the purchase price. Imported wheat receives the highest consumer price subsidy of any of the grains because bread and pasta are important urban wage goods. In addition to the exchange rate subsidy, wheat flour prices have been constant since 1985. Although a complete price series is unavailable, according to the state mill administration, wheat flour prices in 1988 covered only one-half of the costs of production (cost of wheat purchased and milling charges).

Conclusions

Agricultural policy in the Dominican Republic is complex, involving many government institutions and methods of intervention. The main objective is social and economic stability, which, in the government's opinion, requires several elements: low and stable food prices for urban consumers; reduction of producer-consumer marketing margins; provision of aid to poor urban citizens, and attainment of rice self-sufficiency.

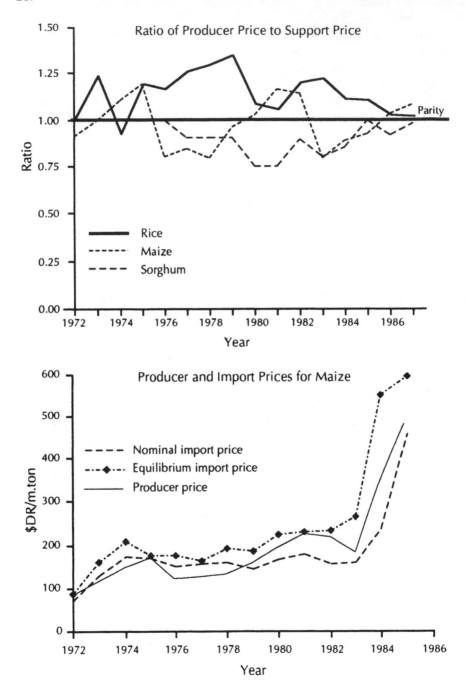

FIGURE 9.5 Grain Producer Price Relationships. *Source:* Data provided by INESPRE; Greene and Roe, 1989.

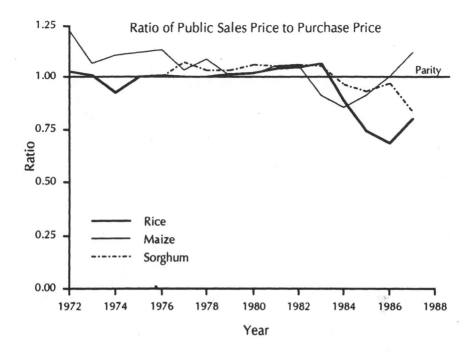

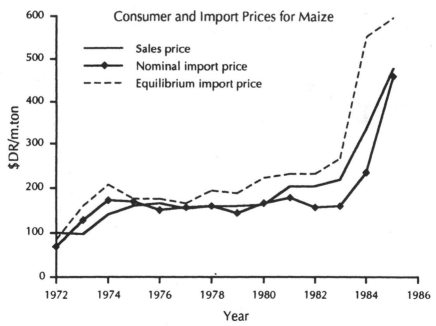

FIGURE 9.6 Grain Consumer Price Relationships. *Source:* Data provided by INESPRE; Greene and Roe, 1989.

However, public grain marketing is unorganized and poorly managed. Several agencies duplicate efforts at the retail level; maize and sorghum procurement is small and uncertain; the marketing agency, INESPRE, has lost credibility and its future role is unclear; agencies, such as INESPRE, BAGRICOLA, and the SEA, compete for funds and influence; and import policy is confusing, because import taxes are inconsistently applied. Existing agencies seem united primarily in their opposition to efforts to curb their control over grain imports, in spite of the uncertainty and confusion for producers and traders created by their intervention. An alternative would be to allow free imports of grain under a better defined and consistent tariff system, such as, for example, a variable levy linked to a price band. This would help to stabilize prices and would be more transparent than the current system (Quezada, 1989).

The experience of the Dominican Republic also demonstrates how macroeconomic policies can either reinforce or neutralize the effects of agricultural policies. For many years, the government emphasized policies that supported industrialization. The provision of cheap food to urban areas was viewed as a way to keep labor costs reasonable. To achieve this, the agricultural sector was implicitly taxed through trade policy instruments, such as tariffs, quotas and prohibitions, and the exchange rate. INESPRE and Molinos Dominicanos played an important role in enforcing this policy through quantity control, while the DGCP used price controls. Although control of staple food prices can be justified on equity grounds, decision-makers face the dilemma of how to furnish affordable food to poor urban consumers, while at the same time providing farmers with sufficient production incentives.

A practical inconsistency exists in the Dominican Republic between macroeconomic and sectoral policies. One objective of agricultural policy is to stimulate the production of grain crops, but macroeconomic policy neutralizes these efforts. By 1989, the real exchange rate had again appreciated to its 1985 level; this provides a strong negative incentive to production through import subsidization. Other authors (Green and Roe, 1989) conclude that macroeconomic policy had a more important effect on rice trade than grain policy itself. Should the policy environment of recent years continue, wheat, maize, and sorghum imports will likely increase at a rate higher than 8 percent per annum. The country is barely self-sufficient in rice, and seems destined to become less so in the future.

Notes

1. The currency of the Dominican republic is the peso ($DR). The peso was at par with the US dollar until 1984, but has subsequently been devalued substantially. In 1988, there were roughly 6.4 pesos to the US dollar.
2. The peso was devalued substantially since 1985. IMF calculations indicate a

depreciation of roughly one-third in the real effective exchange rate over the period 1985-88. This decline helped to reduce the implicit domestic taxes and subsidies created by overvaluation.

References

Banco Agrícola (BAGRICOLA). *Boletín Estadístico.* Santo Domingo: 1984.

Centro Internacional de Agricultura Tropical (CIAT). "Informe de la Sexta Conferencia Internacional de Arroz para América Latina y el Caribe." Cali, Colombia: August 1985.

Food and Agriculture Organization (FAO). *Food Balance Sheets - 1979-81 Average.* Rome: 1984.

_____. *Estudio de Manejo de Sistemas de Riego.* Rome: 1985.

Greene, D. and T. Roe. "Trade, Exchange Rate, and Agricultural Pricing Policies in the Dominican Republic. Volume I: The Country Study" (The Political Economy of Agricultural Pricing Policy). World Bank Comparative Studies. Washington, D.C.: World Bank, 1989.

Instituto de Estabilización de Precios (INESPRE). *Boletín Estadístico.* Santo Domingo: various years.

_____. *Plan Operativo.* Santo Domingo: 1985 and 1986.

Oficina Nacional de Estadísticas (ONE). *Comercio Exterior.* Santo Domingo: 1983-85.

Quezada, N. A. "Ajuste Económico y Sector Agropecuario en República Dominicana." In Instituto Interamericano de Cooperación para la Agricultura (IICA). *Ajuste Macroeconómico y Sector Agropecuario en América Latina.* Buenos Aires, Argentina: IICA, 1988.

_____. *República Dominicana. Bandas de Precios de Importación según el Modelo Chileno para Arroz, Maíz y Aceite de Soya.* North Carolina: Sigma One Corporation, Research Triangle Park, November 1989.

Rogers, B. and A. Swindale. "Determinants of Food Consumption in the Dominican Republic." Medford, Mass.: Tufts University School of Nutrition, April 1988.

Secretaría de Estado de Agricultura (SEA), Departamento de Economía Agrícola. "Impact of Input Price Increases on Agricultural Production Costs." Santo Domingo: SEA, 1984.

_____. *Plan Operativo.* Santo Domingo: SEA, various years.

10

India

Davendra Tyagi

Agriculture's importance in the Indian economy is demonstrated by its contribution to national income, to the supply of industrial raw materials, and to foreign exchange earnings and employment. Although agriculture's share of national income has been declining with industrialization, it still accounts for over 30 percent of gross national product. Agriculturally based industries have increased in number and expanded production. The four main industries, which use jute, cotton, sugarcane and oilseeds as raw materials, accounted for about one-fifth of India's industrial production in the early 1980s. Indian farmers produced almost all the raw material requirements of the sugar, cotton textile, and jute textile industries. Agricultural exports (primarily, tea, coffee, oilseeds, fruits and vegetables, spices, jute, sugar and sugar products, hides and skins, raw wool, animal hair, and vegetable oils) accounted for more than one-third of total export earnings during the decade ending 1980/81. Also important are cotton and jute textiles. According to the 1981 census, over 75 percent of the country's population of 684 million live in nearly 500 thousand villages scattered throughout the countryside. Over 70 percent of the work force was engaged in agriculture and allied activities. Agriculture directly or indirectly continues to be the main source of income for the majority of the Indian populace and is a primary component of the Indian economy.

Foodgrains are by far the most important agricultural commodity in India.[1] About three-quarters of the total cropped area is devoted to their production, and they represent 55 percent of the gross value of agricultural output. Foodgrain production accounts for about 18 percent of gross domestic product. On average, 37 percent of household consumption expenditures are spent on foodgrains,—for low-income consumers the proportion can be as high as 56 percent. Grains also make up the bulk of domestic trade of agricultural commodities—29 percent of the total in the early 1980s. The share of grains in India's foreign trade is not very significant. In 1984/85, they represented less

252

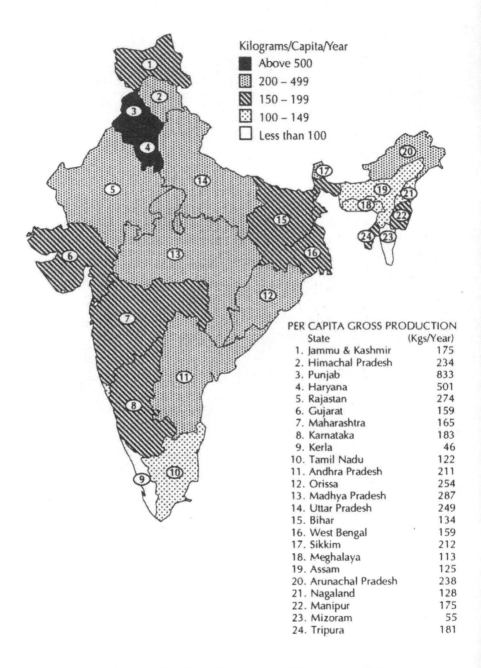

FIGURE 10.1 Statewide Per Capita Production of Foodgrains, 1983-84 (kg).
Source: Institute for Studies in Industrial Development, New Delhi.

than 6 percent of agricultural exports and 15 percent of agricultural imports. However, in drought years, grain imports can increase substantially. For example, in 1965/66—a severe drought year—foodgrain imports accounted for almost 80 percent of total imports of agricultural commodities.

The importance of foodgrains in the economies of the twenty-five states that make up the Indian union varies markedly (Figure 10.1). Grains represent more than 70 percent of the total cropped area of the northwestern states of Punjab and Haryana, but less than 50 percent in the western state of Gujarat, and less than 30 percent in the southern state of Kerala.[2] Because of differences in population and productivity, per capita production of foodgrains varies substantially by state. In 1983/84, 833 kg were produced in Punjab, whereas in Kerala the figure was only 46 kg.

India has a mixed economy, but government planning has had an important role in economic development since independence was achieved from Britain in 1947. The basic objectives of government planning are economic growth, equity, national self-reliance, the alleviation of poverty, modernization, and increased productivity.[3] These objectives are pursued within a democratic framework that allows citizens to own property, to choose an occupation, and to conduct business freely in most areas of economic activity. Because exclusive reliance on free markets and prices is likely to discriminate against those who lack productive assets, the state works to guide the process of development in order to protect disadvantaged groups (Hasim and Singh, 1986). Markets are, however, an integral part of the economy and play a significant role in economic development.

The trade policy regime for India has been extensively documented and analyzed by Bhagwati and Srinivasan (1975). The principles behind India's import policy have traditionally been based on the protection of developing domestic industries from foreign competition and the conservation of scarce foreign exchange. Import substitution was stressed because India's main exports—tea, jute, and raw cotton—in which it has a large market share, are primary products with fluctuating prices and uncertain long-term earning prospects. To an extent, grain trade policies were influenced by a similar unwillingness to rely on the vagaries of international markets and by the overall economic development goals listed above.

The Grain Production System in India

A Land of Smallholders

According to the 1981 agricultural census, there were roughly 89 million farms in India occupying 163 million hectares of land.[4] Average farm size was 1.8 hectares, a decline from 2 hectares in the 1976/77 census. Between the two census years, the number of large holdings (those over 10 hectares) declined by

11 percent, and the average area operated per farm declined by 12 percent. The population of India is increasing at about 2 percent per year. Population pressure is causing the subdivision of holdings, and the number of farms that are too small to be economically viable is on the rise. Marginal farms (less than 1 hectare) accounted for 57 percent of holdings, while small holdings (1 to 2 hectares) made up about 18 percent of the total. Marginal and small farmers together operated only about 26 percent of the total cropped area.[5] Large holdings (10 hectares and above) represented 2 percent of all farms, but controlled nearly 23 percent of the area farmed. Mid-size units—those between 2 and 10 hectares—accounted for 21 percent of the total, but used 50 percent of the agricultural area. In India, most farms are operated by the owner. According to the 1981 agricultural census, 91 percent of holdings were "wholly owned and self-operated," whereas "wholly leased holdings" accounted for only 1 percent of the total.

Small farm size is the primary factor that keeps many Indian farmers at a subsistence level. Most have little opportunity to cultivate more area, which would allow them to generate a marketable surplus. In general, the percentage of area planted to food crops on marginal and small holdings is much higher than on large farms. According to the 1981 agricultural census, farmers owning marginal holdings devoted 72 percent and farmers owning small holdings 69 percent of their farm area to cereals. In contrast, large farms planted only 55 percent of their crop area to cereals.

Variable Climatic Conditions

India's climate is dominated by tropical monsoons. The Indian Meteorological Department distinguishes four distinct seasons: (a) the cold weather season (December-March); (b) the hot weather season (April-May); (c) the rainy season (June-September); and (d) the retreat monsoon (October-November). The southwestern monsoon, which lasts from June to September, accounts for nearly 74 percent of the country's annual rainfall and is the chief source of water in most of peninsular India. The retreat monsoon brings heavy rainfall to the southern portion of the peninsula between October and December. Although it does not usually rain in the cold weather season, some parts of India do receive rainfall sufficient to support another crop. About one-third of the net sown area (total crop area) is in the low rainfall region (below 750 millimeters), while another third is in the high rainfall regions (1,150 mm and above).

A large part of the low rainfall area is in the states of Rajasthan, Gujarat, Maharashtra, and Karnataka. The level of rainfall and its distribution not only affects the per hectare yields of different crops but also restricts crop choice. In the low rainfall regions, only crops that require less rainfall, such as sorghum and pearl millet, can be raised. In high rainfall regions, crops with greater water requirements, such as paddy and sugarcane, can be grown successfully.

Because of the seasonal distribution of rainfall, the principal (*kharif*) harvest occurs in the autumn months (September onwards), while the winter (*rabi*) crop is harvested from April to June. Kharif cereal crops include rice, sorghum, pearl millet, maize, ragi (*Eleusine coracara*), and small millets. Rabi cereals are comprised of wheat and barley, and rabi pulses, including Bengal grams, pigeon pea, green gram, black gram, and lentil. During the period 1969/70 to 1973/74, rice accounted for 41 percent, wheat 23 percent, other cereals 26 percent, Bengal gram 5 percent, and other pulses 6 percent of total foodgrain production. By 1983/84, the mix of grains had changed considerably. Wheat's share rose to 30 percent, other cereals declined to 23 percent, and total pulses dropped to 8 percent. Rice's share declined slightly to 39 percent. The increase in the share of wheat was due to changes in technology and the relative profitability of the crop. These issues are discussed further below.

The Importance of Irrigation

Because of the variability in rainfall, irrigation has long been important in India. As population increased, the need for irrigation grew. Total irrigated area jumped from less than 20 million hectares in 1950/51 to over 60 million hectares in 1983/84. Over the years, the share of this area supplied by tubewells increased at a phenomenal rate (Table 10.1). In 1960/61, tubewells accounted for less than 1 percent of total irrigated area, but by 1970/71, the proportion increased to 14 percent and to 27 percent by 1984/85 (Table 10.1). Although the government helped to stimulate this expansion through rural electrification, the sharp increases in area irrigated were made possible largely because of investments by farmers. Tanks for water storage are mainly owned and maintained by village communities or by individual farmers, with little government assistance or control. Wells are also privately owned, although the government occasionally provides a subsidy to dig wells for irrigation purposes. Government investment in the construction of canals accounted for most of the growth in the area irrigated by this type of system.

The 1981 agriculture census revealed that irrigated area as a percentage of net sown area was as high as 40 percent on marginal holdings, declining as the holding increased in size. For large holdings, irrigated area was as low as 16 percent. Thus, small and marginal farms have better access to irrigation than large farms. Canal irrigation accounted for about 39 percent of total irrigated area in 1980/81.[6] The proportion of area irrigated by canals across farms of different sizes was more or less the same. The percentage of area irrigated by wells grew with an increase in size of holding. However, the percentage of area irrigated by tubewells on various size farms was more or less the same. Water availability is an important factor affecting the level and variability of grain production in India, and the spread of irrigation has been important in allowing farmers to take advantage of new high-yielding varieties (HYVs) of wheat and rice.

TABLE 10.1 Irrigated Area by Type of Irrigation (million hectares)

Source	1950/51	1960/61	1970/71	1975/76	1976/77	1980/81	1983/84	1984/85
Government canals	7.2	9.2	12.0	12.9	13.0	14.5	15.7	15.4
	(34)	(37)	(39)	(37)	(37)	(37)	(38)	(37)
Private canals	1.1	1.2	0.9	0.9	0.8	0.8	0.5	0.5
	(6)	(5)	(3)	(3)	(2)	(2)	(1)	(1)
Tanks	3.6	4.6	4.1	4.0	3.9	3.2	3.8	3.3
	(17)	(19)	(13)	(12)	(11)	(8)	(9)	(8)
Tubewells	—	0.1	4.5	6.8	7.4	9.5	11.0	11.3
	—	(1)	(14)	(20)	(21)	(25)	(26)	(27)
Other wells	6.0	7.2	7.4	7.5	7.6	8.2	8.5	8.7
	(29)	(29)	(24)	(22)	(22)	(21)	(20)	(21)
Other sources	3.0	2.4	2.3	2.4	2.3	2.6	2.4	2.6
	(14)	(10)	(7)	(7)	(7)	(7)	(6)	(6)
Total (net irrigated area)	20.9	24.7	31.1	34.5	35.1	38.8	42.0	41.8
	(100)	(100)	(100)	(100)	(100)	(100)	(100)	(100)

Source: GOI, Directorate of Economics and Statistics, Ministry of Agriculture. Indian Agriculture in Brief. Various issues.
Note: Figures in parentheses are percentages of the total net irrigated area. Detail may not add because of rounding.

Development Policy and Growth in Grain Output

Attaining foodgrain self-sufficiency has been a primary aim of the Indian government since the inception of the planning process. During the first two plan periods (1951/52 to 1960/61), the main goals were to expand irrigated area, to increase the area under cultivation, to bring about institutional changes to secure the rights of tenant farmers, to abolish money lending at excessive rates of interest and improve farmers' access to credit, and to improve the availability of education, health, and other social services to the poor.[7]

These objectives were seen as contributing to increased food production and higher rural welfare. However, as the opportunities to increase food production by extending the area under cultivation declined by the mid-1960s, emphasis shifted to raising the productivity of land already under cultivation. At the same time, new high-yielding, short-duration varieties of wheat (from Mexico) and rice (from the Philippines) released by international research centers became available for commercial cultivation (Dantwala, 1986). Indian planners promptly took advantage of these discoveries, enthusiastically applying their administrative and technological resources toward the rapid adoption of high-yielding varieties. HYVs of wheat and rice quickly became popular, but varieties of sorghum, pearl millet, and corn hybrids, which were also promoted, were found to be susceptible to disease and failed to bring about any major changes in coarse grain production.

The new rice and wheat varieties had enormous potential for increasing yields, but required the use of substantial amounts of fertilizer and other modern inputs. Thus, from 1967/68 onwards emphasis shifted toward increasing the availability of modern farm inputs, such as new seeds and chemical fertilizers, encouraging farmers to use these, and promoting the mechanization of some agricultural operations. The goal of this development strategy was to find methods to increase yields through the use of modern inputs and improved methods of production. Thus, it was explicitly recognized that technology itself is a major agricultural input. The strategy required farmers to invest in technology that allowed them to shift to a higher production possibility curve. For this to occur, favorable price incentives were needed as well as the ready availability of modern farm inputs at reasonable prices. The creation of a favorable price environment was a major objective of government policy for foodgrains.

The Importance of Pricing Policy

The Agricultural Prices Commission, set up to advise the government on agricultural pricing, has played a significant part in the policy process. The Commission bases its advice on the following: (a) the need to provide incentives to encourage the adoption of new technology and the maximization of production; (b) the need to ensure rational use of land and other production resources; and (c) the likely effect of agricultural price policy on the rest of the economy, particularly on the cost of living, wages, and the industrial sector. During the late 1960s, the primary emphasis was on maximizing production because of India's then recent experience with several serious drought-induced shortages of basic agricultural products. By 1980, the position had changed because of substantial and sustained growth in agricultural output. The focus shifted from maximization of production to developing a balanced production pattern suited to national requirements. The activities of the Price Commission and its role in grain policy are described in more detail below.

Minimum support prices fixed by the government, after considering the recommendations of the Price Commission, provide a long-term guarantee to the producer that prices will not fall below a certain level, even in the event of a glut on the market caused by excess production or lack of demand. If the support prices are set sufficiently high, farmers can be induced to make capital investments and to expand input use, thereby increasing output and improving farm productivity and income. Not only do minimum prices need to be fixed, but they must also be enforced. The Food Corporation of India, which was set up at the same time as the Commission, fills this role by purchasing wheat and rice at the support price and distributing these commodities to consumers through "fair price shops," which are administered by the state governments.[8]

Input Subsidies

Quality seed is a critical production input. Seeds of high-yielding and improved varieties of wheat, rice, and millet are supplied to farmers at concessional prices by the government through its own or cooperative seed stores. During 1987/88, 65 thousand tons of seed were made available to farmers at subsidized prices ranging from Rs. 1,500 to Rs. 2,000 per ton.[9] To ensure that farmers obtain fertilizers at reasonable prices, maximum retail prices are fixed by the government. Fertilizer prices are made uniform throughout the country by equalizing freight charges. In 1986/87, the value of fertilizer subsidies (computed as the difference between the price paid by farmers and the farm-gate price of imported fertilizer at the official exchange rate) totalled Rs. 2,560 million.

Irrigation is another area in which implicit subsidies are paid. Although the rates farmers pay for water vary from state to state, they are subsidized by the government. Revenues collected often do not cover fully the costs to operate and maintain irrigation systems, let alone service the capital investment. Expenses to operate and maintain large and medium irrigation systems per hectare of gross irrigated area are estimated at roughly Rs. 137 in 1981/82, while farmers were charged an average of Rs. 63 per hectare (Government of India, 1984). Thus, there is a subsidy on the order of Rs. 74 per hectare of irrigated area. In 1981/82, the total irrigated area of major and minor irrigation systems was approximately 23 million hectares. Thus, the total subsidy spent on irrigation in 1981/82 was approximately Rs. 1.7 billion. Since irrigation is important in the production of foodgrains, this represents an important subsidy to foodgrain producers.

Finally, the average tariff charged for electricity supplied to the agricultural sector is Rs. 0.18 per unit, while the average generating cost was Rs. 0.80 per unit in 1987. The difference represents a further subsidy to the farm sector.

Supply Response and Changes in Production

The share of modern inputs in the total used in Indian agriculture is estimated to have risen from less than 2 percent during 1950/51 to 1959/60 to over 30 percent by 1982/83 (Chakravarty, 1988). As a consequence, supply elasticity estimates based on older historical data can be misleading. In one of the more recent studies, Krishna and Raychaudhuri (1980) calculated output elasticities for rice and wheat that are significantly higher than the acreage elasticities obtained for earlier pre- and post-independent India. At the national level, they found that the price elasticity of rice output is 0.45 percent, and the price elasticity of wheat output is 0.59. For Punjab, an important wheat-producing state, the price elasticities of wheat acreage, yield, and output were 0.28, 0.43, and 0.82 compared with 0.22, 0.34, and 0.59 for the country as a whole. The responsiveness of supply to changes in price has important implications for the balance between supply and demand, and places significant constraints on government pricing policies.

As a result of development efforts, foodgrain production (including pulses) rose from 74 million tons in 1966/67 to over 152 million tons in 1983/84, and in 1988/89 production is estimated to have surpassed 171 million tons. The average yield per hectare rose from 644 kg in 1966/67 to 1,184 kg in 1985/86. Rice production jumped from an annual average of 39 million tons during the triennium ending 1969/70 to 61 million tons during the triennium ending 1985/86. The growth in wheat production was even more spectacular, rising about 2.5 times from an average of 18 million tons to almost 46 million tons between the two triennia. Cereal output grew 3 percent per year between 1967/68 and 1985/86. Rice production increased by 2.5 percent per annum and wheat production at a rate of over 5 percent (Table 10.2).

In India, rice accounts for the largest area under a single crop and is grown during the spring, summer and autumn seasons, although most of the crop is produced during the winter and autumn.[10] On average, the area under rice represents about 40 percent of the area under cereals, about 30 percent of the area under foodgrains, and 25 percent of total cropped area. Although some rice is grown in all states, Andhra Pradesh, West Bengal, Uttar Pradesh, Punjab, Tamil Nadu, and Bihar account for over 50 percent of total output (Table 10.3). Rice accounts for about 45 percent of cereal output and 40 percent of total foodgrain production.

More than 4 thousand varieties of rice are grown in India. Of these, about 700 are commercially significant. All rice varieties except the scented ones are classified into three groups based on the length-breadth ratio: common, fine, and super fine.[11] The scented varieties, which are often preferred by consumers, are primarily classified by the area where they are grown.

Despite seasonal fluctuations, rice area has trended upward during the last two decades—expanding from 36 million hectares in 1967/68 to 41 million hectares in 1978/79 (Table 10.4). After falling slightly in the early 1980s, it has again regained this higher level. The area under rice grew by 0.7 percent per annum between 1967/68 and 1985/86 (Table 10.2). The area planted with high-yielding varieties of rice expanded dramatically, from less than 2 million hectares in 1967/68 to almost 24 million hectares in 1985/86. HYVs now account for more than half of total rice area. The proportion of the area that is irrigated also increased from less than 38 percent to 42 percent over this period.

Rice production, has fluctuated from year to year, primarily because of the effects of weather on yields, but the overall trend in output has been upwards. On average, rice production grew at the rate of 2.5 percent per annum between 1967/68 and 1985/86. Over the same period, yields for India as a whole rose from 1,032 kg to 1,552 kg per hectare. The contribution of increases in yield to the growth in rice output during this period was about 75 percent. However, productivity varies quite markedly by state. Compared with the all-India average of 1,552 kg per hectare in 1985/86, the average yield in Punjab was about 3,179 kg per hectare and 2,797 kg per hectare in Haryana. Conversely, it is only 1,077 and 1,128 kg in Madhya Pradesh and Bihar, respectively.

Next to rice, wheat is the second most important cereal, accounting for roughly 30 percent of total foodgrain production. Wheat is generally sown in November and December and harvested during April-May. It is mainly grown in the northwestern states, with Uttar Pradesh producing over 35 percent of the total, followed by Punjab with about 22 percent. Even before the introduction of HYV wheat to India, of eighteen recognized species, only four—*Triticum*

TABLE 10.2 Growth In Area, Production, and Yield of Foodgrains in India, 1967/68 to 1985/86 (percent per year)

Crop	Area	Yield	Production
Rice	0.6	1.9	2.5
Wheat	2.4	3.2	5.6
Coarse cereals	-0.9	1.6	0.7
Total cereals	0.4	2.5	2.9
Total pulses	0.4	0.1	0.5
Total foodgrains	0.3	2.3	2.6

Source: Calculated from data provided by the GOI.

TABLE 10.3 Average State Shares in the Production of Foodgrains, 1983/84 to 1985/86 (percent)

State	Rice	Sorghum	Pearl millet	Maize	Wheat	Barley	Total cereals	Gram	Total pulses	Total foodgrains
Andhra Pradesh	13	11	5	6	0	0	7	a	4	7
Assam	4	0	0	a	a	0	2	0	1	2
Bihar	9	0	a	12	7	4	7	3	7	7
Gujarat	1	4	22	4	3	1	3	2	4	3
Haryana	2	a	8	1	10	7	5	8	4	5
Himachal Pradesh	a	0	0	7	1	2	1	0	a	1
Jammu and Kashmir	1	0	a	5	1	a	1	0	a	1
Karnataka	4	15	4	6	a	0	5	1	4	5
Kerala	2	0	0	0	0	0	1	a	1	1
Madhya Pradesh	8	16	2	13	7	8	9	28	20	10
Maharashtra	4	40	11	2	2	a	6	4	9	7
Manipur	1	0	0	a	0	0	a	0	0	a
Meghalaya	a	0	0	a	0	0	a	0	0	a
Nagaland	a	0	0	a	0	0	a	0	0	a
Orissa	8	a	a	2	a	0	4	1	8	4
Punjab	8	0	1	7	22	5	12	2	1	11
Rajasthan	a	4	27	13	9	26	5	24	13	6
Sikkim	0	0	0	1	0	0	a	a	0	a
Tamil Nadu	8	5	6	1	0	0	5	a	2	5
Tripura	0	0	0	0	0	0	a	0	0	a
Uttar Pradesh	12	5	14	19	35	46	20	25	21	20
West Bengal	13	0	0	1	2	1	7	1	2	6
Union Territory	a	0	a	a	a	0	a	0	0	a
All India	100	100	100	100	100	100	100	100	100	100

Source: GOI, Directorate of Economics & Statistics. Bulletin on Food Statistics (thirty-fourth issue).
aLess than 0.5. Detail may not add because of rounding.

TABLE 10.4 Area of Grain Planted, Irrigated, and Under High-Yielding Varieties (HYVs)

	1967/68	1970/71	1975/76	1978/79	1983/84	1985/86
— Rice —						
Total area	36.5	37.6	39.5	40.5	41.2	41.1
Area under HYV	1.8	5.5	12.4	17.0	21.7	23.5
Percentage HYV	5	15	32	42	53	57
Area under irrigation	13.8	14.3	15.1	16.9	17.4	17.3
Percentage irrigated	38	38	38	42	42	42
Yield per ha. (tons)	1.03	1.12	1.24	1.33	1.46	1.55
— Wheat —						
Total area	15.0	18.2	20.5	22.6	24.7	23.0
Area under HYV	2.9	6.5	13.5	16.5	19.4	19.2
Percentage HYV	20	36	66	73	79	83
Area under irrigation	6.5	9.9	12.7	14.9	17.9	17.3
Percentage irrigated	43	54	62	66	73	75
Yield per ha. (tons)	1.10	1.31	1.41	1.57	1.84	2.05
— Sorghum (Jowar) —						
Total area	18.4	17.4	16.1	16.1	16.4	16.1
Area under HYV	0.6	0.8	2.0	3.1	5.4	6.1
Percentage HYV	3	5	12	19	33	38
Area under irrigation	0.7	0.6	0.8	0.8	0.6	0.7
Percentage irrigated	4	4	5	5	4	5
Yield per ha. (tons)	0.55	0.47	0.59	0.71	0.73	0.63
— Pearl millet (Bajra) —						
Total area	12.8	12.9	11.6	11.4	11.8	10.7
Area under HYV	0.4	2.0	2.9	2.9	5.4	5.0
Percentage HYV	3	16	25	26	46	47
Area under irrigation	0.4	0.5	0.6	0.5	0.6	0.6
Percentage irrigated	3	4	6	4	6	5
Yield per ha. (tons)	0.41	0.62	0.50	0.49	0.65	0.34
— Maize —						
Total area	5.6	5.9	6.0	5.8	5.9	5.8
Area under HYV	0.3	0.5	1.1	1.3	1.9	1.8
Percentage HYV	5	8	19	23	33	31
Area under irrigation	0.7	0.9	0.9	0.9	1.0	1.0
Percentage irrigated	12	16	16	16	17	18
Yield per ha. (tons)	1.12	1.28	1.20	1.08	1.35	1.15

Source: GOI, Directorate of Economics & Statistics.
Note: Area in million hectares.

vulgare, Triticum durum, Triticum dicoccum, and *Triticum turgidum* —were cultivated. *Triticum vulgare* accounted for over 86 percent of wheat production, followed by *Triticum durum* with about 13 percent, and the remaining two species with about 1 percent. With the introduction of HYVs in the mid-1960s, the importance of *Triticum durum* species declined, while that of the *Triticum vulgare* species increased.

The area under wheat began rising rapidly with the introduction of new agricultural technology. Area increased from 15 million hectares in 1967/68 to over 23 million hectares in 1985/86. Area sown grew by over 2 percent per annum while yields improved by over 3 percent, resulting in an annual increase in wheat production of over 5 percent. Area under HYV wheat grew from roughly 3 million hectares to over 19 million hectares (Table 10.4). The proportion of wheat area irrigated also increased from less than 45 percent in 1967/68 to over 75 percent in 1985/86. The average wheat yield in India increased from 1,100 kg in 1967/68 to 2,046 kg in 1985/86, while yields in Punjab over the same period rose from 1,860 kg to 3,531 kg per hectare, and in Haryana from 1,730 kg to 3,094 kg per hectare.

Coarse cereals, including sorghum (*jowar*), maize, and pearl millet (*bajra*), are largely grown under dryland conditions, resulting in large year-to-year fluctuations in production. Sorghum and maize are grown during both the autumn and winter seasons, but pearl millet is only grown during the autumn.[12] The three biggest pearl millet-producing states are Rajasthan and Gujarat in the northwest, and Uttar Pradesh in north-central India (Figure 10.1). Maharashtra in the peninsular west accounts for over 41 percent of the sorghum produced, but Karnataka and Andhra Pradesh in the south-central part of the country are also important. The main maize-producing states are Uttar Pradesh, Rajasthan, and Madhya Pradesh. In 1985/86, roughly 5 percent of the sorghum and pearl millet areas, and 18 percent of the maize area were irrigated (Table 10.4). Coarse grain HYVs do not cover as high a percentage of the sown area as do rice and wheat HYVs. Barley, another important coarse cereal, is only grown during the winter season.

Between 1967/68 and 1985/86, coarse grain production was roughly constant. Area declined by just under 1 percent per annum and yields rose by about 1.5 percent each year, resulting in a modest increase in production of about 0.6 percent per year. The area planted to high-yielding varieties of sorghum and pearl millet increased steadily, but production fluctuated sharply from year to year because of climatic variability. Sorghum productivity increased impressively in Maharashtra, where the crop is grown during both winter and autumn seasons. Maize yields in the state of Bihar reached an estimated 4-5 tons per hectare during the rabi season. Since high-yielding varieties of wheat were introduced to India, the area under barley has declined. Barley production has fallen steadily from over 3 million tons in 1967/68 to less than 2 million tons in 1985/86.

India is the world's largest producer of pulses. These are a rich source of protein in a country where a significant portion of the total population is vegetarian. India produces about 13 million tons of pulses each year on about 24 million hectares, occupying about one-fifth of the total area under foodgrains. The major pulses are gram, green gram, black gram, pigeon pea, pea, and lentils. Some pulses are grown in both autumn and winter seasons, whereas others, like Bengal gram, are grown only during the autumn season. The increase in pulse production has been modest over the last twenty years. Area expanded by roughly 0.4 percent annually between 1967/68 and 1985/86, yields per hectare rose by only 0.3 percent per annum, while production increased by 0.7 percent annually. As a consequence, annual per capita availability of pulses declined from roughly 25 kg in the early 1960s to about 14 kg in the early 1980s.

Despite the increased use of irrigation, foodgrain production is still highly dependent on timely rainfall, and year-to-year fluctuations in Indian grain production are quite marked. Data in Table 10.5 show that shortfalls in grain production in the 1980s compared to the previous peak year ranged from about 4 to 14 million tons in contrast to 10 million tons in the early 1970s and 22 million tons in 1979/80. The standard deviation in annual output growth rates for 1968-85 at the national level was 14.3 for rice, 11.4 for wheat, 13.4 for coarse cereals, 16.1 for pulses, and 11.4 for total foodgrains, further emphasizing the instability of Indian crop production. The standard deviation in annual production growth rates is estimated at 13.9 for kharif foodgrains and 9.8 for rabi foodgrains (Rao, et al., 1988). The effect of production variability on national food security has had important implications for Indian foodgrain policies, particularly storage and trade policies.

Grain Consumption

Over the years, the per capita availability (apparent consumption) of foodgrains in India has risen, particularly since the introduction of HYVs (Table 10.6). Based on net production (gross production minus 12.5 percent for seed, feed, and waste), per capita availability, which was 395 grams/day in the early 1950s, rose to between 436 and 478 grams/day between 1982-87 (Table 10.6). However, with the increase in yields, seed now accounts for a lower proportion of production. If an adjustment is made for this, per capita availability in the 1980s would actually be between 450 and 492 grams/day.[13]

In 1985, rice accounted for 46 percent and wheat 33 percent of total per capita consumption of cereals, while all other cereals made up the remaining 21 percent. The most notable change in consumption patterns occurred in the case of wheat, whose share rose from just over 20 percent of total cereals in 1963, to 36 percent in 1983. This growth was mainly at the expense of coarse grains. While the consumption of wheat jumped from 79 gm/capita/day in 1963 to 144 gm/capita/day in 1983, consumption of other cereals declined from 118 gm/ capita/day to 83 gm. The share of coarse cereals fell from 31 percent of total consumption to 21 percent. Wheat is now the second most important staple cereal after rice in India.

An insignificant amount of grains are used for livestock feed in India (Government of India, 1986; Techno Economic Research Institute, 1988). During the past ten years, the use of grain for feed in the poultry and dairy industries has increased, but that for draft animals has dropped substantially because of farm mechanization. Consequently, there has been little net growth in feed demand. However, as average per capita income rises in the future, the demand for dairy products is expected to expand, and this will lead to additional requirements for feedgrains.

Only limited use is made of cereals in the brewing/distilling, starch manu-facturing, and malting industries in India. It is estimated that only about 1 percent of total grain output is used for manufacturing purposes.[14] Data on post-harvest losses of grains are limited, and in calculating net availability, an allowance of 2.5 percent is usually made for waste. The Directorate of Marketing and Inspection (Government of India, 1986) estimates that less than 3 percent of wheat production is lost in storage at the producer level in an average year. Additional losses at the post-producer level are estimated to be less than 0.1 percent. The Committee on Handling of Foodgrains, set up in 1974 by the Food Corporation of India, reported that from 1969/70 to 1972/73 transit and storage losses accounted for between 1.03 and 1.09 percent of all stored grain. Recent studies (Techno Economic Research Institute, 1988) also indicate that losses in storage are low.

TABLE 10.5 Variability in Foodgrain Production

Year	Peak (million tons)	Trough (million tons)	Shortfall of trough from previous peak (percent)	(percent)
1970/71	104.8			
1972/73		97.0	11.4	10.5
1975/76	121.0			
1976/77		111.2	9.8	8.1
1978/79	131.9			
1979/80		109.7	22.2	16.8
1981/82	133.3			
1982/83		129.5	3.8	2.9
1983/84	152.4			
1987/88		138.2	14.2	9.3

Source: Calculated from data from the GOI, Directorate of Economics & Statistics.

TABLE 10.6 Net Availability of Foodgrains

Year (1)	Population (mill.) (2)a	Cereals Net prod. (mill. tons) (3)bc	Cereals Net imports (mill. tons) (4)d	Change in gov't. stocks (mill. tons) (5)	Net avail.a (tons) (6)e	Net availability of foodgrains f (mill. gr.) (7)	Per capita net avail. of foodgrains/day (8)g
1951	361.2	40.1	4.8	0.6	44.3	52.4	395
1956	397.3	50.4	1.4	-0.6	52.4	62.6	431
1961	442.4	60.9	3.5	-0.2	64.6	75.7	469
1966	493.2	54.6	0.3	0.1	64.8	73.5	408
1971	551.3	84.5	2.0	2.6	84.0	94.3	469
1976	617.2	94.5	6.9	10.7	90.7	102.1	453
1980	675.2	88.5	-0.5	-5.8	93.8	101.4	410
1981h	690.1	104.1	0.5	-0.2	104.9	114.3	454
1982h	705.2	106.6	1.6	1.3	106.8	116.9	455
1983h	720.4	103.0	4.1	2.7	104.4	114.7	436
1984h	735.6	122.0	2.4	7.1	117.4	128.6	478
1985h	750.9	116.9	-0.4	2.7	113.9	124.3	454
1986h	766.1	119.9	-0.1	-1.6	121.5	133.2	476
1987h	781.4	115.3	-0.4	-9.5	124.4	134.6	472
1988h	796.6	111.5	1.9	-5.4	118.7	128.4	441

Source: GOI, 1989. Detail may not add because of rounding.

Notes:
a Population figures from 1971 to 1980 are based on the latest projections made by the Registrar General of India. Estimates from 1981 onwards are based on the Expert Committee's population projections as approved by the Planning Commission.
b Production figures relate to the agricultural year ending in June.
c Net production is equal to 87.5 percent of gross production; the remainder, 12.5 percent, is used for feed and seed requirements or wasted. During 1987, 1.7 million tons of wheat in Punjab were lost because of unseasonal rains.
d Net imports from 1981 onwards are based on imports and exports by the Government of India only.
e The level of private stocks is unknown. Estimates of net availability are, therefore, not strictly equivalent to consumption. Net availability = Col. (3 + 4 - 5).
f Includes pulses.
g Calculated by using net imports and stock position as of 12 January.
h Provisional.

Demand Elasticities

Demand elasticities for the various cereals differ, as do elasticities in rural and urban areas. Based on data collected by the National Sample Survey Organization on consumer expenditure in rural and urban areas, the National Commission on Agriculture (Government of India, 1976a) estimated income elasticities of demand for grains. For rice, income elasticities were 0.41 for rural areas and 0.18 in urban areas. For wheat, the estimates were 0.67 for rural areas and 0.37 for urban areas, while for other cereals, elasticities were calculated to be less than 0.01 for rural areas and -0.47 for urban areas. For pulses, the quantity elasticities were much higher at 0.85 for rural areas and 0.66 for urban areas. Thus, as incomes rise in rural areas, demand for wheat and rice, which are preferred over coarse grains, will likely expand. Further, demand for pulses may also grow appreciably. The response of cereal consumption to price changes is high in India. In the case of people below the poverty line, in rural areas the price elasticity is estimated at -0.73 and in urban areas at -0.66. For those above the poverty line, it is estimated at -0.30 in rural areas and at -0.04 in urban areas (Alagh, 1985).

Much of the grain grown in India is retained by the producer for home use, seed, and to pay agricultural laborers. The quantity kept for personal use depends on several factors, such as family size, number of hired laborers and the way they are paid, and total grain production. Because of the generally small surplus, it is often uneconomical for farmers to transport produce to nearby markets. Village merchants, itinerant merchants, agents of large traders, and cooperative organizations collect small lots of grain from farmers, although whenever their marketed surplus is large enough, farmers transport their own grain to market.

Grain Markets

Agricultural markets in India can be broadly classified into three types: (a) *hats* and *shandies*, (b) wholesale markets or *mandies*, and (c) retail markets. Hats and shandies are primary markets held weekly or biweekly, serving a radius of 8-15 kilometers. Haggling is the common method used to fix prices. Mandies are daily wholesale markets that draw supplies from a wider area of 15-60 kilometers and provide some facilities for storage and banking, etc. Retail markets are scattered throughout towns and villages and market all types of goods. These markets receive supplies, not only of grains, but also of other farm products from middlemen in addition to farmers.

At mandies, produce is generally sold by farmers or village merchants through agents known as *katcha arhatias*, who operate on a commission basis (Figures 10.3 and 10.4). These agents provide all the services needed by farmers to sell produce. They may also lend money to farmers for planting and other farm expenses with the stipulation that the producer market all his output through the agent. The buyers, or *pucca arhatias*, may be individuals or firms.

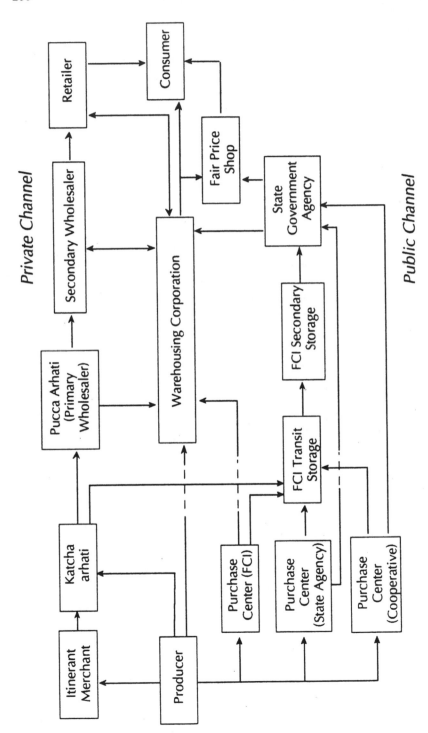

FIGURE 10.3 The Wheat Marketing System in India

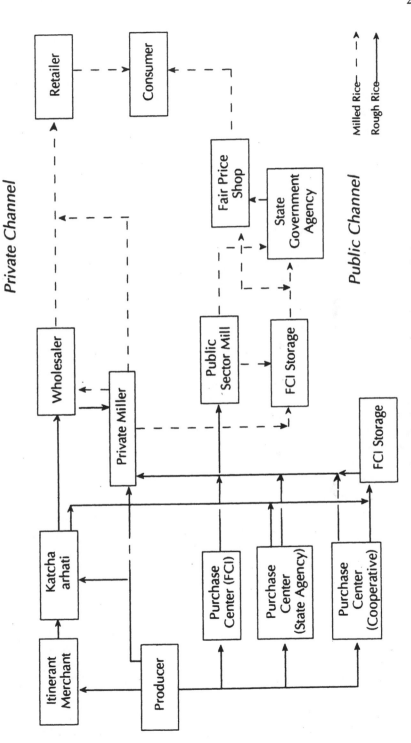

FIGURE 10.4 The Rice Marketing System in India.

They generally use their own capital to buy produce in the assembling markets either for themselves or on behalf of clients in more distant markets. Pucca arhatias usually do not deal directly with the seller (Figure 10.3). The buyer charges his clients a commission for obtaining the grain and also recovers expenses incurred for weighing, processing, packing, and transport.

Over the last forty years, the government has established regulated markets, to solve problems, such as short weights, excessive charges, unauthorized deductions and allowances made by commission agents, and the absence of a dispute settlement machinery for sellers and buyers. The principal objectives of market regulation are to safeguard producer interests and to raise standards in local markets. Farmers, traders, sellers' cooperatives, and representatives of the state governments sit on market committees, which typically have a producer majority. All of the functionaries in the regulated markets are licensed, and business transactions are conducted under rules and laws framed and administered by the market committees. Charges, allowances, and deductions are fixed by the market committee, and anyone charging more is likely to have their license revoked. Disputes between sellers and buyers are submitted for arbitration to a disputes subcommittee. Regulated markets now exist in all important foodgrain-producing states.

Marketing Channels

In India, there are two important marketing channels for grain. In the first, grain moves directly from the producer to the consumer through various middlemen, while in the second, public sector agencies handle grain sales (Figures 10.3 and 10.4). In the private channel, the producer sells grain to *katcha arhatias* either directly or through village merchants, and it is then transported to a wholesaler. The wholesaler may store the grain in a warehouse or merchandise it to another wholesaler in the secondary market. Retailers purchase supplies from wholesalers in the secondary market. Grain that moves through the public sector marketing channel is either sold directly by producers to the Food Corporation of India (FCI) or to state agencies purchasing on behalf of FCI. This grain is transported to the Food Corporation of India's transit stores and then to the FCI's depots. On the advice of the Central Government, the FCI releases grain stocks to state agencies for distribution. The state agencies in turn send grain to fair price shops, where consumers may purchase it at prices fixed by the government.

The Food Corporation of India was established on 1 January 1965 as the sole agency of the central government to purchase, store, transport, and distribute foodgrains. Until the late 1970s, the FCI handled all cereals including coarse grains, but beginning in 1980, its responsibilities were reduced to wheat and rice. Coarse cereals are handled by the National Agricultural Cooperative Marketing Federation (NAFED). The FCI maintains offices in most states and also at the district level. The FCI also owns godowns (warehouses), silos, and some processing plants. At the end of 1987, the FCI had 151 district offices, 19 regional offices, four port operation offices, and four zonal offices, with a staff of over 71 thousand. In addition, the FCI owns 21 rice mills equipped with modern drying and milling facilities. On 30 March 1988, the FCI owned 12.7 million tons of storage capacity and rented another 8.6 million tons from state warehousing corporations and private parties. In some markets, the FCI buys directly from producers, whereas in others, state government agencies operate on its behalf. During 1987/88, the number of purchase centers/mandies (markets) for paddy procurement run by the FCI and state government agencies was over 4,400. Of these, roughly 2,400 were directly operated by the FCI. For wheat, the total number of purchase centers exceeded 7,800, of which less than 1 thousand were managed directly by the FCI.

As indicated earlier, the FCI purchases foodgrains at prices fixed by the government. Procurement by the FCI largely depends on the volume of production and the difference between open market prices and government procurement prices. During times of scarcity, open market prices are higher than the government-declared procurement price, and FCI purchases of grain are small. When supplies are abundant, its purchases are large (Table 10.7). Until 1977, to maximize procurement, the government imposed restrictions on private movement of grains among states (Kahlon and Tyagi, 1983). This lowered prices in the states that produced a surplus and narrowed the difference between open market and procurement prices. As a consequence, the FCI was able to procure substantial amounts of foodgrains. However, with increasing production, FCI purchases since 1978/79 have primarily been made to provide producer price support. For wheat, FCI or its agents purchase directly from farmers in the regulated market by participating in auctions. In some states, the state government exercises the right of pre-emption in market purchases. Rice is procured by levying millers, who are required to sell a specified percentage of the rice that they mill to FCI at a price equal to the procurement price for paddy plus a milling allowance. However, when the price of paddy reaches the procurement price, the FCI also purchases paddy and mills this in its own facilities or in private mills.

In a good crop year when prices are low, FCI purchases may be substantial, whereas in a poor crop year when market prices are high, they may be small. In 1985/86, the FCI purchased 10.5 million tons of wheat and 9.8 million tons of rice, or about 23 percent of total wheat production and 15 percent of total rice production. Assuming that 40 percent of the wheat crop and 35 percent of rice output is marketed each year, FCI purchases accounted for about 56 percent of all wheat and 43 percent of all rice sold in the country.

Each state government has an administrative department of "civil supplies" or its equivalent. Its secretary is responsible for implementing the state's policy on foodgrain pricing, procurement, and distribution. Thus, general control over the public distribution systems rests with the civil supplies/food depart-

TABLE 10.7 Net Availability, Procurement, and Public Distribution of Foodgrains (million tons)

Year (1)	Net production of foodgrains (2)	Net imports (3)	Net availability of foodgrains^a (4)	Procurement (5)	Public distribution (6)	Col. 3 as % of col. 4 (7)	Col. 5 as % of col. 2 (8)	Col. 6 as % of col. 4 (9)
1951	48.1	4.8	52.4	3.8	8.0	9.2	8.0	15.3
1961	72.0	3.5	75.7	0.5	4.0	4.6	0.7	5.3
1967	65.0	8.7	73.9	4.5	13.2	11.7	6.9	17.8
1970	87.1	3.6	89.5	6.7	8.8	4.0	7.7	9.9
1971	94.9	2.0	94.3	8.9	7.8	2.1	9.3	8.3
1972	92.0	-0.5	96.2	7.7	10.5	-0.5	8.3	10.9
1973	84.9	3.6	88.8	8.4	11.4	4.0	9.9	12.9
1974	91.6	5.2	97.1	5.7	10.8	5.3	6.2	11.1
1975	87.4	7.5	89.3	9.6	11.3	8.4	10.9	12.6
1976	105.9	6.9	102.1	12.9	9.2	6.8	12.1	9.0
1977	97.3	0.1	99.0	10.0	11.7	0.1	10.2	11.8
1978	110.6	-0.6	110.3	11.1	10.2^b	-0.5	10.0	9.2
1979	115.4	-0.2	114.9	13.9	11.7^b	-0.2	12.0	10.2
1980	96.0	-0.3	101.4	11.1	15.0^b	-0.3	11.6	14.8
1981^c	113.4	0.7	114.3	13.0	13.0^b	0.6	11.4	11.4
1982^c	116.6	1.6	116.9	15.4	14.8^b	1.4	13.2	12.6
1983^c	113.3	4.1	114.7	15.7	16.2^b	3.5	13.8	14.1
1984^c	133.3	2.4	128.6	18.7	13.3^b	1.8	14.0	10.4
1985^c	127.4	-0.4	124.3	20.1	15.8^b	-0.3	15.8	12.7
1986^c	131.6	-0.1	133.2	19.7	17.6^b	d	15.0	13.2
1987^c	125.5	-0.4	134.6	15.7	18.4^b	-0.3	12.5	13.6
1988^c	121.1	1.9	128.4	14.1	18.3^b	1.5	11.6	14.3

Source: GOI, 1989. Detail may not add because of rounding.
a Net availability = net production - net imports - change in government stocks.
b Includes quantities released under the Food for Work Program.
c Provisional.
d Less than 0.05 percent.

ments in each state government. The stocks for the fair price shops are also channelled through the state governments. The civil supplies departments estimate their monthly foodgrain requirements and, by the third week of the month, request grain supplies from the Government of India for the coming month. The Department of Food of the Government of India allocates foodgrains among the states. Later, following the instructions of the Department of Food, the regional offices of the FCI, in consultation with the state civil supplies departments, allocate quotas to be issued from each depot. These departments then issue permits to the fair price shops, which collect the foodgrains from FCI or state government godowns.

The government distributes subsidized imported and internally procured foodgrains to meet the needs of deficit areas and the poorer members of the population. Fair price shops were first instituted during the Second World War, when the government opened grain stores in various areas to distribute specified quantities of subsidized foodgrains to laborers. Since 1951, the ever expanding public distribution of foodgrains reflects the growing ability of the government to procure and, in bad harvest years, to import large quantities of foodgrains (Table 10.7). The level of demand for grains through the fair price shops is related to the difference between subsidized and open market prices, and the number of shops, which has increased. At the beginning of 1987, there were more than 333 thousand fair price shops.[15] The grain procured by the FCI is distributed through the shops at fixed prices. Each year, the government sets a uniform issue price for wheat and rice, which is usually much lower than the cost of the procured grain (procurement price plus procurement and distribution costs). Hence, the government pays a consumer subsidy on the distribution of foodgrains. Thus, in India, there is a dual market for foodgrains, a free market where prices are determined by demand and supply, and a managed market where both the prices paid by consumers and received by producers are fixed by the government.

Foreign Trade in Grains

The central government monopolizes foreign grain trade. The government, acting through the FCI, decides which grains are to be imported and from which country. Wheat accounts for the majority of imports; rice and coarse grain imports are relatively insignificant. Foodgrain imports vary inversely from year to year with domestic output. Exports of grains are generally insignificant.

Grain Imports

India lost substantial rice-producing areas in East Bengal and wheat-producing areas in West Punjab when Pakistan became a separate country at independence in 1947. Consequently, imports of both wheat and rice of about 1 million tons each became necessary. By the middle 1950s, the domestic availability of foodgrains deteriorated even further. Subsequently, India signed an agreement with the United States to import 16 million tons of wheat and one million tons of rice under PL-480 (see the chapter on the United States in this volume) over a four-year period to meet increased consumption requirements and to aid in rebuilding reserve stocks. However, the situation continued to deteriorate with serious droughts in the crop years 1965/66 and 1966/67. In 1966, over 10 million tons of foodgrains were imported, accounting for 15 percent of total availability (Table 10.6). Most of these imports were wheat, while rice accounted for less than 10 percent (Table 10.8). Wheat was imported from the USA, Canada, and Australia, while most rice imports were supplied by Burma.

After HYVs were introduced in 1967/68, foodgrain output and wheat production, in particular, grew substantially. Breakthroughs in the development and dissemination of new wheat seeds, combined with the expectation of similar developments in the production of rice and other cereals, led to a decision by the government to halt concessional imports by the end of 1971. Since then, most grain imports have been on cash terms or on a deferred payment basis. Between 1967/68 and 1970/71 India was gradually able to reduce imports (Table 10.8).

During the 1970s, foodgrain imports fluctuated from year to year. Total imports declined from about 3.6 million tons in 1970 to 0.5 million tons in 1972 but rose to 7.4 million tons in 1975 before falling again in 1976 to 6.5 million tons, mainly because of increases in domestic output. During 1978 and 1979, there were no imports at all. Wheat accounted for about 90 percent of foodgrain imports during this period. The United States was the main supplier, followed by Canada and Australia. As compared with the 1960s, rice constituted a much lower proportion of imports of foodgrains in the 1970s, despite the fact that rice is preferred by the majority of Indians and that rice production increased at a slower pace than wheat. The reason is that rice is relatively more expensive than other grains on the world market, especially in comparison to the domestic price of rice. Burma and Thailand continued to be the most important rice suppliers.

During the 1980s, imports of foodgrains were smaller on average than in the 1970s, fluctuating between almost negligible quantities in 1979/80, 1980/81, and 1986/87 to 4.2 million tons in 1983/84 (Table 10.9). Wheat continued to account for an overwhelming proportion of total imports, with the United States, Australia, Canada, and Argentina as the major suppliers. Rice imports remained largely the same. Thus, over the years, the composition and sources of grain imports remained unchanged, but the share of imports in total domestic supply declined drastically.

TABLE 10.8 Indian Cereal Imports by Principal Countries of Origin (million tons)

Year	Total cereals	Rice USA	Rice Thailand	Rice A.R. and Egypt	Rice Burma	Rice Total[a]	Wheat Canada	Wheat USA	Wheat Australia	Wheat Total[a]
1960	5.2	0.3			0.3	0.7	b	4.0	0.3	4.4
1961	3.5	0.2			0.2	0.4	0.2	2.5	0.4	3.1
1962	3.6	0.2			0.2	0.4	b	2.8	0.4	3.3
1963	4.5	0.3	b		0.2	0.5	b	3.9	0.2	4.1
1964	6.3	0.3	0.2		0.2	0.6	0.1	5.3	0.3	6.6
1965	7.5	0.3	0.2		0.2	0.8	0.2	6.0	0.3	5.6
1966	10.4	0.1	0.2		0.4	0.8	0.9	6.6	0.3	7.8
1967	8.7	b	0.2		0.2	0.5	0.8	4.5	0.8	6.4
1968	5.7		0.1		0.1	0.4	0.6	3.9	0.2	4.8
1969	3.9	0.1	b		0.2	0.5	0.6	2.2	0.1	3.1
1970	3.6		0.1		0.1	0.2	0.7	2.6	0.1	3.4
1971	2.1		0.1		0.1	0.2	0.5	1.2	0.1	1.8
1972	0.5					0.1	0.2	0.1	b	0.3
1973	3.6						0.6	1.4		2.4
1974	4.9						0.3	1.9	0.2	4.2
1975	7.4		0.1			0.1	0.4	4.6	0.5	7.0
1976	6.5	0.1	b	11		0.1	0.3	4.1	1.0	5.8
1977	0.6					b	b	0.1	0.3	0.5
1978	0.0									
1979	0.0									
1980	0.1									
1981	0.9							0.8		0.8
1982	2.1							1.3	0.8	2.1
1983	4.1		0.2		0.1	0.3	0.1	3.6		3.8
1984	2.4		0.2		0.2	0.5	0.4	0.8		1.8

Source: GOI, Bulletin on Food Statistics, various issues. Detail may not add because of rounding.
a Includes imports from other sources.
b Less than 0.05 million tons.

Grain Exports

Since 1956, cereal exports have generally been prohibited. However, exports of scented, long grain rice, certain pulses, and some coarse cereals were permitted from time to time, and small quantities of wheat were exported in some years. Cereal exports are relatively unimportant but have tended to increase on average such that India became a net exporter in the late 1970s. During the 1980s, it was a net exporter in 1980, 1985, 1986, and 1987 (Table 10.9). Rice accounts for the bulk of exports, with basmati (scented, long grain rice) exported even in years when ordinary rice is imported. Basmati normally commands a price on world markets that is three to four times the price of ordinary, long grain rice. Wheat is occasionally exported, and recently barley exports have increased. The main market for basmati rice is the Middle East, while parboiled rice is exported to the USSR, Sudan, and the United Kingdom.

TABLE 10.9 Imports and Exports of Rice and Wheat by Indian Government (000 tons)

Year	Exports			Imports		
	Rice	Wheat	Total	Rice	Wheat	Total
1977/78	41	494	534	0	179	179
1978/79	12	778	790	0	0	0
1979/80	397	641	1,038	0	0	0
1980/81	165	58	223	0	49[a]	49
1981/82	333	0	333	78[a]	2,114	2,192
1982/83	361	104	467	0	1,952	1,952
1983/84	0	24	24	466[b]	3,739	4,204
1984/85	0	32[c]	32	381	690	1,070
1985/86	0	337[d]	337	10[e]	0	10
1986/87	0	287[f]	287	0	0	0

Source: GOI, Bulletin on Food Statistics, thirty-fourth issue.

[a]Received from Bangladesh in repayment for a wheat and rice loan given by India in April 1979 under the Indo Bangladesh Agreement.

[b]Includes 100 thousand tons of rice received from Bangladesh in repayment for a rice loan given in 1979.

[c]Aid to African countries given through the World Food Program.

[d]Includes 67.5 thousand tons of wheat sent as aid to African countries.

[e]Received in payment for a rice loan given to Vietnam in 1980.

[f]Commodity Loan to the World Food Program.

Detail may not add because of rounding.

Grain Policies

Grain policies in India are an integral part of the national food policy developed over the last 40 years. Basic goals include: (a) increasing foodgrain production by offering to purchase grains at a stipulated minimum price, thereby encouraging farmers to use more inputs; (b) safeguarding consumer interests, particularly those of more vulnerable sections of the community, by selling foodgrains at a reasonable and uniform price through the public distribution system, and in the process, helping to stabilize market prices; and (c) holding adequate stocks of foodgrains to ensure food security between harvests and to meet emergencies caused by crop failures. These goals broadly coincide with three basic food security goals pursued by many countries; that is, the attainment of desirable levels of food production, the assurance of access to food supplies by all levels of society, and the stabilization of grain supplies.[16] A summary of the major instruments used to achieved the policy objectives is given in Table 10.10.

Setting Grain Prices

To attain these objectives, farmers need to receive a fair price for their grain. The Commission for Agricultural Costs and Prices (CACP) plays a major role in this by advising the government on grain price policy. The CACP obtains factual information on the crop economy and solicits opinions on price policy from the states and concerned public and private sector agencies. Based on the information gathered and the advice of its own specialists, the CACP formulates recommendations, which are submitted to the central government and to various economic ministries for comment. These are then considered before a final recommendation is sent to the Cabinet for its decision.

When working on recommendations for support/procurement prices, the CACP uses several criteria: cost of production and changes in input prices; trends in market prices and the demand and supply situation; intercrop price relationships; the likely effect of price changes on manufacturing industry and the cost of living; the international price situation; and the relationship between prices paid and prices received by producers. The prices fixed by the government provide a guaranteed floor, thus reducing producers' risk.

Although the Commission does not rely solely on international market prices in making its recommendation, it is not insensitive to international developments. When world market wheat prices are higher than domestic prices, there is a tendency for those representing the interests of the farming community to demand that domestic prices be adjusted upward to reflect international market prices. In response to such pressure, the APC (now CACP) stated in one of its reports (1975) that

> a facile argument for a sizeable increase in the procurement price for wheat is often built on the ground that the world market price for

TABLE 10.10 Major Grain Policy Instruments in India

Instrument	Commodity		
	Coarse grains	Rice	Wheat
PRODUCTION/CONSUMPTION			
Producer guaranteed price	X	X	X
Deficiency payments			
Government purchases	X	X	X
Production quota			
Input subsidies			
- credit	X	X	X
- fert./pest.	X	X	X
- irrigation	X	X	X
- machinery/fuel			
- seed	X	X	X
Crop insurance			
Controlled consumer price			
TRADE			
Imports			
- tariff			
- quota			
- subsidies			
- licensing			
- state trading	X	X	X
Exports			
- taxes			
- restrictions		X[a]	X[a]
- subsidies			
- licensing			
- state trading			
OTHER			
Marketing subsidies			
- storage			
- transport			
- processing			
State marketing	X	X	X
Buffer stock		X	X

[a] No exports except for government exports.

the cereal is high. In so arguing, it is forgotten that what sustains the high price of wheat outside is the purchasing capacity of the consumer in the affluent economies. The appropriateness of an administered price for the grain in the Indian context cannot be detached from the paying capacity of the vast mass of the low income consumers in the country. (emphasis added) (GOI, 1976:9).

Similarly, when world market prices drop below domestic market prices, there is a tendency for some groups to suggest that domestic prices be lowered. The volatility of international market prices and the disruptive effects of reductions in procurement prices on farm investment and production are two important reasons why world prices are not used as the primary guide in domestic pricing policies in India.

Over the longer term, however, trends in world market prices have an influence on decisions. For example, wheat competes with rapeseed/mustard (an oilseed crop) for land. If more mustard is planted, some wheat production is foregone. In recent years, India's imports of vegetable oils have been growing. One response could be to increase producer prices of edible oils relative to those of wheat, resulting in the production of more domestic edible oils and reduced imports. However, this could mean more wheat imports. Relative prices could also be kept in favor of wheat production, and edible oils could continue to be imported. The Indian government chose the path of importing more edible oils rather than wheat until the early 1980s. However, recently prices were adjusted to favor mustard, even though crop productivity is low in comparison to wheat (Kahlon and Tyagi, 1989).

Changes in Relative Prices

Adjustments that are made in domestic prices in response to changing market conditions can be illustrated by considering the course of events since the mid-1960s. There were severe droughts in India in 1965/66 and 1966/67 that depressed foodgrain production. Foreign exchange constraints and the curtailment of concessional supplies by the United States limited the availability of foreign wheat and pushed domestic prices to high levels. The wholesale price index for wheat rose from 100 in 1961/62 to 178 in 1966/67 and 214 in 1967/68. To stimulate wheat production, the procurement price was increased from Rs. 540 per ton in 1966/67 to Rs. 700 in 1967/68 and Rs. 760 in 1968/69. The procurement price was kept at this level until 1973/74, despite phenomenal growth in wheat output. To sustain this price, the government curtailed cereal imports and used a large part of remaining imports to build reserve stocks.

With the increase in production, the Food Corporation of India began to assume a major price support role during the early 1970s. It procured 2.4 million tons in 1969/70, 3.2 million tons in 1970/71, and 5.1 million tons in 1971/72 to maintain the market price. Public wheat distribution in 1971 totalled 4.5 million tons, and the remainder was put into the buffer stock. Had the FCI not intervened in the market, wheat prices would probably have fallen below the support price.

The history of pricing policies for rice differs to some extent from that for wheat. Large increases in rice production did not occur until the mid-1970s when new HYVs were released on a commercial scale. State procurement prices for paddy were fixed at between Rs. 460 and Rs. 525 per ton in 1967/68. These prices remained almost unchanged until 1972/73. The procurement price was raised to Rs. 700 per ton in 1973/74, to Rs. 770 per ton in 1977/78, and to Rs. 1,050 in 1980/81. These prices were maintained by purchasing rice at the procurement price and allocating part of the surplus to the buffer stock. At the end of 1979, government rice stocks amounted to over 9 million tons, almost entirely amassed out of domestic procurement purchases made between 1971 and 1979.

Procurement prices for wheat rose 71 percent between 1972/73 and 1980/ 81, whereas rice prices increased by 81 percent over the same time period. Further, the price realized by farmers for paddy during the 1970s grew at a faster rate than the wheat price (Figure 10.5). The differential increase in prices was intentionally maintained to sustain the rapid growth in rice production. Whereas in the early phases of the HYV revolution price policies favored wheat producers, after the mid-1970s, price policies were tilted toward rice producers.

Price Stabilization

In addition to influencing long-term production trends with price levels and price relativities, government stock management has consciously been used to stabilize domestic market prices. In 1971/72, wheat production increased 11 percent over the previous year. Given an estimated price elasticity of demand of .44, wheat prices should have declined by about 25 percent. However, the government purchased all of the wheat offered at the fixed price of Rs. 740 per ton, and prices actually increased by 7 percent (Table 10.11). Again, in 1982/83 when wheat production increased by more than 14 percent over the previous year, prices should have declined by 32-33 percent. However, because of government purchases, prices rose by 2 percent. Consumer prices were also stabilized with government grain price policies. In 1972/73, wheat production declined by 7 percent over the previous year, and total cereal production fell by 19 percent. Given the demand elasticity for wheat, this decline should have caused a 14 percent rise in prices. In actuality, prices rose by less than 2 percent (Table 10.11). Similarly, in 1979/80, wheat production declined by 10 percent and cereal production by 16 percent, yet wheat prices increased only 10 percent rather than the 24 percent rise that would have occurred without government intervention.

In conclusion, price policies in conjunction with policies to stimulate the adoption of new technologies have had a major impact on the foodgrain

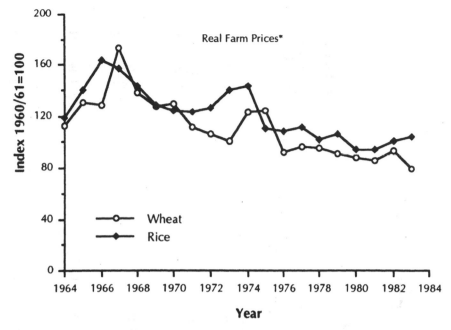

*Farm price deflated by weighted average of prices paid for inputs, final consumption, and capital formation.

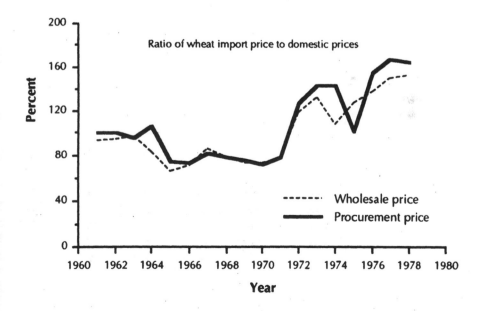

FIGURE 10.5 Trends in Indian Grain Prices. *Source:* Based on data from Kahlon and Tyagi, 1983; and Krishna and Chhibber, 1983.

TABLE 10.11 Fluctuations in Crop Production and Changes in Prices

Year	Wheat			All cereals			
	% change in production over the previous year	Expected change in prices[a]	Actual change in prices[b]	% change in production over the previous year	Expected change in prices[c]	Actual change in prices[b]	% change in all commodities index[c]
1971/72	+10.8	-24.5	+7.0	-2.6	+6.5	+1.9	9.9
1977/78	+4.4	-21.4	-1.7	+14.7	+36.8	-2.7	0.0
1980/81	+14.1	-32.0	+8.7	+17.6	-44.0	+12.4	9.2
1982/83	+14.3	-32.5	+1.8	-3.4	+8.5	+9.7	9.4

Note: Price changes are for the corresponding marketing years.
a Percentage change multiplied by the price flexibility coefficient.
b Percentage change in the annual average of wholesale price indices.
c Change in the all commodities index reflects the inflation rate.

situation in India. Under stable and remunerative prices, wheat area increased from 15 million hectares in 1967/68 to 25 million hectares by 1983/84. During this period, the area of high-yielding varieties increased from 3 to 19 million hectares, irrigated area from 6 to over 16 million hectares, and the per hectare application of plant nutrients increased significantly in almost all wheat-growing regions. Wheat production rose at a rate of over 5 percent per annum, with yields increasing at 3 percent. Today, India is marginally surplus in wheat. In 1970, 13 percent of average per capita income was required to buy 100 kg of wheat, but by 1984/85, the figure had fallen to only 6 percent. A similar decline occurred for rice.

National food policy has also been designed to safeguard consumer interests by making grains available throughout the country through the public distribution system. In the 1950s and early 1960s, consumers received top priority in the formulation of domestic food policy. Not only was the issue of gross availability addressed, but infrastructure was also improved to deliver grain to the remote and inaccessible areas where a large portion of the population lives. Apart from subsidizing the sale of rice and wheat, the government focused on improving the public distribution system. The number of fair price outlets increased from about 240 thousand in March 1979 to over 330 thousand by 1987. When new outlets are opened, special emphasis is given to remote and rural areas. Outlets are established using the target of one shop per 2 thousand people. While the number of items distributed through these shops varies from state to state, there are seven that the central government considers essential: wheat, rice, sugar, imported edible oils, controlled cloth, kerosene, and soft coke. Sales made through the public distribution system increased from roughly 18 million tons in 1979 to over 27 million tons in 1983, while the value of sales grew from Rs. 26.7 billion to Rs. 65.1 billion during the same period. Generally, publicly distributed rice is allocated to rice-consuming areas and wheat to wheat-consuming areas. However, along with rice, small quantities of wheat are also made available in rice-consuming areas, and similarly, very small amounts of rice are distributed to wheat consumers. In 1985, of the total 15.8 million tons distributed through the public distribution system, 7.2 million tons were rice and 8.5 million tons were wheat. Other grains accounted for only 91 thousand tons.

The final element of the national foodgrain policy is food security, ensured by maintaining adequate grain stocks. The Food Corporation of India is responsible for procuring foodgrains and managing buffer stocks, which vary throughout each year and among years. Recently, stocks within a year have ranged between 16 and 21 million tons; wheat stocks have varied between 7 and 13 million tons; and rice stocks from 7 to 11 million tons. The FCI, which holds stocks for the government, is paid a subsidy, which changes with the tonnage held.

TABLE 10.12 Imported and Domestic Wheat Costs (Rs/100 kg)

	Pooled price for imported and indigeneous	1971/72 Imported	1971/72 Indigeneous	1972/73 Indigeneous
Average procurement price	70.61[a]	55.02	75.71	76.00
Port clearance/ procurement changes	9.91	6.64	11.02	11.02
Storage, movement, and distribution costs	7.29	7.16	7.40	6.91
Total cost excluding carrying costs	87.84	68.82	94.13	93.96
Carrying costs of buffer stock	6.34	1.18	11.01	7.94
Total cost including carrying costs of buffer stock	94.18	70.00	105.14	101.90

Source: GOI, Ministry of Agriculture, 1972.
[a] Based on different purchase prices for sound and rain-damaged wheat.

TABLE 10.13 Foodgrain Subsidies Paid by the Indian Government (million rupees)

Year	Consumer subsidy Rice	Consumer subsidy Wheat	Carrying cost of buffer stocks	Total
1975/76	-14	2,110	765	2,861
1976/77	31	1,453	2,926	4,410
1977/78	272	2,441	2,630	5,343
1978/79	411	2,535	2,628	5,574
1982/83	3,270	4,330	1,840	9,440
1983/84	4,665	3,551	2,697	10,912
1984/85	4,326	4,222	4,979	13,527
1985/86	5,084	8,650	5,176	18,910
1986/87	6,865	9,361	5,101	21,326
1987/88	7,840	11,403	2,038	21,282

Source: Food Corporation of India.
Detail may not add because of rounding.

Some Emerging Problems

The policy of providing remunerative prices to farmers and making foodgrains available at reasonable prices to consumers requires that grains be distributed at a price lower than the economic cost (procurement price plus procurement and distribution costs). When India was importing large quantities of wheat and world market prices were lower than the domestic procurement price, part of the subsidy needed to meet distribution costs was obtained from the surplus earned on imported wheat. From 1963 to 1971, import prices were lower than domestic wholesale prices and the procurement prices fixed by the government (Figure 10.5). As a consequence, the cost of imported wheat (Rs. 700 per ton during 1971/72) was lower than that for indigenous wheat (Rs. 1,050 per ton during 1971/72) (Table 10.12). In 1972, the price at which wheat was released for consumption (the issue price) was Rs. 780 per ton. Thus, the issue price was higher than the cost of imported wheat but was substantially lower than the cost of domestically procured wheat. As long as cheap, imported wheat constituted the bulk of grains distributed through the public distribution system, the subsidy burden was small. However, when world wheat prices rose sharply after 1971, subsidy costs climbed quickly. By the middle 1970s, a subsidy on the distribution of rice also became inevitable because of increases in domestic procurement prices. Consequently, the bill for grain subsidies rose rapidly in the 1970s and now accounts for a substantial portion of the government's budget. During the fiscal year 1985/86, total subsidies, including those on foodgrains, made up almost 14 percent of the government's expenditures. Food subsidies alone accounted for over 5 percent of expenditures. In addition to the distribution subsidy, the costs of maintaining buffer stocks are quite large (Table 10.13), and if current trends persist, will continue to grow. It may be difficult for the government to continue to pay these escalating costs without adversely affecting long-term development plans.

Conclusion

During the mid-1960s, India faced a critical supply situation with drought-reduced crops, escalating consumer prices, and increasing reliance on food aid shipments of grains. India has placed a major priority on increasing foodgrain production. Investments in infrastructure, the use of input subsidies, favorable pricing policies, and the spread of new high-yielding varieties of wheat and rice have had a dramatic effect. Foodgrain output increased from 74 million tons in 1966/67 to over 150 million tons by 1983/84. Wheat production in particular improved rapidly, growing at over 5 percent per year. Significantly, over 56 percent of the growth in wheat output can be attributed to improvements in yields. As a result, in the 1980s India is essentially self-sufficient in foodgrains.

Because of the instability in foodgrain production caused by the vagaries of the monsoon, food security concerns continue to dominate Indian grain policy. The government not only purchases foodgrains from farmers at a fixed price, but also distributes these grains at set prices through the public distribution system. A large public buffer stock is also maintained. Government intervention helps the producer in years of bumper harvests by preventing large declines in prices, and it aids the consumer, particularly in years of crop shortfall, by limiting price increases and ensuring the availability of supplies. As a result, more Indians than ever before can now afford to buy an adequate supply of foodgrains.

The linkage between India's domestic foodgrain market and international grain markets is strictly controlled. In the 1950s and early 1960s, large imports of foodgrains were allowed under the United States' PL-480 program. Exports have usually been limited to specialty grains, such as scented rice varieties. International grain prices are considered when domestic grain prices are set, but are not given undue weight. As the level of self-sufficiency has increased, the relationship between the domestic foodgrain market and the international market has become a more important issue. One view is that India should export rice and import wheat, given the proportionally higher price of rice. However, rice exports have been allowed only in years of large domestic supplies. With continued increases in productivity in the Indian grain economy, serious consideration must be given to the country's position in the international grain market to permit efficient management of the domestic grain economy.

Notes

1. In India, pulses are a major part of the diet. They are normally included in cereal production and consumption statistics. For this reason, "foodgrains" include pulses in this chapter. The principal pulses are green gram, black gram, pigeon pea, peas, and Bengal gram.

2. Gross cropped area is defined as the net area sown plus area sown more than once, i.e., if two crops are raised in the same field in the same year, the area is counted twice.

3. To help provide guidance to the development in India, a planning commission was set up, and the first five-year plan was launched in April 1951. The planning model adopted in India was different from the one followed in centrally planned economies. Planning has been characterized as "instrumental interference" rather than an alternative to the market system (Chakravarty, 1988).

4. The first agricultural census in India was completed in 1970/71 (July 1970 to June 1971). The information discussed here is based on the third census completed in 1980/81. In India, July to June covers a complete crop year. Therefore, most dates on agricultural production, inputs, etc., refer to the July-June agricultural year. See Government of India, 1987.

5. There can be differences between area owned and operated. A farmer may lease land from another farmer to supplement holdings or lease land to another farmer. Operated area is defined as the area that the farmer actually cultivates.

6. Before independence, planners had to prove that new irrigation projects would

yield a positive rate of return before the government would sanction their construction. Many post-independence schemes included storage dams and channels in undulating terrain. As a result, they were expensive to construct, and some did not meet financial criteria. Consequently, the government subsidizes canal irrigation.

7. Community development efforts included a comprehensive program involving cooperation between the government and community members to improve general living conditions. Principal responsibility for enacting the program rested with the community, but the government was expected to guide and assist by supplying technical advice and needed inputs such as seed, fertilizer, and irrigation facilities.

8. In November 1964, a bill was passed in Parliament that established the Food Corporation of India to trade in foodgrains and other agricultural commodities, and on 1 January 1965, the FCI came into being. For more details, see Birla Institute, 1980.

9. The Indian currency is the rupee. During the 1970s, there were between 7 and 9 rupees to the US dollar. During the 1980s, the currency depreciated and there were roughly 15 rupees to the dollar in 1988. What constitutes a subsidy is a point of controversy, and it is difficult to estimate subsidies in an economy where there are many distortions in market prices. The subsidies discussed here are essentially overt; that is, they involve public expenditures.

10. In some states, rice is grown during the summer season. It is planted between November and February, and the crop is harvested during the summer months-March to June.

11. Scented varieties are only grown in a few select areas and command prices on the domestic market that are four times those of coarse or regular rice. Even in the world market, scented varieties are three to four times the price of ordinary rice. The yield per hectare is, however, very low, and these varieties require special environmental conditions that limit where they can be grown.

12. In 1983/84, sorghum accounted for 35 percent, pearl millet 22 percent, and maize 23 percent of total coarse grain production.

13. For a discussion of the need to revise the seed requirement component in the netting factor in light of recent technological change, see Tyagi (1982).

14. According to input-output transaction (commodity by industry) tables prepared by the Central Statistical Organization for 1973/74, industrial uses of food crops accounted for less than 1 percent of total domestic use (CSO, 1981).

15. The shops are licensed by the civil supplies departments and are required to distribute foodgrains to ration card holders at fixed prices. They are generally privately owned and are paid a commission based on how much they sell.

16. Although food security has always been an integral part of the government's foodgrain policy, the decision to accumulate a buffer stock for foodgrains was only made in the early 1970s.

References

Alagh, Y.K. "Issue of Planning for Food Security." Paper presented at the Seminar on the Role of Foodgrain Agencies in Food Security in Asia and the Pacific, Seminar papers Volume II. New Delhi: Food Corporation of India, 1985.

Bhagwati, J.N. and T.N. Srinivasan. *Foreign Trade Regimes and Economic Development: India*. New York: National Bureau of Economic Research, 1975.

Birla Institute of Scientific Research. *State of Foodgrains Trade in India: A Study of Policies and Practices of Public Distribution System*. New Delhi: Economic Research Division, 1980.

Central Statistical Organization (CSO). *National Accounts Statistics: 1970-71 - 1978-79*. New Delhi: Department of Statistics, Ministry of Planning, 1981.

Chakravarty, S. *Development Planning, the Indian Experience*. Delhi: Oxford University Press, 1988.

Dantwala, M.L. *Indian Agricultural Development Since Independence*. New Delhi: Oxford IBH, Publishing Company, 1986.

Government of India (GOI). *Report of the Eighth Finance Commission*. New Delhi: Government of India, 1984.

_____, Directorate of Economics and Statistics, Ministry of Agriculture. *Bulletin on Food Statistics*, various issues.

Government of India (GOI), Directorate of Economics and Statistics, Ministry of Agriculture. *Indian Agriculture in Brief*. Various issues.

Government of India (GOI), Ministry of Agriculture. *Report of Agricultural Prices Commission on Price Policy for Rabi Foodgrains for the 1972-73 Season*. New Delhi: Ministry of Agriculture, 1972.

_____. *Report of the Agricultural Prices Commission on Price Policy for Rabi Foodgrains for the 1975-76 Season*. New Delhi: Ministry of Agriculture and Irrigation, 1976.

_____ a. *Report of the National Commission on Agriculture: Part III, Demand Supply*. New Delhi: Ministry of Agriculture and Irrigation, 1976.

_____. Marketable Surplus and Post-Harvest Losses of Wheat in India. Faridabad (Haryana): Directorate of Marketing & Inspection, Ministry of Agriculture, 1986.

_____. *All India Report on Agricultural Census 1980-81*. New Delhi: Department of Agriculture and Cooperation, Ministry of Agriculture, 1987.

_____. *Economic Survey 1988-89*. New Delhi: Ministry of Finance, Economic Division, 1989.

Hasim, S.R. and P. Singh. *Growth Rates in Agriculture and Industry: Their Implications for the Economy*. Paper presented at VIIIth World Economic Congress of I.E.A. (Dec. 1-5, 1986), New Delhi.

Kahlon, A.S. and D.S. Tyagi. *Agricultural Price Policy in India*. New Delhi: Allied Publishers Private Ltd., 1983.

_____. *Agricultural Price Policy in India*. Second Edition. New Delhi: Allied Publishers Pty. Ltd., 1989.

Krishna, R. and A. Chhibber. *Policy Modelling of A Dual Grain Market: The Case of Wheat in India*. Report No. 38. Washington, D.C.: International Food Policy Research Institute, 1983.

Krishna, R. and G.S. Raychaudhuri. *Some Aspects of Wheat and Rice Price Policy in India*. World Bank Staff Working Paper No. 381. Washington D.C.: World Bank, 1980.

Pisharoty, P.R. "Water Management and Waste Land Development." *Waste-land News* 3(1)(1987):3-11.

Rao H.C.H., S.K. Ray, and K. Subbarao. *Unstable Agriculture and Droughts*. New Delhi: Institute of Economic Growth, 1988.

Techno Economic Research Institute. "Report on Feed, Seed and Wastage Rates of Foodgrains." New Delhi: Mimeograph, J-7, Saket, 1988.

Tyagi, D.S. "How Valid are the Estimates of Trends in Rural Poverty." *Economic and Political Weekly*, Review of Agriculture 26(1982):A-54 - A-62.

11

Indonesia

Roley Piggott and Kevin Parton

Introduction

Indonesia is an archipelago consisting of more than 13,000 tropical islands (Figure 11.1). It is the fifth most populous country in the world, with a 1990 figure of 179 million inhabitants (Hull 1991). Just below 60 percent of the population lives on the island of Java, which comprises only 7 percent of the land area of Indonesia. There are many reasons for this concentration, perhaps the most important of which is Java's strategic location on colonial trade routes. Further, the soils on Java are naturally fertile because of the presence of abundant volcanic ash. Java accounts for most of the country's agricultural and industrial output. GNP per capita was $US 440 in 1988. Economic growth has proceeded fairly rapidly, with an average annual rate of growth of GNP per capita of 4.3 percent between 1965 and 1988 (World Bank, 1990). However, the average growth rate during the 1980s has been considerably slower than during the 1970s.

The main macroeconomic targets have been accelerated growth and minimal inflation. Since a hyperinflationary period in the early 1960s, inflation has been controlled with tight fiscal and monetary policies. Government expenditures have been kept in line with revenues (the latter including foreign capital inflow). However, as government revenues from oil have fallen in recent years, so have development expenditures in the agricultural sector.

Although its share of GDP has fallen, agriculture (including forestry and fishing) is still the largest sector (Table 11.1). Also, since 1980, agriculture's share has declined more slowly as the output of food crops has expanded following price policy stimuli, technological change, and the relative decline in the value of oil exports. The main food crops (rice, maize, cassava, soybean, groundnut, and sweet potato) continue to provide more than 60 percent of agricultural output (Table 11.2). Food production is characterized by small farm holdings (most are less than .3 hectares), which occupy more than 80

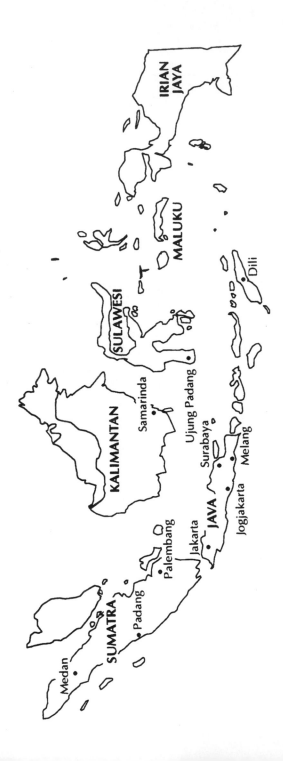

FIGURE 11.1 Map of Indonesia

TABLE 11.1 Major Sectors Contributing to Gross Domestic Product at Current Prices (percent)

Sector	1960	1970	1980	1986[a]	1987[a]	1988[ab]	1989[ab]
Agriculture, forestry, and fishing	53.9	48.6	24.0	24.2	23.3	24.1	23.5
Mining and quarrying	3.7	5.3	23.0	11.2	13.8	12.1	13.1
Manufacturing	8.4	9.0	13.0	16.7	17.0	18.5	18.4
Electricity, gas, and water supply	0.3	0.5	0.5	.06	0.6	0.6	0.6
Construction	2.0	3.1	5.3	5.2	4.9	5.1	5.3
Trade	14.3	16.6	15.0	16.7	16.9	17.2	17.0
Transport and communications	3.8	3.0	4.5	6.2	6.0	5.7	5.5
Public administration and defense	4.5	5.6	6.6	8.1	7.1	6.7	6.7
Other	9.1	8.3	8.1	11.1	10.4	10.0	9.9
Total	100.0	100.0	100.0	100.0	100.0	100.0	100.0

Sources: 1960: Booth and Glassburner, 1975; 1970: United Nations, 1983; 1980: EIU, 1986; 1986-89: BPS, 1990a.
[a]In 1989, BPS revised the national accounts for the period 1983-88. Consequently, the previous data are not comparable with the new series.
[b]Preliminary.

percent of the agricultural land (Kasryno, Budianto and Birowo, 1982). With more than 75 percent of the population located in rural areas (Birowo and Sanusi, 1982), agriculture is also a major source of employment (Table 11.3). On the consumption side, about 63 percent of consumer expenditure is devoted to food (BPS, 1987).

Until the early part of the 1980s, mining was the second largest sector of the Indonesian economy (Table 11.1). The previously-predominant position of oil as an export earner meant that events in the international oil market were the main determinant of the fortunes of the sector as a whole. The years 1974 and 1979 were peak years for mining's share of GDP because oil prices reached their highest levels. After 1982 as the oil market began to weaken, the continuing oil price declines brought Indonesian production to a ten-year low in 1985.

Indeed, the accelerated oil price decline of 1986 reduced the sector's contribution to GDP to less than half its total at the beginning of the decade.

The manufacturing sector is dominated by state enterprises in the processing industries, but private firms feature prominantly in all other areas of the sector (Hill, 1988). Initially, the sector was developed largely to produce substitutes for imports. Its share of GDP grew from 8.4 percent in 1960 to just over 13.0 percent in 1980. After a period of sluggish growth in the early 1980s, a series of policy changes attempted to deregulate and restructure the economy. The earlier highly-protectionist attitude was re-oriented to a more export-led strategy and resulted in a growth rate of the manufacturing sector of 10.4 percent per annum between 1984 and 1989. The revised series of national accounts indicate manufacturing growing from 16.7 percent of GDP in 1986 to 18.4 percent in 1989 (Table 11.1). Should the current growth rate be sustained, manufacturing in the early 1990s will overtake agriculture as a major component of GDP.

Three other sectors of the economy deserve mention. These are trade, which, since 1960, has accounted for between 14 and 17 percent of GDP; transport and communications, which accounts for around 6 percent of GDP; and construction, which was briefly a growth sector before relative stagnation ensued during most of the early and mid-1980s. At the end of the decade, the more buoyant economic conditions have assisted with some sectoral recovery.

Since 1969, five Repelitas (five-year plans) have been instituted by the Indonesian government. Through these plans, the government attempted to control the relative growth of the various economic sectors by strategic placement of public investment, which accounted for more than half of total investment. The general approach of the government has been to expand rice production (the major food staple), while at the same time attempting to foster

TABLE 11.2   Food and Non-food Crops Contributing to Agricultural Output Measured in Current Prices (percent)

Product	1960	1970	1980	1986[a]	1987[a]	1988[ab]	1989[ab]
Farm food crops	63.8	61.0	52.0	60.7	60.2	61.8	62.0
Non-food crops, livestock products,forestry, and other	36.2	39.0	48.0	39.3	39.8	38.2	38.0
Total	100.0	100.0	100.0	100.0	100.0	100.0	100.0

Sources: 1960: Booth and Glassburner, 1975; 1970: United Nations, 1983; 1980: EIU, 1986; 1986-89: BPS, 1990a.
[a]In 1989, BPS revised the national accounts for the period 1983-88. Consequently, the previous data are not comparable with the new series.
[b]Preliminary.

TABLE 11.3 Employment by Sector (percent of the employed population aged ten years and over)

Sector	1961	1971	1980	1985
Agriculture, forestry, and fishing	73.6	64.2	56.3	54.7
Mining and quarrying	0.0	0.2	0.8	0.7
Manufacturing	—	6.5	9.1	9.3
Electricity, gas, and water supply	7.8[a]	0.1	0.1	0.1
Construction	—	1.6	3.2	3.4
Trade	—	10.3	13.0	15.0
Transport and communications	8.9[b]	2.3	2.9	3.1
Public administration, defense, and other	9.7	14.8	14.6	13.7
Total	100.0	100.0	100.0	100.0

Sources: 1961: Dapice, 1980; 1971: EIU, 1986; 1980 and 1985: BPS, 1987: 174.
[a]Manufacturing, electricity, gas, water supply, and construction.
[b]Trade, transport, and communications.

rapid growth in the manufacturing and mining sectors. Self-sufficiency in rice was achieved in 1985, but is proving difficult to maintain. Initially less spectacular, the growth of the manufacturing sector has improved markedly since the deregulation and diversification commenced in the mid-1980s. The mining sector has undergone several dramatic swings and downturns, mainly in response to price changes in the international oil market.

Through the first four Repelitas, the food-producing sector received special attention. At first this was because it dominated the economy, contributing more than 50 percent of GDP, making agricultural growth an essential component of overall economic growth. More recently, as agriculture's share of GDP has declined, attention given to the food-producing sector has been designed to aid achievement of self-sufficiency and to ensure stability in agriculture as the basic employment-generating sector of the economy. Agriculture is projected to grow at 3.6 percent annually during the fifth Repelita (which began on April 1, 1989), a rate slightly faster than that achieved during the 1980s (Booth, 1989:5).

The two grain crops, rice and maize, together with cassava, account for both

TABLE 11.4 Relative Importance of Rice and Other Food Crops in Indonesia, 1988

	Area ('000' ha)	Production ('000' tons)	Apparent per capita consumption[ac] (kg/yr)	Equivalent rice calories available per capita[b] (kg)
Milled rice	10,138	28,340	150	150
Maize	3,406	6,652	31	31
Cassava	1,303	15,471	51	15
Soybean	1,177	1,270	9	—
Sweet potato	248	2,159	11	3
Wheat flour	—	—	7	7

Sources: BPS, Buletin Ringkas, 1989; Food Balance Sheet in Indonesia, 1988.
[a]Calculated as production converted from paddy, less seeds and losses, plus imports, less increases in stock. Population assumed to be 174 million (Preliminary figures).
[b]Conversion ratios used to calculate rice equivalents are maize 1.0, cassava 0.3, sweet potato 0.27, and wheat 0.97.

the bulk of food crop production and calories consumed in Indonesia (Table 11.4). Various government schemes have attempted to encourage the production of other secondary food crops, but with only partial success. Many producers are unwilling to grow other crops because rice has provided a stable income over many years. Also, an especially difficult problem is that secondary food crops are considered inferior to rice by many consumers.

On the consumption side, rice, maize, and imported wheat are most important. While rice and wheat flour are consumed mainly by humans, about a quarter of maize produced is used for animal feed, particularly for poultry. In 1984, 63 percent of expenditure was devoted to food, and of this, 31 percent was spent on cereals (BPS, 1987: 142-44). That is, cereals accounted for more than 19 percent of consumer expenditure and more than 11 percent of GDP (BPS, 1986a: 53). One outcome of this is that prices for these commodities, which are regulated by the government, have considerable influence on the macroeconomy through their impact on inflation.

Turning to the relative importance of grain trade in agricultural trade and total trade, it is important to note that, since 1960, Indonesia has maintained a positive balance of trade (Table 11.5). This has provided an important source of funds for economic development and is the net result of a large surplus in oil and gas products and a smaller deficit in non-oil products. There has also been a continuing surplus provided by broadly defined agricultural and forestry products, including rubber, logs (until 1985) and wood products, coffee, and tea exports. However, Indonesia is a net importer of grain (Table 11.6).

Since the decline in oil prices, and hence export revenues, the government has both encouraged export diversification into non-oil commodities and attempted to reduce imports. Rice, wheat, and maize imports have been strictly controlled by the government. In the case of rice, Indonesia has moved from being the world's largest importer in 1980, when payments for rice accounted for 6.5 percent of Indonesia's import bill (EIU, 1986), to that of a rice exporter in 1985, 1986, and to a lesser extent in 1987 and 1989. This is widely attributed to domestic agricultural policies that promoted rice production (ROI, 1988a: 1; Rosegrant, et al., 1987: 1; Tabor, et al., 1988: 171).

Production

General Background

The Indonesian archipelago is subject to a variety of topographic and climatic influences, with two distinct monsoons. The major northwest monsoon commences in October/November and spreads eastward across the islands. Java, the main food-producing island, is dominated by an east-west chain of volcanic mountains, which means the northern part of the island benefits from this monsoon. A southeast monsoon, which begins in May/June, brings little or no useful rainfall to the eastern islands, but does provide West Java, Sumatra, and East Sulawesi with another rainy season. North Sumatra benefits from year-round rainfall because of the two monsoons.

TABLE 11.5 Value of Exports and Imports ($US million)

	Total		Excluding petroleum and gas	
	Exports	Imports	Exports	Imports
1960	841	578	620	552
1970	1,108	1,002	662	987
1975	7,103	4,770	1,792	4,516
1980	23,950	10,834	6,169	9,086
1981	25,165	13,272	4,501	11,550
1982	22,328	16,859	3,929	13,314
1983	21,146	16,352	5,005	12,206
1984	21,888	13,882	5,870	11,185
1985	18,587	10,262	5,869	8,988
1986	14,805	10,718	6,528	9,632
1987	17,136	12,370	8,580	11,302
1988	19,219	13,249	11,537	12,339
1989	22,160	16,360	13,480	15,385

Sources: 1960, 1970, 1975: BPS, 1985; 1980-89: BPS, 1990c.

TABLE 11.6 Indonesian International Trade in Rice, Wheat, and Maize ('000' tons)

Year	Rice[a] imports	% of world rice exports	Wheat[b] imports	% of world wheat exports	Maize exports	% of world maize imports	Maize imports	% of world maize exports
1960	962	13.8	—	—	0	0.00	0	0.00
1970	956	12.0	467	0.8	50[c]	0.20	0	0.00
1975	693	8.0	730	0.9	51	0.10	0.3	d
1980	2,012	15.4	1,488	1.5	15	0.02	34	0.04
1981	538	4.1	1,421	1.3	5	0.01	2	d
1982	310	2.6	1,501	1.4	0.5	d	76	0.10
1983	1,169	10.2	1,746	1.6	18	0.03	28	0.04
1984	414	3.3	1,448	1.2	160	0.20	59	0.09
1985	34	0.3	1,338	1.3	4	d	50	0.07
1986	28	0.2	1,623	1.7	4	d	58	0.10
1987	55	0.4	1,697	1.5	5	d	221	0.34
1988	33	0.3	1,601	1.3	37	0.06	63	0.10
1989	397	2.7	1,921	1.8	234	0.30	40	0.05

Source: FAO, Trade Yearbook, 1987 and various issues. BPS, Buletin Ringkas, 1990b and previous issues.
[a]Imports include glutinous rice not produced domestically.
[b]Wheat plus wheat flour in wheat equivalents (conversion factor based on an extraction rate of 72 percent).
[c]FAO estimate.
[d]Figures .008 or less.

As a result of these influences, an extremely wide range of agricultural commodities can be grown in Indonesia (Birowo and Sanusi, 1982). Farm enterprises range from intensively-cropped small farms, to shifting cultivation, to larger estates that specialize in industrial crops. While export crops like palm oil and tea are produced on plantations, food crops, including rice and maize, are grown mainly by smallholders. On Java, less than 0.7 million hectares are under estate cultivation, compared with more than 5 million hectares occupied by smallholdings (Birowo and Sanusi, 1982).

Rice

Rice is by far the predominant staple grown by the smallholder (Table 11.7), usually in conjunction with one or more of maize, cassava, sweet potato, groundnut, and soybean (the *palawija*). The relative importance of these secondary crops varies according to location. This is both because soil types and climate are different, and various cultures and traditions lead to preference to grow secondary crops.

In 1989, 95 percent of the rice crop was wet (*sawah*) or lowland rice, of which nearly 60 percent was grown on Java and Madura. Sawah rice areas are scattered throughout Java, with the exception of mountainous areas and parts of the southern coastline, where limestone ridges result in dry and infertile soils. Some terracing occurs in the hilly areas of West and Central Java, but the alluvial northern plains, crossed by a number of north-flowing rivers, are the dominant sawah-producing areas.

Off-Java, sawah rice tends to be grown in fertile river valleys and on smaller plains, or on the terraces characteristic of the small volcanic islands of Bali and Lombok. In 1989, Sumatra accounted for almost 20 percent of sawah production, while other islands accounted for smaller proportions (e.g., Sulawesi, 10 percent; Bali, 2 percent; and Kalimantan, 4 percent). The drier climate of the eastern islands (e.g., East Nusa Tenggara), where the dry season may last up to six months, restricts sawah plantings. Palawija, especially maize and cassava, are more important in these areas.

Dry, or upland rice (*ladang*), is seeded directly into the ground and grown under rainfed conditions. In 1989, Java accounted for 36 percent, Sumatra 36 percent, and Kalimantan 17 percent of upland rice production. Of the areas where upland rice is grown, those regions with a fairly predictable rainfall (i.e., West Java, Sumatra, and Kalimantan) tend to dominate production. In drier regions (e.g., East Java), or in particularly dry seasons, only palawija may be grown. On some of the outer islands, upland rice may be planted in areas of shifting cultivation.

Before the initiation of the government's rice intensification programs (see later discussion), sawah rice was planted during the wet season, followed by maize and/or legumes during the dry season, unless irrigation allowed a second rice crop. The upland areas generally have less fertile soils and, for the

TABLE 11.7 Rice and Maize Area, Yield, and Production

	Rice (milled)			Maize		
	Area (m.ha)	Yield (tons/ha)	Production (m.tons)	Area (m.ha)	Yield (tons/ha)	Production (m.tons)
1970	8.14	1.62	13.14	2.94	0.96	2.83
1971	8.32	1.65	13.72	2.63	0.99	2.61
1972	7.90	1.67	13.18	2.16	1.04	2.25
1973	8.40	1.74	14.61	3.43	1.08	3.69
1974	8.51	1.80	15.28	2.67	1.13	3.01
1975	8.49	1.79	15.18	2.44	1.19	2.90
1976	8.37	1.89	15.84	2.10	1.23	2.57
1977	8.36	1.90	15.88	2.57	1.22	3.14
1978	8.97	1.95	17.52	3.02	1.33	4.03
1979	8.80	2.03	17.87	2.59	1.39	3.61
1980	9.00	2.24	20.16	2.73	1.46	3.99
1981	9.38	2.38	22.29	2.96	1.53	4.51
1982	8.99	2.54	22.84	2.06	1.57	3.23
1983	9.16	2.62	24.01	3.00	1.69	5.09
1984	9.76	2.66	25.93	3.09	1.71	5.29
1985	9.90	2.68	26.54	2.44	1.77	4.33
1986	9.99	2.71	27.01	3.14	1.88	5.92
1987	9.92	2.75	27.25	2.63	1.96	5.16
1988	10.14	2.80	28.34	3.41	1.95	6.65
1989	10.53	2.89	30.41	2.94	2.10	6.19

Sources: Rice 1970-81: Mears, 1984: 126; 1982-83: Hobohm, 1987: 24; 1984-89: BPS, 1990b and previous issues. Maize 1970-83: BULOG, 1985: 7-9; 1984-89: BPS, 1990b and previous issues.

most part, rely upon rain for moisture. Traditionally, upland rice was planted during the wet season, intercropped with maize/cassava/legumes.

The high-yielding varieties program (HYV) was confined to rice crops grown on the lowland plains because the new varieties need an assured water supply to make use of increased fertilizer applications. However, the construction and renovation of irrigation projects extended the area in which it was possible to grow rice. When improved water supplies were combined with an increased use of pesticides and the availability of better credit and extension services through the various intensification schemes, the scene was set for an expansion of rice production, both through higher yields and an increase in double and triple cropping. A floor price for rice provided a further incentive to farmers (see later discussion). However, the growth in production was often attained at the expense of palawija crops, at least in the early years, as additional rice crops replaced lowland palawija. At the same time, the pattern of foodgrain production in the uplands was not drastically changed by the new intensification schemes for rice and palawija, although some progress was made in improving maize yields, particularly after the introduction of two new varieties in 1979.

Between 1960 and 1977, the growth in rice output was less than 3 percent per annum. Poor support services and the limited availability of fertilizers combined with drought and insect damage to constrain expansion. From 1978 to 1985, the growth in output exceeded 5.5 percent per annum because of a rise in the adoption of new varieties, more widespread use of fertilizers and, not least, because of a stable price policy regime of subsidized input prices and output prices that rose in real terms. Between 1985 and 1987, there was a reduction in the growth of area and yield, reducing output growth to 1.5 percent per annum. Producer prices declined in real terms, and drought and insect damage reduced output. In addition, runoff from excessive fertilizer applications in some areas resulted in a significant pollution problem, and overuse of pesticides led to the development of resistant strains of brown planthopper. An Integrated Pest Management Scheme emphasizing use of natural predators was instituted in late 1986 (ROI, 1988a: 6-7). From 1975 to 1986, three-quarters of the increase in rice output can be attributed to yield improvement and the remainder to increases in harvested area. In the future, the contribution of HYVs to growth in yields is expected to decline (World Bank, 1983).

The 1989 crop has been officially estimated at 30.4 tons (BPS, 1990b), increasing the growth rate between 1985 and 1989 to about 3.4 percent per annum. The estimated population growth rate for the decade 1980/90 is 1.96 percent per annum (Hull, 1991:140), so that recent rice production has managed to keep pace with population and income growth estimates. This was achieved despite two poor crop years (1986 and 1987), and obviously bolstered by the 1988 and 1989 crops which brought production increases of 4 and 7 percent, respectively. A 5 percent expansion in sawah area, probably induced by favourable climatic conditions, contributed to the bumper 1989 crop, of which BULOG (Badan Urusan Logistik), the government agency, procurred 2.52 million tons, just exceeding the previous procurement peak of 1984 (Conroy and Drake, 1990:17).

In contrast, the 1990 crop is expected to be only 30.3 million tons (FAO, 1991:23), just below the previous year, but indicating that production has actually declined for the first time since the mid-1970s (Table 11.7). A fall in harvested area, poor rainfall, insect pests and a slight decrease in the rice/ fertilizer price ratio all seem to have contributed to the production slowdown. The target growth rate for rice in Repelita V has been projected at 3.2 percent per annum (ROI, 1988b: 159), but this may be difficult to attain given the gap now emerging between existing and predicted yields, as well as the gap between yields on- and off-Java.

Maize

While maize is grown throughout Indonesia, it is produced mainly on Java and Madura, chiefly under rainfed upland conditions as part of traditional intercropping or monocrop systems. During the 1980s, Java and Madura accounted for about 70 percent of total maize production, with Sulawesi and East Nusa Tenggara contributing about 20 percent. Some areas (e.g., Central Java) concentrate on producing white maize, which is consumed by humans. Yellow maize is preferred by the poultry industry because of its high carotene levels. This covers about 65 percent of the maize produced. Maize is the principal crop in the upland areas of East Java, where some 40 percent of total production is grown, because it tolerates relatively marginal soils and can survive uncertain climatic conditions.

Maize production systems are extremely diverse, and it is important to distinguish between sawah (covering about 21 percent of planted area) and upland systems, as well as the cropping frequency and productivity of each system (Mink, Dorosh and Perry, 1987). Irrigated sawah, growing one or two high-yielding crops, is more productive than rainfed sawah. The latter usually produces two crops, planted before and after the main rice crop. Under this system, the first maize crop is usually intercropped.

The upland system is divided between single and multiple crop land. Under the former system, maize is usually intercropped with longer-duration HYV rice. The major cropping system, the multiple crop upland system, covers about 55 percent of total planted area and may produce three crops per year. The main (first) crop is usually a HYV, the second crop may be a HYV, while the third crop is often a short-duration, traditional variety (lower yielding).

The HYV maize, Harapan 6, was introduced in late 1978 to overcome downy mildew disease, which devastated crops during the mid-1970s. Arjuna and other resistant varieties followed in the early 1980s. These varieties tend to be slower maturing and are more suited to monocropping than the tighter schedule of intercropping. Since 1983, hybrids have been adopted by some farmers, especially in East Java, but other farmers have not found them to be worth the effort (Mink, 1987). The major constraint to the expansion of improved seed use is the inadequacy of the seed production system. However, physical, social, and economic constraints also contribute. Increased use of new varieties and fertilizers tends to shift farmers to monocropping, which makes less demands on moisture levels. Monocropping does, however, appear to be more labor intensive, especially in the case of hybrids (Djuahari, Djulin and Soejono, 1988: 19).

While two and sometimes three maize crops are grown each year, the main crop is the "winter" crop, which is harvested during the wet season in the two to three months prior to the main rice harvest. The main crop accounts for about 60 percent of East and Central Java production, but between 75 and 93 percent in other areas (Mink and Dorosh, 1987). The timing and volume of the main harvest exacerbate the problems associated with drying maize. Farmers are usually limited to drying small quantities by the kitchen fire, while traders, who purchase wet maize on the cob, must find sufficient rainless time to sun dry it on cement aprons (Mears, 1978:57). To obtain the floor price set by BULOG (see later discussion), maize must be dried to 14 percent moisture content, the same standard as that set for rice. If maize is not properly dried before storage, it is particularly susceptible to the development of aflatoxins, which contaminate feedgrain rations.

Since 1975, the annual production, yield, and area of maize has been more volatile than that of rice (Table 11.7). Downy mildew was a problem in 1976, and during 1979, which was a wet year, farmers planted more rice in place of maize. Nevertheless, the trends in production, yield, and area have been upward, with production almost doubling in the last decade. Despite the improvement in yield, new maize varieties have not been adopted to the same extent as in the case of rice, and maize yields are below those of other comparable countries (e.g., 2.0 tons/ha in Indonesia compared to 2.6 tons/ha in Thailand).

Wheat

There have been sporadic attempts to grow wheat dating back to the Dutch colonial era, and trials still continue (ROI, c1987: 16-17). However, Indonesia's climatic conditions are not conducive to wheat production. As a result, all wheat used in Indonesia is imported.

Marketed Surplus

Only a proportion of rice and maize production is marketed because of home consumption requirements. Estimates of the marketed surplus for rice are variable in quality and magnitude. Mears (1987: 143) suggests that marketings now account for about 50 percent of the total, more than double his estimate made in the late 1950s. Amat (1982: 151) estimates that 40 percent of the rice crop is marketed. He attributes growth in the percent marketed to farmers' needs for more cash income and some feeling of confidence that, if the proportion of retained production subsequently proves insufficient to meet household needs, additional supplies can be purchased in the market at "comparatively reasonable" prices, given the price stabilization activities of BULOG. World Bank economists (1987) estimate that 40-45 percent of the rice crop is marketed.

The proportion of the maize crop marketed varies according to the production system (Mink, Dorosh and Perry, 1987: 63). Around 40 percent of rainfed sawah production is marketed, while 50-80 percent of irrigated sawah is sold. The marketed surplus for single-crop upland maize varies from about 15 percent in low-productivity areas to 60-85 percent in high-productivity areas. The marketed surplus for multiple-crop upland maize is similarly broad, ranging from 40 percent in low-productivity areas to 30-80 percent for high-productivity areas. Timmer (1987: 201) argues that about 50 percent of maize is marketed, which is reasonably consistent with the World Bank (1987: 12) estimate of 46 percent.

302

TABLE 11.8 Rice, Maize, and Wheat Annual Per Capita Consumption by Region, 1981 (kg)

	East Java	Central Java	West Java	North Sumatra	Other Sumatra	South Sulawesi	Other Sulawesi	Other Indonesia	Indonesia
Rice	91.64	93.77	140.84	132.65	131.57	123.51	115.11	114.78	115.94
Rural	90.60	94.21	151.88	143.15	139.51	124.99	113.53	116.83	118.76
Urban	95.27	92.62	116.17	115.17	115.88	120.55	118.16	111.88	110.28
Maize	21.43	12.17	1.12	0.83	1.69	14.05	27.46	19.23	12.53
Rural	26.50	16.55	1.20	1.11	1.88	19.84	35.34	30.57	17.43
Urban	3.73	0.49	0.95	0.36	1.32	2.52	12.25	3.21	2.73
Wheat and wheat products	1.50	2.54	3.37	1.23	3.38	2.96	7.73	5.40	3.58
Rural	1.20	2.29	2.91	0.65	3.18	3.16	4.72	4.62	2.93
Urban	2.77	3.25	4.47	2.53	3.81	2.54	1.77	6.57	5.08

Source: Rosegrant, et al., 1987, tables 3.31, 3.34, and 3.46.

Consumption

Rice and Maize

Rice is the preferred staple food for most of the Indonesian population. More than 60 percent of the carbohydrate-based calorie intake of the average Indonesian in 1984 was derived from rice (BPS, 1987: 121). However, there are some regional differences. Annual per capita consumption of rice, maize, and wheat (Table 11.8) is higher on the outer islands, while that of cassava and soybean is greater on Java. Rice consumption in urban areas was slightly higher than in rural areas during the 1970s, but the position is now reversed.

While rice is consumed almost exclusively by humans, maize has three main uses: (a) a staple food; (b) an animal feed (mainly poultry); and (c) an industrial raw material for noodles, oil, starch, and glue. As the main maize harvest precedes the main wet-season rice harvest, maize provides a relatively cheap source of calories (Dixon, 1984:73), especially for the poorer people, at a time of year when the price of rice tends to be high. Maize consumption is greatest during this time and declines during the remainder of the year. Inhabitants of East and Central Java consume the largest quantity of maize, but on a per capita basis, maize is a much more important staple food in East Nusa Tenggara, North Sulawesi, and Southeast Sulawesi (Monteverde and Mink, 1987: 113). In these three regions, per capita maize consumption is three to four times greater than the national average. Monteverde and Mink (1987: 110) estimate that some three-quarters of total maize production is consumed as a staple food, predominantly in rural areas.

As per capita income rises, it is likely that the poultry industry will expand, resulting in an increased demand for maize. Mink (1987: 147) estimates that livestock feed use already accounts for over 20 percent of production. However, a year-round supply is required to meet livestock demand effectively, and (excluding exports) this will require technical improvements to overcome problems involved in drying and storing maize. Some white maize is used in the noodle industry, but overall, industrial use is small, accounting for less than 1 percent of production in the mid-1980s (Pearson, 1987: 176).

As incomes have risen over time, so has rice consumption. However, the production-consumption gap has been gradually closing, and Indonesia became self-sufficient in 1985. In the case of maize, growth in production has just kept pace with increases in demand. In some years Indonesia is a small net importer and in others a small net exporter, although exports during the 1980s were influenced by higher world prices and devaluations.

Wheat

Wheat is consumed almost exclusively by humans in the form of wheat flour, bread, cookies, crackers, and noodles. From 1974-83, wheat consumption

TABLE 11.9 Expenditure Elasticities of Demand Estimates for Rice, Maize, and Wheat

Author	Type	Elasticity value		
		Rice	Maize	Wheat
Mears (1981)	Review	0.25 to 0.69	—	—
Mears (1981)	Time series of per capita private expenditure against per capita rice availability (1968-79)	0.32 (but considerable uncertainty)	—	—
Mears, Rachman, Sakrani (1981)	Time series (1968-79)	0.32 to 0.50	—	—
Magiera (1981)	Time series (1951-79)	—	—	0.79
Teken and Soewardi (1982)	Review	0.45	-0.31	—
World Bank (1983)	Cross-section (1976)	0.53	-0.55	—
World Bank (1983)	Time series (1968-79)	0.20	-0.21	0.99
Teklu and Johnson (1986)	Time series	0.15 to 0.56	—	—
Rosegrant, et al. (1986)	Cross-section (1981)	0.15 to 0.45	-0.20 to -0.90	0.30 to 0.60
Rosegrant, et al. (1987)	Cross-section (1981)	0.05 to 0.30	-0.10 to -0.30	0.30 to 0.60

almost doubled to 1.7 million tons. Consumption declined to 1.3 million tons in 1985 because the consumer price for rice fell relative to that for wheat, before increasing to 1.8 million tons by 1989.

Although per capita consumption of wheat and wheat products remains small (6.86 kg/capita/yr in 1987), a high income elasticity of demand (see Table 11.9) suggests that as the economy grows, wheat consumption should increase. The urbanization of the population could reinforce such an expansion, given that wheat consumption participation rates and per capita consumption levels are higher in the cities. Wheat consumption rates are also higher off-Java, and this is believed to be the result of Dutch dietary influences and relative affluence (Magiera, 1981: 63).

Supply and Demand Elasticities

The issue of which values are most appropriate for supply and demand elasticities has received considerable attention in economic analyses of grains in Indonesia (Mitchell, 1988). It is clear that there is considerable uncertainty about the accuracy of available estimates. As Hedley (1987) points out, the values of a few key elasticities can alter policy conclusions. For example, in a number of contemporary studies, estimates are provided for the elasticity of the response of rice output to fertilizer price (Rosegrant, et al., 1987; Tabor, et al., 1988; World Bank, 1987). The estimates range from -0.03 to -0.22, and each estimate could be used to arrive at different conclusions about the impact of changes in fertilizer subsidies. Bearing this proviso in mind, follow some brief comments on available estimates.

Supply Elasticities

Much less work has been carried out in Indonesia on estimating supply parameters than on estimating demand parameters. Mears (1981: 43) reviewed studies based on data from 1950 to 1976 and reported a range of elasticity values for the response of rice area to rice price from zero to 0.15. Pitt (1977) was one of the first to examine cross-price effects. He found a significant response in rice area to the relative profitability between rice and other crops. The World Bank (1983: 63) depended on supply elasticity estimates from comparable countries of between 0.10 and 0.30 for its Indonesian modeling work. Likewise, Timmer (1986) used a value of 0.10 for the price elasticity of supply of rice, apparently based on his previous work (Afiff, Falcon and Timmer, 1980: 411).

Given the rudimentary nature of the estimates that have been made, the work of an IFPRI team (Rosegrant, et al., 1987) to achieve a consistent set of area, yield, and supply elasticities for the significant food crops is particularly important. For rice, they estimated the short-run elasticity of supply with respect to own price to be 0.20, and with respect to fertilizer price to be -0.16. Comparable elasticities for maize are 0.31 for own price, -0.24 for rice price, and -0.03 for fertilizer price. Another significant result of their analysis was that rice supply responded strongly to the use of irrigation.

Other recent studies have been undertaken by the Indonesian Ministry of Agriculture (Tabor, et al., 1988) and the World Bank (1987). Mitchell (1988) compares the elasticity estimates presented in these two studies and Rosegrant, et al. (1987). Apart from the differences in the estimates of the response of rice output to fertilizer price alluded to above, the significant differences are: (a) Tabor, et al., obtain a much higher own-price elasticity of supply for maize (0.61 on-Java; 0.74 off-Java); (b) there is a strong cross-price effect (-0.42) from rice price to supply of maize in the World Bank model; and (c) both the World Bank and Tabor, et al. show a small effect (elasticity = -0.06) of maize price on the supply of rice.

The various elasticity estimates suggest a limited overall short-term response to changes in output prices. However, the supply response of rice and maize is greater than other food commodities. Indeed, changes in rice and fertilizer prices in recent seasons resulted in significant and observable changes in rice output.

Demand Elasticities

Some estimates of expenditure elasticities are shown in Table 11.9. Rice has a positive expenditure elasticity, but this declines in cross-sectional observations as income rises. Maize has a negative expenditure elasticity in the sample shown in the table. Some researchers (e.g., Rosegrant, et al., 1987; Mears, 1981) suggest that a more appropriate value would be closer to zero, especially if demand for maize from the poultry and processing sectors is included. (Rosegrant, et al., 1987, estimate an income elasticity of animal feed demand for maize of 1.50.) Wheat appears to have the highest expenditure elasticity among the foodgrains.

Table 11.10 contains a summary of recent own-price elasticity of demand estimates. Once again, there is an element of considerable uncertainty about these estimates. Overall, however, the demand for rice seems more price inelastic than that for either maize or wheat.

Grain Marketing Systems and Institutions

Marketing Channels

There is considerable regional variability in the way rice marketing is handled in Indonesia, and it is difficult to make generalizations about marketing channels (Mears 1981: 119). Regional variability is a function of such factors as the proportion of the crop marketed, location of the farmer relative to milling facilities, and whether a particular region is generally rice deficit or surplus.

TABLE 11.10 Own-Price Elasticities of Demand Estimates for Rice, Maize, and Wheat

Author	Type	Elasticity value		
		Rice	Maize	Wheat
Timmer and Alderman (1979)	Cross-section (1976)	-1.10	—	—
Mears (1981)	Review	-0.40	—	—
Magiera (1981)	Time series (1951-79)	—	—	-1.78
World Bank (1983)	Time series (1968-79)	-0.15	-2.27	-0.53
Dixon (1982)	Cross-section (1976)	-0.48 to -0.84	—	—
Rosegrant, et al. (1986; 1987)	Cross-section (1981)	-0.12 to -0.40	-0.30 to -0.60	-0.20 to -0.50
Tabor, et al. (1988)	Time series and cross-section	-0.17	-0.19	—
World Bank (1987)	Time series (1969-85)	-0.20	-0.43 to -0.60	—

308

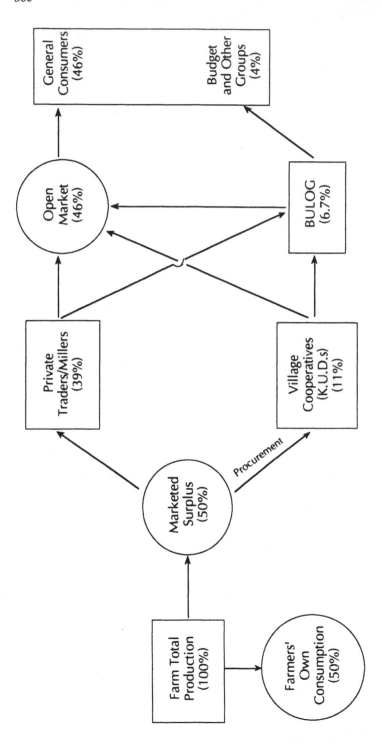

FIGURE 11.2 Indonesia Rice Marketing Channels and Percentage Shares, 1981-82. *Source:* Mears (1987). Adapted and modified from Figure 6.2.

However, most observers make a distinction between two channels: government and private. The agents working in these channels are described briefly in subsequent subsections. Mears (1987: 143) reports that in 1981-82, the government channel (BULOG) handled about 8.7 percent of production, of which the "Budget" group (army and civil service) received an allocation of 4 percent as part of their salary (Figure 11.2). In addition, the government channel is also responsible for imports and exports. The remainder of rice marketed, some 46 percent of total production, was handled in the private channel (Figure 11.2).

Because of the diversity of the maize production system, the variation in the size of marketings and the isolation of many of the smaller producers (especially those on the outer islands) from potential markets, the maize marketing system serves some producers more efficiently than others. Large commercial maize farmers in East Java, for instance, have ready access to maize marketing systems (Timmer, 1987: 207). Good local transportation and communications also contribute to more efficient market integration in East Java. On the other hand, small, scattered, and irregular marketings lead to a higher volume of transactions and a generally less efficient and competitive system. In addition, there is a significant lack of integration between the markets for white and yellow maize. Timmer (1987: 204-5) argues that, although less than one-quarter of maize produced in East Java is destined for Surabaya or Jakarta feed mills, the demand from these mills is the key to price formation, largely because they have access to imported maize at internationally competitive prices through BULOG.

Private Agencies

Commonly called KUDs (Koperasi Unit Desa), cooperatives in Indonesia cannot be equated with those found in Western countries. Rather, they operate more like semi-governmental institutions than private ones. Initiated in 1973 to aid BULOG with rice procurement and marketing, they were also closely associated with the rice intensification programs (Daroesman, 1981). KUDs play an important role in BULOG's rice price policy by purchasing rice from farmers at the official floor price for resale to BULOG (or, in some cases, private merchants). They also assist in making purchases as part of the floor price scheme for maize (and other palawija products that have floor prices), but this marketing role is much less significant. For example, the World Bank (1987:33) indicates that at no time has BULOG's maize procurement exceeded 3 percent of production. Other services provided by some KUDs include storage of rice and fertilizers, small-scale rice milling, and grain drying.

Small traders (sometimes on bicycles) buy paddy from farmers for resale to KUDs or small mills, while large traders operate mainly at the provincial level, purchasing from smaller traders and millers or from the wholesale market to which the latter also contribute. BULOG buys only a small part of the marketed rice crop, while the balance (80 percent or more) is handled through private marketing channels. Mears (1987: 143) indicates that private traders tend to either purchase high-quality rice at prices above the floor price, or low-quality rice at "sacrifice prices". They also prefer to concentrate on rapidly completed transactions that involve little storage.

BULOG Marketing

BULOG purchases paddy (*gabah*, dried to 14 percent moisture level) or medium-quality milled rice through both the KUDs and private rice traders (mostly millers) to supply rice rations to the "Budget" group and operate the rice price stabilization scheme. Rice (and much smaller amounts of maize and other crops) acquired for stabilization purposes is stored at regional DOLOGs (Depot Logistik) or district depots for later sale through private traders at the wholesale or retail level. As rice procurement levels rose during the good harvests in 1984 and 1985 (existing stocks were still high from previous harvests), BULOG was forced to find extra storage capacity and to build more warehouses. BULOG is constrained by the fact that its operations are based on the level of the floor price and not on the level of its stocks. In 1985, it was forced to tighten quality standards in order to defend the floor price.

The dwindling gap between the floor price and the ceiling price means that inter-seasonal storage is now unprofitable for most private traders. As a result, BULOG must bear the costs of long-term storage which, in 1985, threatened it with insolvency (Falcon Team Report, 1985).

Grain Trade: A Drive for Self-Sufficiency

The main trends in Indonesia's grain trade reflect the government's drive for foodgrain self-sufficiency. Until 1980, Indonesia was a major rice importer, accounting for about 15 percent of total world imports (Table 11.6) and an even higher percentage in the late 1970s. As a result of increased domestic production, since 1980 Indonesia first reduced and then finally eliminated rice imports (except for small amounts of glutinous rice, which is not grown domestically). During the first half of the 1980s, rice imports constituted, on average, only about 4 percent of world imports (less than 0.5 million tons/yr). Indonesia's dwindling imports helped lead to reductions in international rice prices during the 1980s. The implication of Indonesia's declining imports for an already-thin rice market are outlined in Barker, Herdt, with Rose (1985).

In 1985, Indonesia exported a little over 300,000 tons of rice and 600,000 tons more in 1986. This was generally low-quality rice, with about 25 percent or greater brokens, which sells for one-quarter to one-third less than high-quality (5 percent brokens) rice (Barker, Herdt, with Rose, 1985: 190). Indonesia is in a rather precarious situation, therefore, with respect to the targeting of production levels. Production beyond domestic requirements sells on international markets for substantially less than the domestic procurement price. This implies that the Indonesian government is providing a large export subsidy, which was in the order of $US100 per ton for exports made in 1986 (World Bank, 1987: 22).

Indonesia's international trade in maize is highly seasonal, highly erratic, and rarely free of government restriction (Dorosh, 1987). Between 1970 and 1974, Indonesia exported small amounts of maize until an overvalued exchange rate made exports uncompetitive. In 1976, Indonesia both imported and exported maize, largely because of high storage costs and limited storage capacity (Timmer, 1987). This pattern continued until 1980, but in all years Indonesia was a net importer because of increased demand from the poultry industry. BULOG's maize imports reached a record high of 76,000 tons in 1982 because of drought. In 1984, following the 1983 devaluation, domestic prices were roughly similar to world prices allowing for transport costs, and 160,000 tons were exported. However, 1984 exports were something of a special case. World prices rose following bad weather and a one-year US special program that cut production by one-half (Dorosh 1987: 244, 248). Exports are restricted to those countries accepting bagged grain, as Indonesia lacks bulk-handling facilities.

Indonesia's maize imports comprise an insignificant proportion of world maize trade. On a regional basis, however, it is a significant importer in some years. The relative importance of Southeast Asian countries as importers and exporters of maize within the region fluctuates considerably because of climatic variability (this is manifest partially in the substitution of maize production for rice production where seasonal conditions are not conducive to the latter) and the transport costs and storage problems associated with maize.

As stated previously, attempts to grow wheat in Indonesia have so far been unsuccessful, and trade in wheat is restricted entirely to imports (Table 11.6). Wheat flour was imported from the beginning of the 1950s and, between 1951 and 1961, averaged 182,000 tons per annum. Imports declined during the political troubles of the early 1960s, but then recovered substantially because the New Order government used controlled flour prices to contain inflationary pressure.

In 1971, following a considerable investment in milling capacity (which resulted from the deterioration of flour in humid tropical conditions and a desire to create jobs), Indonesia began to increase imports of wheat grain and to decrease imports of wheat flour. Extra milling capacity was added until, by 1977, over 1 million tons of wheat was imported. Imports of 1.7 million tons were recorded in 1983, after which imports declined somewhat until rising again to 1.8 million tons in 1989.

Most of the early wheat imports were made on a concessional basis (P.L. 480 Title I), with financing on the softest possible terms (Magiera, 1981: 49), but in the 1970s imports shifted to the commercial market. The United States dominates the Indonesian wheat import market, with about 60 percent of the total, followed by Australia with about 30 percent. Canada, the Netherlands, and France share the remainder of the market. Some of the US imports are still made on concessional terms, and a $US 30 million food-aid grant was awarded in mid-1986.

Since 1972, BULOG has controlled all of the wheat trade, with the objective of stabilizing flour prices and synchronizing grain shipments relative to mill and stock needs (Magiera, 1981: 53). However, because Indonesian imports have never exceeded 2 percent of world trade, the supply of wheat to Indonesia probably can be considered to be perfectly price elastic, unlike the supply of rice.

Grain Policy: Background

There is a long history of government involvement in the rice economy of Indonesia. Concise but comprehensive accounts are available in Timmer (1975) and Rasahan (1983). Government involvement occurred for both political and economic reasons and subsequently spread to secondary food crops. In this section, attention is focused on three matters: the policy-making process, the objectives of foodgrain policy, and the various policy instruments used to pursue the objectives.

The Creation of Foodgrain Policies

Glassburner (1978: 32) describes policy making in Indonesia as an "obscure process". It has never been comprehensively documented, and this observation itself is suggestive of some of the characteristics of the policy-making environment. These include a general lack of public debate and the probability that policy is determined ultimately by a small group of key individuals. Those who do have an intimate knowledge of how policy is formulated do not make this knowledge public. Also, it is likely that some policy decisions are made in response to crises during which there is no time for public debate.

Liddle (1987a: 206-208) argues that there are three interest groups competing "for the President's ear". Depending on the prevailing economic climate, the President will seek advice from any one of these three groups. The most important and presently most influential group is made up of mainly Western-trained economists and technocrats with a market-oriented outlook. Another group supports policies that protect government-owned enterprises and the *pribumi* (Indonesian as opposed to ethnic Chinese) business class. The third group consists of politicians and seems to be concerned primarily with fostering the popularity of the political "party" in power.

When policy is decided, it is often not implemented because of coordination problems among ministries and agencies, as well as inadequate planning (Liddle, 1987b: 129; Stone, 1987: 20ff). Policies are often rescinded because of political costs or the impossibility of implementation (McCawley, 1972: 3; Liddle, 1987a: 211). No doubt, problems with policy formulation (Liddle, 1987b: 129-36) and implementation also occur because of inadequate and unreliable information.

The current government frequently turns to foreign advisers for assistance when crises occur. There is a substantial number of foreign advisers and agencies involved in providing economic advice in Indonesia, particularly in relation to staple food production, marketing, and price policy. However, Glassburner (1986: 27) emphasizes that foreign advisers do not make policy decisions but are there only to provide counsel. Their presence reflects, in part, a scarcity of professionally-trained domestic policymakers.

BULOG (established in 1967) reports directly to the President. In addition, it has a strong influence over the KUDs. (The BULOG Chairman is also the Minister of Cooperatives). Although BULOG's primary role is one of policy implementation, it is also represented on key committees responsible for formulating recommendations on food price policy.

In summary, it is difficult for outsiders to gain an appreciation of the precise details of how food price policy in Indonesia is determined. However it seems that there is a "...general belief in consensus in decision making, with open opposition to policies being seen as socially incorrect." (MacAndrews, 1986:31) Communication within the bureaucracy has been described as "...predominantly oral, with day-to-day decision making based on oral discussion and only formally finalized, if required, in a letter." (MacAndrews 1986: 31)

Objectives of Grain Policies

During the 1970s and early 1980s, there were four main objectives underlying Indonesian grain policies (World Bank, 1983: vi): self-sufficiency, especially with respect to rice; higher farm incomes; reasonable and stable food prices for urban consumers; and containment of the expenses required to achieve these goals. As in other Southeast Asian countries (e.g., Malaysia and the Philippines), the first objective, self-sufficiency, was pursued the most vigorously. In recent years, the government has reassessed its food policy objectives. Some factors that contributed to this were: (a) the success of the rice expansion program; (b) historically low real world prices for rice and other food commodities; and (c) budgetary constraints brought about in part by a decline in the price of oil (Rosegrant, et al., 1987: 6.1). In a recent report on Indonesia, the World Bank (1987: 41) summarizes broad agricultural policy objectives as follows: (a) to reduce subsidies for agriculture while obtaining greater returns from available resources; (b) to sustain rice self-sufficiency while diversifying the rural economy and furthering import substitution of other food crops; (c) to improve the standard of living of farmers and the poorer segments of society; (d) to promote more equitable regional development; and (e) to increase foreign exchange earnings and savings.

There is no doubt that rice still dominates the foodgrain policy agenda. Maintaining rice self-sufficiency is, perhaps, the principal foodgrains policy objective. In the case of maize, the primary objective seems to be the maintenance of stable prices for feedgrain maize. Policy with respect to wheat is aimed at providing wheat flour to urban consumers at stable prices.

Instruments Used to Enforce Grain Policies

Since the early 1960s the government has promoted rice self-sufficiency through a series of intensification programs. The first of these, the BIMAS program, was introduced in 1963, and the current program is called SUPRA-INSUS. Other programs included BIMAS GOTONG ROYANG, PERFECTED BIMAS, INMAS, and INSUS. The BIMAS programs were designed to increase production through the use of improved seeds, fertilizers, pesticides, water management, improved cultural practices, and the development of farmer cooperatives. The parallel scheme, INMAS, aided farmers with capital (or access to it) and a keen desire to use modern inputs. In the case of the INSUS scheme, about 50 to 100 farmers with contiguous plots were encouraged to make joint decisions about seeds, planting times, and crop choices in addition to rice.

TABLE 11.11 Indonesian Grain Policy Instruments[a]

Instrument	Commodity		
	Rice	Maize	Wheat[b]
Output prices			
Floor price	X	X[c]	
Ceiling price	X		
Storage cost subsidy	X	X	
Input subsidies			
Fertilizer	X	X	
Pesticides	X	X	
Credit	X	X	
Irrigation	X	X	
Research/extension	X	X	
HYV seeds	X	X	
Machinery/fuel	X	X	
Trade measures			
Export subsidy (implicit in 1985)	X		
Export licensing		X	
BULOG import monopoly	X	X	X

[a]Excluded are an array of regulations and decrees that the government uses to influence cropping patterns (World Bank, 1987: 37).
[b]Wheat not produced domestically.
[c]Floor price applies to yellow maize.

The government also realized that adequate price incentives would help encourage farmers to adopt new technologies. The first stage in the development of price incentives was the announcement of the *rumus tani* (Farmer's Formula) in 1968, which linked the price of rice to the price of imported urea fertilizer (Arndt, 1968: 11-12). While this marked the introduction of a floor price policy, it was not effective until fertilizer distribution was also improved (Mears, 1981: 390). Following the 1972 rice crisis, the introduction of the cooperatives (KUDs) gave BULOG the opportunity to implement the pricing formula at the farm level. There were problems of interpretation and implementation (Afiff and Timmer, 1971; Timmer, 1975), but a rice price policy based on a Mears and Afiff model (Mears, 1981: 391) was gradually put in place, became fully operational by 1974, and continues through to the present time.

The main price-policy instruments used in the grain sector are summarized in Table 11.11. In the case of rice, the key instruments are floor prices, which are used to maintain a minimum market price for rice delivered to the KUDs; various input subsidies; ceiling prices, which limit the amount that consumers must pay to purchase rice; and storage-cost subsidies. The introduction of implicit subsidies on rice in 1985 added a new policy instrument to the list. While there is a floor price policy for maize, BULOG intervention in the maize market is relatively slight compared with the rice market. This is, in part, because of the greater political importance of rice and the fact that marketing channels for maize are less clearly defined. Control over imports and exports is the principal means by which the government influences the maize market. BULOG is the sole importer of wheat and issues processing contracts to one firm in the private sector.

In summary, the Indonesian government uses a mixture of price-policy instruments to influence grain production and marketing, and BULOG is the key implementer of these policy decisions. In addition, the government devotes considerable resources to the development and dissemination of improved technologies. Some policy instruments can be viewed as supply shifters (e.g., fertilizer subsidies and technological innovations), while others encourage output expansion along supply functions (e.g., producer price supports).

The fertilizer subsidy accounted for over 3 percent of the Development Budget during the first half of the 1980s, reaching 7.4 percent in 1984/85. The government attempted to reduce its share in the 1987/88 budget, but actual expenditure on the subsidy exceeded 9 percent of the Development Budget for that year. For subsequent budgets, the government decided in early 1988, to make only "nominal" provision (1 or 2 percent of the Development Budget or less) for fertilizer subsidies, and shift the remaining expenditures off-budget, funding it through bank credit (see Booth 1988: 10, 14).

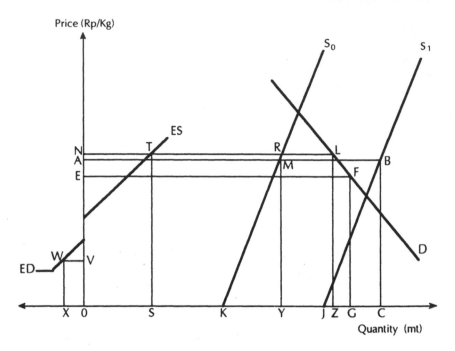

FIGURE 11.3 Comparison Between the 1985 Policy and a Free-Market Equilibrium. *Source:* Adapted from Timmer (1986:36).

Public Policy: A Partial Evaluation

In this section, a partial evaluation of foodgrain price policy is provided. The section is divided into three parts. First, a description is given of the scenario in 1985 with respect to rice in a conventional, partial equilibrium, demand and supply framework. Second, a partial evaluation is made of 1985 rice price policy, using a competitive market situation as the base. Third, two marginal changes in rice price policy are evaluated giving emphasis to the effects on secondary food crop markets.

The Policy Situation in 1985

A partial equilibrium description of Indonesia's rice policy in 1985 is shown in Figure 11.3, which is amended from Timmer (1986: 36). The analysis is performed at the wholesale level, assuming linear demand and supply functions with base-point elasticities (adapted from Rosegrant, et al., 1987 and Timmer, 1986: 35) of -0.6 (domestic demand) and 0.3 (domestic supply). The domestic demand is labelled D, the actual domestic supply is labelled S_1, and the domestic supply that would appear in the absence of the fertilizer subsidy is labelled S_0. Also shown is the demand for Indonesian rice on the world market, labelled ED, and the supply of imported rice, labelled ES. Indonesia exports a relatively low-grade rice, so there is some price discounting. The export demand function eventually becomes perfectly elastic as rice substitutes for other feedgrains.

The equilibrium (import parity) international price level is shown as ON (=332 Rp/kg). However, domestic producers received a price OA (329.47 Rp/kg), while domestic consumers paid a price of OE (=314.36 Rp/kg). Consumers received an implicit subsidy because BULOG did not pass on the full cost of its intra-seasonal storage activities. Domestic production was OC (=24.2 million tons net), consumption was OG (=23.55 million tons), and exports were OX (=OC - OG = 650 thousand tons net). The price received for the exported rice was OV (=165.45 Rp/kg) (EIU, 1986).

The three components of rice price policy are clear from the figure. The floor/ceiling price scheme had the effect of stimulating supply, while still providing a consumer subsidy. The fertilizer subsidy scheme shifted the rice supply curve to the right (S_0 to S_1), while the export subsidy removed excess supply from the domestic market and dumped it on the international market. Exports received a subsidy per unit of OA minus OV (=164.02 Rp/kg).

An important distinction is that the fertilizer subsidy policy and producer price support (floor price) acted to influence supply in the preharvest period, whereas the consumer (ceiling) price and export subsidies helped to manage supply once it had entered the marketplace. Thus, the fertilizer subsidy and floor price policy can be used interchangeably or in tandem when the goal is to influence production, while the ceiling price and export subsidy act to influence consumption.

Partial Evaluation of the 1985 Policies

In this section, the effects of the 1985 policy on efficiency are examined. The comparison is against a competitive market situation in which all the policy measures described in the previous section are removed.

A policy change to a competitive situation would result in the domestic price for both consumers and producers increasing to the import parity level, ON (=332 Rp/kg) (Figure 11.3). Domestic output would fall from OC (=24.20 million tons) to OY (=20.74 million tons), while domestic consumption would drop from OG (=23.55 million tons) to OZ (=22.76 million tons). Hence, Indonesia would become a net importer of some 2 million tons of rice, rather than a net exporter.

Using Figure 11.3 to compare the situations with and without the 1985 interventions, the net social benefits can be calculated. The budgetary cost of the interventions per unit of output is (a) the difference between the average producer price of rice and the weighted average of consumer and export prices, plus (b) the cost of the fertilizer subsidy. The gain in consumer surplus derived

TABLE 11.12 Estimates of Components of Net Social Cost of Rice Price Policy

	Rp. billion	$US million
Gain in producer surplus	864	783
Gain in consumer surplus	408	370
Budgetary cost	1,507	1,366
Net social cost	235	213
Foreign exchange savings	778	705

from the intervention is represented by the area NLFE, while the increase in producer surplus is measured by the difference between the areas MBJK and NRMA. There is a saving in foreign exchange from the interventions of OSTN (i.e., the amount that BULOG would need to spend on rice imports if interventions were removed) plus OVWX (i.e., the amount received for exports).

Overall, the budgetary cost of the intervention measures was in excess of Rp 1,500 billion (Table 11.12). This expenditure yielded a gain in producer surplus of Rp 864 billion and an increase in consumer surplus of Rp 408 billion. Hence, there was a net social cost of Rp 235 billion. However, because of the interventions, about $US 705 million was added to foreign exchange earnings.

It should be noted that this calculation of the net social cost may result in an underestimate of the total, because government intervention in the rice market may cause spillover effects on secondary food crops. Because of cross-price relationships between rice and secondary food crops, both on the supply side and the demand side, intervention in the rice market would have resulted in a decrease in production and consumption of secondary crops. Moreover, when calculating the net social cost estimate, some positive effects are also ignored. First, foreign exchange is conserved. In Indonesia's case, this might be considered important given the shortfall in oil revenues caused by the decline in the world price for crude oil. Second, the production of a small rice surplus in normal years provides a buffer against shortfalls in years with poor weather, thus reducing the worries associated with year-to-year instability in production.

Effects of Marginal Changes in Rice Price Polices

In this section, the effects of marginal changes in rice price policy on markets for rice and secondary food crops is demonstrated. There has been some

TABLE 11.13 Response to a 10 Percent Reduction in Real Producer Price of Rice (percentage change from 1985 values)

Commodity	Price	Consumption	Production
Rice	-10.00[a]	-0.01	-2.99
Maize	0.00[b]	-0.01	1.21
Cassava	0.00[b]	-0.04	0.80
Sugar	-0.21	0.14	0.14
Groundnut	-0.08	0.07	0.07
Soybean	0.00[b]	-0.02	0.20
Sweet potato	-0.21	0.05	0.05
Wheat	0.00[b]	-0.01	0.00

Trade Results
  Rice: exports decrease from 644 kt to imports of 78 kt.
  Maize: imports decrease from 34 kt to exports of 22 kt.
  Cassava: exports increase from 750 kt to 870 kt (16%).
  Soybean: imports decrease from 278 kt to 276 kt (<1%).
  Wheat: imports decrease marginally. Crop not produced domestically.

[a]Indicates a change in the producer wholesale price. Consumer wholesale price fixed at 329 Rp/kg.
[b]Price determined exogenously.

concern in Indonesia that the emphasis on the expansion of rice production has come at the expense of secondary food crop production and consumption. If this is the case, it is an outcome inconsistent with the government's objective of diversifying the food base.

The results were obtained by using a simultaneous model of demand and supply covering eight commodities that account for 90 percent of caloric intake: rice, maize, cassava, sugar, groundnut, soybean, sweet potato, and wheat. Details of the model are given in the Appendix.

The marginal changes in policy reported here are designed to encourage increased production and consumption of secondary food crops, because this is a stated government objective. Increased production of secondary crops can be encouraged in a variety of ways. Output price policy alternatives include, for example, price supports for secondary crops, reduced price supports for rice (since, from the supply side, it competes with secondary food crops), or a combination of the above. Given the fact that policy instruments designed to control the price of secondary food crops are less developed than those for rice and fertilizer prices, attention here is directed toward the effect of reductions in the real producer price of rice and the fertilizer subsidy.

The estimated effects of a 10 percent reduction in the real producer price of rice compared with its 1985 value are shown in Table 11.13. It is clear that rice production declines, while the production of all other crops rises, with maize and cassava production growing the most rapidly. On the consumption side, the effects are small and occur as a result of induced decreases in the prices of sugar, groundnut, and sweet potato following alterations in production. The prices of these commodities are determined endogenously by domestic demand and supply forces, while the consumer prices of the other commodities, including rice, are set by BULOG. Growth in the consumption of sugar, groundnut, and sweet potato reflects a decrease in their relative prices, while reduced consumption of the other crops is caused by increases in their relative prices.

A 10 percent rice price reduction is more than sufficient to cause Indonesia to become a net importer of rice and a net exporter of maize. There is also a significant percentage increase in cassava exports. Although not shown in the table, the effects of the 10 percent reduction in the real producer price of rice gave rise to about a 12 percent reduction in the producer surplus resulting from rice production, but the effects on the producer and consumer surpluses for other commodities were generally less than one percent in magnitude.

If the farm price of fertilizer is increased by 20 percent, the output of domestically-produced crops falls (Table 11.14). Maize, rice, and soybean are most affected, presumably because they derive the most benefit from the fertilizer subsidy. The effects on consumption are generally smaller than those on production. The consumption of those commodities with endogenously-

TABLE 11.14 Response to a 20 Percent Increase in Fertilizer Price (percentage change from 1985 values)

Commodity	Price	Consumption	Production
Rice	0.00[a]	0.15	-3.10
Maize	0.00[a]	0.16	-4.13
Cassava	0.00[a]	0.43	-1.04
Sugar	3.07	-2.08	-2.08
Groundnut	1.52	-1.40	-1.40
Soybean	0.00[a]	0.32	-3.03
Sweet potato	2.12	-0.50	-0.50
Wheat	0.00[a]	0.10	0.00

Trade Results
Rice: exports decrease from 644 kt to imports of 141 kt.
Maize: imports decrease from 34 kt to exports of 229 kt (573%).
Cassava: exports decrease from 750 kt to 541 kt (28%).
Soybean: imports increase from 278 kt to 306 kt (10%).
Wheat: imports increase from 1333 kt to 1334 kt (<1%). Crop not produced domestically.

[a]Price determined exogenously.

determined prices declines because their relative prices rise following decreased production. Consumption of the other commodities increases because their relative prices fall (absolute prices remain fixed).

A reduction in the fertilizer subsidy would force Indonesia to become a net rice importer, while maize imports would increase six-fold. There would also be a substantial decline (28 percent) in cassava exports, and increases in soybean and wheat imports. Producer and consumer surplus effects (not shown in the table) are generally small in percentage terms.

The magnitudes of the results of the two marginal policy changes examined here are, of course, dependent upon assumptions made about price and income elasticities. However, the qualitative effects of these marginal policy changes are robust under the alternative elasticity assumptions used in the modelling work. It can be safely assumed that changes in price incentives for rice production do have cross-commodity effects, and if the government is concerned about the production and consumption of secondary food crops, these deserve consideration. It also seems clear that decreasing the price incentives for rice production will result in production and consumption increases for some secondary food crops, but declines for others. That is, attempting to diversify the food base through rice price policy alone will have only limited success. In addition, also needed are policies specifically designed to affect the prices of secondary crops.

Conclusions

Rice is the dominant grain (and food) crop in Indonesia. Moreover, it is also used to supplement salary payments to government employees. Rice production increased markedly in the course of the last two decades in response to a government drive for rice self-sufficiency. This growth was caused by the government's emphasis on dissemination of new technologies, improved infrastructure (e.g., irrigation), and a combination of price policies that favored rice production over other commodities. The price policies involved a floor and ceiling price arrangement implemented by BULOG and subsidization of key inputs, especially fertilizer. As a result of these policies and a build-up of stocks, Indonesia had to sell rice on the international market at a loss over its procurement price. Hence, a certain measure of export subsidization was involved. Indonesia's exit as a major rice importer, in itself, helped to depress international rice prices.

In terms of its rice self-sufficiency status, the Indonesian agricultural economy is "delicately poised." The treasury cost to attain the 1985 surplus was large and will be expensive to maintain given a population growth rate of nearly 2 percent per annum and sustained income growth. Too, the marginal reductions in the level of price incentives for rice examined in this paper indicate a return to net importer status. Aberrant weather conditions, disease, and insect problems could also cause a resumption of rice imports. The international rice market is sufficiently "thin" that Indonesia's switch from net importer to net exporter status will result in significant fluctuations in the international price.

The marginal changes in policies toward rice examined in this paper do affect the production and consumption of secondary food crops. However, in percentage terms, the impact is relatively small. This is not to deny, however, that the total incentive package, which favors rice, may have resulted in large distortions in the production and consumption of secondary food crops. The main effects on trade of secondary crops appear to be with respect to maize, and while these effects are small in terms of world maize trade, they are much more significant when considered on a regional basis.

World oil prices act as an important constraint on policy formation with respect to foodcrops in Indonesia. Oil revenues play an essential part in financing incentives for increased agricultural output. The renewed emphasis on the budgetary costs associated with rice price policy no doubt reflects, in part, the pressures created by low oil prices compared to their levels during the 1970s and early 1980s.

Notes

The authors express their appreciation to Elaine Treadgold for her assistance at all stages in the preparation of this chapter.

References

Afiff, S. and C.P. Timmer. "Rice Policy in Indonesia." *Food Research Institute Studies* 10(1971): 131-59.

Afiff, S., W.P. Falcon, and C.P. Timmer. "Elements of a Food and Nutrition Policy in Indonesia." In G.F. Papanek (ed.), *The Indonesian Economy*. New York: Praeger, 1980.

Amat, S. "Promoting National Food Security: The Indonesian Experience." In A.H. Chisholm and R. Tyers (eds.), *Food Security: Theory, Policy and Perspectives from Asia and the Pacific Rim*. Lexington: Lexington Books, 1982.

Arndt, H.W. "Survey of Recent Developments." *Bulletin of Indonesian Economic Studies* 10(1968): 1-28.

Barker, R., R.W. Herdt, with B. Rose. *The Rice Economy of Asia*. Washington, D.C.: Resources for the Future, 1985.

Biro Pusat Statistik (BPS). *Statistik Indonesia 1985*. Jakarta: 1985.

_____. *Buletin Ringkas*. Jakarta: 1986b.

_____. *Buletin Ringkas*. Jakarta: 1989.

_____. *Buletin Ringkas*. Jakarta: 1990a.

_____. *Buletin Ringkas*. Jakarta: 1990b.

_____. *Food Balance Sheet in Indonesia*. Jakarta: 1989.

_____. *Indikator Ekonomi*. Jakarta: 1990c.

_____. *Indikator Kesjahteraan Rakyat*. Jakarta: 1987.

_____. *National Income of Indonesia 1983-1985.* Jakarta: 1986a.

_____. *National Income of Indonesia 1984-1989.* Jakarta: 1990a.

Birowo, A.T. and N.A. Sanusi. "Introduction and Overview." In Mubyarto (ed.), *Growth and Equity in Indonesian Agricultural Development.* Yayasan Agro Ekonomika. Yogyakarta: Gadjah Mada University Press, 1982.

Booth, A. "Survey of Recent Developments." *Bulletin of Indonesian Economic Studies* 24(1988): 3-36.

Booth, A. "Repelita V and Indonesia's Medium Term Economic Strategy." *Bulletin of Indonesian Economic Studies* 25(1989): 3-30.

Booth, A. and B. Glassburner. "Survey of Recent Developments." *Bulletin of Indonesian Economic Studies* 11(1975): Appendix A, 29-31.

BULOG. *Bulletin Statistik Jagung 2.* Jakarta: August 1985.

Conroy, J. D. and P. J. Drake. "Survey of Recent Developments." *Bulletin of Indonesian Economic Studies* 26(1990): 5-42.

Dalton, B. *Indonesia Handbook.* Rev. Edn. Chicago: Moon Publications, 1985.

Dapice, D.O. "An Overview of the Indonesian Economy." In G.F. Papenek (ed.), *The Indonesian Economy.* New York: Praeger, 1980.

Daroesman, R. "Survey of Recent Developments." *Bulletin of Indonesian Economic Studies* 17(1981): 1-41.

Djauhari, A., A. Djulin, and I. Soejono. "Maize Production in Java: Prospects for Improved Farm-Level Production Technology." CGPRT (Centre for Coarse Grains, Pulses, Roots and Tubers). No. 13. Bogor: 1988.

Dixon, J.A. "Use of Expenditure-Survey Data in Staple-Food-Consumption Analysis: Examples from Indonesia." In A.H. Chisholm and R. Tyers (eds.), *Food Security: Theory, Policy and Perspectives from Asia and the Pacific Rim.* Lexington: Lexington Books, 1982.

_____. "Consumption." In W.P. Falcon, W.O. Jones and S.R. Pearson (eds) *The Cassava Economy of Java.* Stanford: Stanford University Press, 1984.

Dorosh, P.A. "International Trade in Corn." In C.P. Timmer (ed.), *The Corn Economy of Indonesia.* Ithaca, NY: Cornell University Press, 1987.

Economist Intelligence Unit (EIU). *Country Profile Indonesia 1986-87.* London: 1986.

Falcon Team Report. *Rice Policy in Indonesia, 1985-1990: The Problems of Success.* Report presented to BULOG. Jakarta: 1985.

Food and Agriculture Organization (FAO). *Monthly Bulletin of Statistics.* 1986 and various issues.

_____. *Quarterly Bulletin of Statistics.* 1991 and various issues.

_____. *Trade Yearbook.* 1987 and various issues.

Glassburner, B. "Political Economy and the Soeharto Regime." *Bulletin of Indonesian Economic Studies* 14(1978): 24-51.

_____. "Survey of Recent Developments." *Bulletin of Indonesian Economic Studies* 22(1986): 1-33.

Hedley, D. "Proposal for a Workshop on Issues in Research Methodology for Indonesian Agriculture." Jakarta: Bureau of Planning, 1987.

Hill, Hal. "Foreign Investment and Industrialization in Indonesia." East Asian Social Science Monographs. Singapore: Oxford University Press, 1988.

Hobohm, S.O.H. "Survey of Recent Developments." *Bulletin of Indonesian Economic Studies* 23(1987): 1-37.

Hull, T. H. "Population Growth Falling in Indonesia: Preliminary Results of the 1990 Census." *Bulletin of Indonesian Economic Studies* 27(1991): 137-42.

Kasryno, F., D. Budianto, and A.T. Birowo. "Agriculture - Non Agriculture Linkages and the Role of Agriculture in Overall Economic Development." In Mubyarto (ed.), *Growth and Equity in Indonesian Agricultural Development*. Yayasan Agro Ekonomika. Yogyakarta: Gadjah Mada University Press, 1982.

Liddle, R.W. "Indonesia in 1986: Contending with Scarcity." *Asian Survey* 27(1987a): 206-18.

_____. "The Politics of Shared Growth: Some Indonesian Cases." *Comparative Politics* (January 1987b): 127-46.

MacAndrews, C. "The Structure of Government in Indonesia." In C. MacAndrews (ed.), *Central Government and Local Development in Indonesia*. Singapore: Oxford University Press, 1986.

McCawley, P. "Survey of Recent Developments." *Bulletin of Indonesian Economic Studies* 8(1972): 1-32.

Magiera, S. "The Role of Wheat in the Indonesian Food Sector." *Bulletin of Indonesian Economic Studies* 17(1981): 48-73.

Mears, L.A. "Problems of Supply and Marketing of Food in Indonesia in Repelita III." *Bulletin of Indonesian Economic Studies* 14(1978): 52-62.

_____. *The New Rice Economy of Indonesia*. Yogyakarta: Gadjah Mada University Press, 1981.

_____. "Rice and Food Self-Sufficiency in Indonesia." *Bulletin of Indonesian Economic Studies* 20(1984): 122-38.

_____. "BULOG: National Food Authority — Indonesia." In J.C. Abbott (ed.), *Agricultural Marketing Enterprises for the Developing World*. Oxford: Cambridge University Press, 1987.

Mears, L.A., A. Rachman, and Sakrani. "Income Elasticity of Demand for Rice in Indonesia." *Ekonomi dan Keuangan Indonesia* 29(1981): 81-90.

Mink, S.D. "Corn in the Livestock Economy." In C.P. Timmer (ed.), *The Corn Economy of Indonesia*. Ithaca, NY: Cornell University Press, 1987.

Mink, S.D. and P.A. Dorosh. "An Overview of Corn Production." In C.P. Timmer (ed.), *The Corn Economy of Indonesia*. Ithaca, NY: Cornell University Press, 1987.

Mink, S.D., P.A. Dorosh, and D.H. Perry. "Corn Production Systems." In C.P. Timmer (ed.), *The Corn Economy of Indonesia*. Ithaca, NY: Cornell University Press, 1987.

Mitchell, M. "Elasticity and Growth Rate Comparisons." Paper presented at a Workshop on Issues in Research Methodology for Indonesian Agriculture. Jakarta, January 20-21, 1988.

Monteverde, R.T. and S.D. Mink. "Household Corn Consumption." In C.P. Timmer (ed.), *The Corn Economy of Indonesia*. Ithaca, NY: Cornell University Press, 1987.

Pearson, S.R. "Prospects for Corn Sweeteners." In C.P. Timmer (ed.), *The Corn Economy of Indonesia*. Ithaca, NY: Cornell University Press, 1987.

Pitt, M.M. "Economic Policy and Agricultural Development in Indonesia." Ph.D. dissertation, University of California, Berkeley, 1977.

Rasahan, C.A. "Government Intervention in Food Grain Markets: An Econometric Study of the Indonesian Rice Economy." Ph.D. dissertation, University of Minnesota, 1983.

Republic of Indonesia (ROI). "5 Years of Agricultural Research 1981-1986." Agency for Agricultural Research and Development (AARD), Ministry of Agriculture. Jakarta: c1987.

————. National Development Information Office. *Indonesia Development News.* Jakarta: 8/12 (1985): 1; 11/6 (1988a): 6-7.1.

————. "Repletia V - Pertanian, Bukul." Departmen Pertanian, Jakarta: 1988b.

————. Direktorat Bina Usha Petani. *Yearbook of Palawija Prices.* Jakarta: 1986.

Rosegrant, M.W., F. Kasryno, L.A. Gonzales, C. Rasahan, and Y. Saefudin. *Assessment of Food Demand/Supply Prospects and Related Strategies for Indonesia.* Washington, D.C. and Bogor: International Food Policy Research Institute and Center for Agro Economic Research, October 17, 1986. Preliminary Draft Final Report.

————. *Price and Investment Policies in the Indonesian Food Crop Sector.* Washington, D.C. and Bogor: International Food Policy Research Institute and Center for Agro Economic Research, August 14, 1987.

Stone, A. "Management Problems in Irrigated Java: Beyond Rice Self-Sufficiency." Discussion Paper No. ARU63. World Bank, Agricultural and Rural Development Department, Washington, DC, March 1987.

Tabor, S., et al. "Supply and Demand for Food Crops in Indonesia." Jakarta: Ministry of Agriculture, 1988.

Teken, I.B. and H. Soewardi. "Food Supply and Demand and Food Policies." In Mubyarto (ed.), *Growth and Equity in Indonesian Agricultural Development.* Yayasan Agro Ekonomika. Yogyakarta: Gadjah Mada University Press, 1982.

Teklu, T. and S.R. Johnson. "Preliminary Analysis of Demand Parameters for Indonesia." Working Paper for Indonesia. Ames, IA: Food and Agricultural Policy Research Institute, Iowa State University, 1986.

Timmer, C.P. "The Political Economy of Rice in Asia: A Methodological Introduction." *Food Research Institute Studies* 14(1975): 191-96.

————. "Food Policy in Indonesia." Graduate School of Business Administration, Harvard University, Boston, 1986. Draft.

————. "Corn Marketing." In C.P. Timmer (ed.), *The Corn Economy of Indonesia.* Ithaca, NY: Cornell University Press, 1987.

———— and H. Alderman. "Estimating Consumption Parameters for Food Policy Analysis." *American Journal of Agricultural Economics* 61(1979): 982-87.

United Nations. *Yearbook of National Account Statistics.* 1983.

World Bank. "Indonesia: Policy Options and Strategies for Major Food Crops." Report No. 3686b-IND. Washington, D.C.: April 4, 1983.

————. "Indonesian Agricultural Policy: Issues and Options." Washington, D.C.: Asia Regional Office, July 1987.

————. *World Development Report 1990.* New York: Oxford University Press, 1990.

Appendix

A simultaneous demand and supply model incorporating rice and the seven secondary food crops was used for the policy experiments reported in the text. Demand and supply functions were assumed to be linear, although the results from a log-linear specification were found to be quite similar to those from the linear specification.

There are eight exogenous variables in the model composed of income and seven prices set by the government: producer and consumer prices for rice and the prices of maize, cassava, soybean, wheat, and fertilizer. No trade is permitted in the case of sugar, groundnut, and sweet potato. Rather, the prices of these commodities vary such that domestic demand equals domestic supply. The endogenous variables in the model are the prices of sugar, groundnut and sweet potato, eight domestic demand quantities, seven domestic supply quantities (all wheat is imported), and five net import quantities (23 in total).

The model was made operational by: (a) choosing values for supply and demand elasticities based on the authors' judgement and other authors' empirical estimates (shown in Table A.1 and Table A.2); and by (b) solving for implied constant terms for the linear supply and demand functions using 1985 base values for prices, quantities, and income (shown in Table A.3).

TABLE 11A.1 Demand Elasticities Used for Policy Analysis

	Rice	Maize	Cassava	Sugar	Gr.nut	Soybean	Sw.Pot.	Wheat	Income
Rice	-0.60	0.04	0.03	0.02	0.03	0.03	0.02	0.04	0.36
Maize	0.34	-0.80	0.28	0.02	0.02	0.03	0.03	0.04	0.01
Cassava	0.10	0.15	-0.55	0.02	0.05	0.04	0.14	0.02	0.01
Sugar	0.04	0.02	0.02	-0.70	0.02	0.04	0.02	0.02	0.50
Gr.nut	0.05	0.05	0.05	0.02	-1.00	0.10	0.03	0.02	0.65
Soybean	0.08	0.04	0.04	0.03	0.12	-0.90	0.02	0.03	0.50
Sw.Pot.	0.01	0.05	0.10	0.01	0.02	0.01	-0.25	0.01	0.01
Wheat	0.12	0.04	0.01	0.02	0.01	0.02	0.01	-0.80	0.55

Note: Numerical entries in the table show the percentage change in quantities supplied of commodities listed in the left-hand column resulting from a one percent change in prices of commodities shown in the top row.

TABLE 11A.2 Supply Elasticities for Policy Analysis

	Rice	Maize	Cassava	Sugar	Gr.nut	Soybean	Sw.Pot.	Fertilizer
Rice	0.30	-0.05	-0.01	-0.02	-0.01	-0.02	-0.01	-0.15
Maize	-0.12	0.50	-0.02	-0.02	-0.02	-0.02	-0.02	-0.20
Cassava	-0.08	-0.01	0.20	0.00	0.00	0.00	-0.02	-0.05
Sugar	-0.02	-0.02	0.00	0.30	0.00	0.00	0.00	-0.15
Gr.nut	-0.01	-0.02	0.00	0.00	0.40	-0.02	0.00	-0.10
Soybean	-0.02	-0.02	0.00	0.00	-0.02	0.40	0.00	-0.15
Sw.Pot.	-0.01	-0.02	-0.02	0.00	0.00	0.00	0.25	-0.05

Note: Numerical entries in the table show the percentage change in quantities supplied of commodities listed in the left-hand column resulting from a one percent change in prices of commodities shown in the top row.

TABLE 11A.3 Base Prices and Quantities for Cross-Commodity Analysis

Commodity	Wholesale price (Rp/kg)	Demand quantity (kt)	Supply quantity (kt)	Imports (kt)
Rice (milled)[a]	329.47	23,551	24,195	-644
Maize[a]	158.98	4,590	4,556	34
Cassava[a]	125.00	13,593	14,343	-750
Sugar[a]	437.90[b]	1,780	1,780	0
Groundnut	1041.25[c]	494[d]	494[d]	0
Soybean[a]	519.36	1,080	802[d]	278
Sweet potato	86.67[c]	1,876[d]	1,876[d]	0
Wheat	220.43[e]	1,333[f]	0	1,333

[a]Rosegrant, et al., 1986.
[b]Factory price.
[c]ROI, Direktorat Bina Usaha Petani, 1986.
[d]Private communication with D. Hedley, Bureau of Planning, 1986.
[e]BPS, 1986b.
[f]FAO, 1986: 9(6) June.
Note: In addition to the base values shown in the table, the 1985 value for income was set at the aggregate GDP figure for that year, namely, 96 thousand billion rupiah (BPS, 1986a), and the price of fertilizer was set at 100 Rp/kg (Rosegrant, et al., 1986).

12

Mexico

Jaime Matus-Gardea and Ralph Bierlen[1]

Mexico, with an area of almost 2 million km², is the thirteenth largest country in the world. It is bordered on the east by the Gulf of Mexico, on the west by the Pacific Ocean, on the north by the United States, and on the south by Guatemala and Belize (Figure 12.1). The Sierra Madre Occidental Mountains run along the western edge of the country, while the Sierra Madre Orientals occupy the eastern edge. Between these two mountain ranges is a plateau that increases in elevation from north to south, reaching 2,200 meters at the capital, Mexico City. Broadly speaking, southern and coastal Mexico are tropical, the center is temperate, and the north is arid or semi-arid, necessitating irrigation for the growth of crops.[2]

In 1989, the population numbered roughly 80 million people, 80 percent of whom resided in urban areas,[3] making Mexico the eleventh most populous nation in the world. Mexico is an upper-middle income country with a 1987 gross domestic product of $US 139 billion and a per capita income of roughly $US 1,700. The country is a federal republic and has enjoyed political stability since the 1910 Revolution led to establishment of the current system of government.

The Mexican Economy

Until recently, the Mexican government pursued an inward-looking economic development strategy, using a combination of measures to protect the domestic economy from foreign influences. These measures included tariffs on imports and food subsidies. The government also aggressively promoted industrial exports, particularly petroleum, which provide a large share of Mexico's foreign exchange. Following a financial crisis in 1982 (when Mexico was unable to meet its repayments on foreign debt) and the sharp drop in world petroleum prices during the early 1980s, the government gradually reformed many of its economic policies, liberalizing trade and deregulating domestic markets. Monetary reforms have included a realignment of the exchange rate

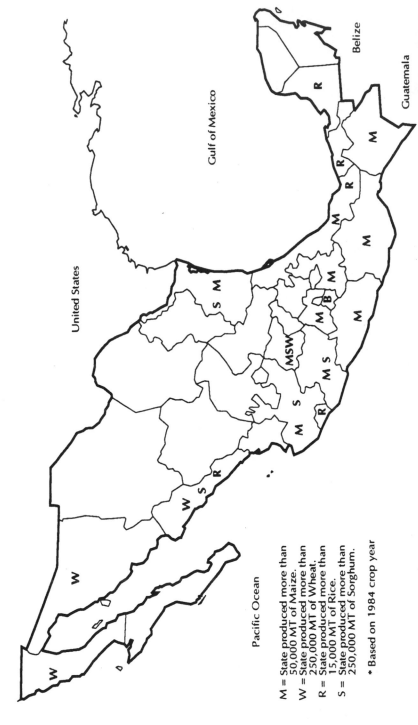

FIGURE 12.1 Major Grain-Producing Areas in Mexico. *Source:* Secretaría de Agricultura y Recursos Hidráulicos and Bassols. Batalla, Angel. Geografía Economica de México (Mexico City: Trillas, 1980).

M = State produced more than
    50,000 MT of Maize.
W = State produced more than
    250,000 MT of Wheat.
R = State produced more than
    15,000 MT of Rice.
S = State produced more than
    250,000 MT of Sorghum.

* Based on 1984 crop year

TABLE 12.1 Basic Statistics for the Mexican Economy, 1970-87

Year	Gross National Product (billion 1970 pesos)			Exports (million 1970 pesos)			Imports (million 1970 pesos)		
	Total	Agricultural	Grains	Total	Agricultural	Grains	Total	Agricultural	Grains
1970	444	51	13	16,121	8,426	19	29,104	1,487	760
1971	463	54	12	17,410	7,870	307	28,435	1,498	195
1972	502	54	11	20,296	9,398	274	32,645	2,060	966
1973	544	56	14	20,411	9,522	33	37,592	2,505	2,129
1974	578	58	13	22,891	7,717	38	48,647	7,295	3,825
1975	610	59	16	19,882	6,070	72	45,660	5,984	3,858
1976	636	59	15	31,128	6,393	21	54,513	3,056	943
1977	658	64	16	33,839	13,659	7	45,269	6,757	2,214
1978	712	68	15	42,776	12,748	109	55,811	7,243	2,246
1979	777	66	12	50,972	12,778	a	69,765	7,359	2,575
1980	842	71	18	71,090	8,340	0	89,279	12,999	4,954
1981	909	75	19	82,092	6,363	0	102,388	13,672	3,966
1982	904	74	15	117,145	6,976	0	83,165	7,706	1,382
1983	856	77	19	134,040	10,296	0	54,104	11,194	6,644
1984	888	79	21	121,205	8,817	0	59,049	10,222	3,641
1985	912	82	23	104,151	7,291	0	66,162	10,067	3,623
1986	878	80	20	110,759	15,613	5	82,342	10,270	1,825
1987	862	79	20	124,429	9,295	0	76,871	6,674	2,525

Source: Data obtained from the Mexican Government.
Note: Agricultural figures include livestock.
aLess than 1.

to reduce the overvaluation of the domestic currency, the peso, and increases in nominal interest rates to provide a real return on domestic savings.[4] Trade policy changes, which have a direct bearing on the grain economy, included reductions in tariffs and nontariff barriers; a process which accelerated with Mexico's accession to the GATT (General Agreement on Trade and Tariffs) in 1986. However, inflationary pressures in 1988 and 1989 forced the government to slow liberalization of trade policies (Mielke, 1989). From 1940 through the early 1980s, economic growth was rapid and remarkably steady, with GDP increasing at 6 percent annually between 1970 and 1982 (Table 12.1). Since 1982, Mexico's economic performance was more variable, although growth has again replaced the stagnation of the mid-1980s.

Although the importance of agriculture in the Mexican economy has declined gradually since the end of the Second World War, agricultural exports continue to be an important source of foreign exchange, and the government places heavy emphasis on food self-sufficiency, particularly for grains, in the formation of agricultural policy. In the 1985-87 period, the average value of agricultural output was roughly 80 billion pesos (1970 pesos) or 9 percent of GNP (Table 12.1). Grains (maize, wheat, rice, and sorghum) accounted for 21 billion pesos (1970 pesos) or about 2 percent of GNP during this period. Grains are grown on over 10 million hectares, of which more than 80 percent lacks irrigation.

From 1970-82, Mexico consistently had a merchandise trade deficit. Since 1982, mostly because of the depreciation of the peso, the trade balance has been positive. In the five years following 1982, the surplus averaged $US 10 billion (current dollars). Exports for 1983-87 averaged $US 21 billion, of which 7 percent were agricultural products (Table 12.1). In the early 1970s, small quantities of wheat and maize were exported, but Mexico is now a major grain importer. Thus, grain is not an important component of agricultural exports, which are dominated by coffee, fresh fruits and vegetables, cotton, and cattle. Total imports in 1983-87 averaged $US 11 billion, with agricultural products accounting for 13 percent of the total (Table 12.1). Agricultural imports are primarily made up of grains, live animals and animal products, and oilseeds. In the late 1980s, grain imports accounted for about 36 percent of total agricultural exports, but this percentage varies markedly from year to year because of weather-induced variability in domestic production.

From 1970-79, Mexico's agricultural trade surplus averaged $US 800 million annually. Rising imports of grain and oilseeds during the 1970s helped to swing the agricultural trade balance into the red by 1980 (Table 12.1). Increased demand for grain created by population and income growth, plus only modest increases in grain production, kept imports high, despite substantial protection for domestic farmers through price supports, input subsidies, and trade barriers. In 1983-85, the agricultural trade deficit averaged $US 375 million per year.

Grain Production

Grains, including maize, wheat, rice and sorghum, are produced throughout the country and account for over half of the roughly 20 million hectares of cultivated land.[5] Less than 20 percent of grain area is irrigated in an average year, but irrigated area is expanding rapidly. Because the Mexican climate is generally dry, the improvement of water supplies plays an important role in sustaining and increasing crop production. Maize, grown on small peasant farms, is usually produced under dryland conditions. However, about 73 percent of the wheat crop is irrigated, and wheat occupies 20 percent of total irrigated area. During the 1985-87 period, total grain production averaged 24 million metric tons, of which more than half was maize, 25 percent sorghum, and 20 percent wheat. Rice is a minor crop and represents only 3 percent of total grain output. The increased use of irrigation and modern inputs has resulted in higher grain yields, particularly for wheat. Between 1970 and 1987, grain output grew at an annual rate of 1.6 percent.

Maize

Maize, the most important grain in Mexico, occupies the largest area and is grown by the greatest number of farmers. Production is concentrated in central and southern Mexico, primarily on peasant or *campesino* farms. Of the 2.5 million farms in Mexico, 2.2 million are small farms operated by peasants, who control 57 percent of Mexico's cultivated area. Campesino farmers rely on family labor, use little fertilizer or hybrid seed, and seldom employ machines. Much of the maize produced by campesino farms is consumed at home, with the surplus sold in local markets. Maize is a low-value crop, and less than 10 percent of maize area is irrigated.

The maize crop, along with the sorghum and rice crops, is planted and harvested in the spring/summer season, which runs from February to October. During the 1980s, harvested maize area averaged over 7.5 million hectares (Table 12.2), 10 percent higher than in the early 1960s. Yields, though low, have grown and, at 1.8 metric tons/ha, were 75 percent above their 1960-64 levels (Table 12.2). The combination of increased area and yields has resulted in a doubling of maize production since the early 1960s to about 13 million metric tons. Much of the 6 percent annual growth in production can be attributed to increased yields. Improvements in yields likely occurred because of the use of improved seeds and additional fertilizer.

Wheat

In contrast to maize, wheat is grown primarily by large commercial farms using modern technology. Mexico has about 47 thousand such farms, which rely on hired labor and employ high levels of mechanization, along with hybrid seeds and fertilizers. These farms account for over 20 percent of the total

TABLE 12.2 Grain Area, Yield, and Production in Mexico, 1970-87

Year	Total grain	Irrigated grain	Total maize	Irrigated maize	Total wheat	Irrigated wheat	Total rice	Irrigated rice	Total sorghum	Irrigated sorghum
Area ('000' hectares)										
1970	9,397	1,370	7,440	458	886	513	150	63	921	336
1975	9,175	1,874	6,694	683	778	541	257	125	1,445	525
1980	9,161	2,063	6,766	880	724	573	127	76	1,543	534
1985	10,178	2,614	7,498	1,036	1,224	937	220	58	1,891	604
1987	11,797	2,570	8,513	989	981	973	195	32	2,108	576
Yield (m. tons per hectare)										
1970	1.6	3.4	1.2	2.6	3.0	3.7	2.7	4.3	3.0	3.8
1975	1.8	3.5	1.3	2.6	3.6	4.4	2.8	4.1	2.9	3.6
1980	2.2	3.4	1.8	2.5	3.8	4.3	3.5	4.2	3.0	4.0
1985	2.6	4.0	1.9	3.0	4.3	4.7	3.7	4.6	3.5	4.3
1987	2.1	3.7	1.7	3.0	4.1	4.1	3.0	4.4	2.9	4.3
Production ('000' m. tons)										
1970	14,708	4,610	8,879	1,199	2,676	1,876	405	272	2,747	1,262
1975	16,091	6,564	8,449	1,780	2,799	2,362	717	509	4,126	1,913
1980	20,294	7,103	12,374	2,199	2,785	2,453	445	321	4,689	2,130
1985	26,524	10,466	13,957	3,079	5,207	4,441	809	271	6,641	2,601
1987	24,893	9,626	14,100	2,983	4,009	4,002	578	139	6,206	2,502

Source: Data provided to the authors by the Mexican Government.
Notes: Totals include maize, wheat, rough rice, and sorghum; minor grains are excluded. Figures for rice are rough rice.

cultivated area in the country and dominate wheat production.

During the 1960s, the advent of Green Revolution technology and large public investment in irrigation resulted in a shift in the wheat-producing area from rainfed central Mexico to the semiarid but irrigated northeast. At present, three-quarters of Mexico's wheat production comes from the northeastern states of Sinaloa, Sonora, North and South Baja California, and Chihuahua. Field sizes in Sinaloa, Sonora and Baja California Norte average more than 20 ha. In addition to irrigation, wheat farmers use other purchased inputs, such as pesticides, herbicides, and machines. They grow modern, high-yielding varieties engineered to exploit the input package. Soft winter wheat accounts for the majority of production. It is planted and harvested in the fall/winter season, which runs from October to February. Essentially all wheat land is fertilized, and over 70 percent is irrigated. The use of certified seed supplied by PRONASE (Productora Nacional de Semillas), the government seed company, is widespread.

As a result of irrigation and modern technology, Mexican wheat yields are high by world standards (70 percent higher than average yields in the United States, where wheat is grown under dryland conditions). Wheat area declined slightly in the 1960s and 1970s as production shifted from central to northeastern Mexico, but rebounded in the 1980s. In recent years, wheat area averaged roughly 1 million hectares (Table 12.2), almost 25 percent above levels in the early 1960s. Average yields more than doubled during the same period and now exceed 4 metric tons per hectare. Although area and yields fluctuate from year to year, increases have led to a sustained growth in wheat production since 1960.

Rice

Rice is a relatively minor crop in Mexico compared to maize, wheat and sorghum, and is grown primarily in the northwestern, south, and central parts of the country. Plots average about 3 hectares in size, although some exceed 15 hectares, and most rice is produced by large commercial farms with access to modern inputs. The harvested area of rice is only about 180 thousand hectares (Table 12.2) and fluctuates from year to year. Approximately 30 percent is irrigated. Average yields have climbed steadily, and are approaching 4 metric tons of paddy per hectare; roughly 70 percent higher than in the early 1960s. Yields on irrigated land are past 4 metric tons per hectare. Yield improvements have resulted in a steady growth in rice production of roughly 1 percent per year since the early 1960s. Annual production now exceeds 500 thousand metric tons. ·

Sorghum

Sorghum is a relatively new crop in Mexico, but has gained rapidly in importance since the 1950s because of its ability to thrive in Mexico's climate and because of rising demand for livestock feed. Initially, sorghum was produced in irrigated areas, but in more recent years the crop has been planted on rainfed land, often displacing maize (Barkin and Suarez, 1982). The central pacific and northern regions now account for almost 90 percent of sorghum production. Sorghum is generally grown by farmers with access to larger plots of land. Farms greater than 5 ha in size account for more than 50 percent of production. Machines are used to prepare and harvest over 90 percent of the area, and the use of purchased inputs is commonplace (Roberts and Mielke, 1986).

During the 1980s, the area under cultivation averaged roughly 1.6 million hectares (Table 12.2), an eight-fold increase since the early 1960s. Sorghum yields have steadily improved, and now average over 3 metric tons per hectare; more than 50 percent higher than in the early 1960s. The rapid expansion of area and the growth in yields have resulted in annual production increases of about 15-20 percent. In recent years, production has totalled 5-6 million metric tons.

Price Responsiveness of Production

The estimation of supply response is still in the formative stages in Mexico. Analysis is hampered by a general lack of data, as well as quality problems with available data. Some recent estimates of price response elasticities for grains are indicated in Table 12.3. Most of these are derived from simple equations incorporating own-price, the price of competing crops, the price of fertilizer, and rainfall and/or irrigation water availability as explanatory variables.

The available estimates suggest that the price elasticity of maize supply at 0.75 is slightly inelastic. Wheat elasticities fall between 1.21 and 1.39. Given the dynamism of the wheat sector in the last forty years, this seems reasonable. Fernandez's (1986) rice elasticity of 1.05 suggests that rice supply is also elastic. A sorghum elasticity of 0.61 shows only moderate response to price changes. Given sorghum's recent production growth, this estimate suggests that factors other than price have caused the rapid growth in output. Given the fact that sorghum is grown for the market, relative price is the main determinant of changes in production. However, because sorghum is more resistant to drought than maize, there is much less risk of crop loss. Maize and sorghum own-price elasticities reflect that these grains are mainly grown in rainfed areas, while wheat is produced in irrigated areas, where competition with other crops is more pronounced. Thus, price increases in maize or sorghum will not necessarily lead to a large growth in production. On the other hand, a fall in wheat prices would lead to a considerable decrease in wheat production.

Grain Consumption: A Case of Inequality

Per capita grain consumption (defined as production plus imports, less exports) varies markedly by grain type in Mexico. Per capita consumption of maize (Table 12.4) grew slowly and erratically over the past twenty years, while

TABLE 12.3 Grain Elasticities

Price Supply Response

	Maize			Wheat		Rice	Sorghum
Elasticity	0.77	0.73	1.21	1.39	1.23	1.05	0.61
Author[a]	HERNANDEZ	HERNANDEZ	AVILA	SALCEDO	HERNANDEZ	FERNANDEZ, P.	AGUIRRE
Data period	1961-78	1961-78	1960-78	1960-79	1960-78	1965-84	1965-82
Elas. period	AVG.	AVG.	AVG.	AVG.	LAST 5 YRS	AVG.	LAST 5 YRS
Form[b]	LOG.	LINEAR	NERLOVE	LOG LIN.	LINEAR	LINEAR	LINEAR
Estimated by[c]	2SLS	2SLS	OLS	OLS	OLS	3SLS	2SLS
R2[d]	65.74	65.74	78.83	69.1	78.8	78.97	86.0

Income Demand Response

	Maize			Wheat	Rice
Elasticity	1.16	1.01	0.97	0.24	-0.19
Author[b]	AGUIRRE	AGUIRRE	HERNANDEZ	SALCEDO	FERNANDEZ, P.
Data period	1965-82	1965-82	1961-78	1960-79	1965-84
Elast. period	LAST 5 YRS	LAST 5 YRS	AVG.	AVG.	LAST 5 YRS
Form	LINEAR	LINEAR	LOG LINEAR	LOG LINEAR	LINEAR
Estimated by	2SLS	2SLS	9LS	2SLS	3SLS
R2	71.0	71.0	76.14	63.7	78.97

Price Demand Response

	Maize			Wheat	Rice	Sorghum
Elasticity	-0.07	-0.13	-0.02	-0.33	-0.21	-0.24
Author[b]	AGUIRRE	HERNANDEZ	HERNANDEZ	SALCEDO	FERNANDEZ, P.	AGUIRRE
Date period	1965-82	1961-78	1961-78	1960-79	1965-84	1965-82
Elas. period	LAST 5 YRS	AVG.	AVG.	AVG.	LAST 5 YRS	LAST 5 YRS
Form	LINEAR	LINEAR	LOG LINEAR	LOG LINEAR	LINEAR	LINEAR
Estimated by	2SLS	2SLS	2SLS	OLS	3SLS	2SLS
R2	71.0	72.69	76.14	63.7	78.97	88.0

[a]See References for complete sources.
[b]Log refers to a double logarithm equation.
[c]Estimation methods include OLS (ordinary least square); 2SLS (two stages least square); and 3SLS (three stages least square).
[d]R2 stands for the determination coefficient.

TABLE 12.4 Apparent Grain Consumption in Mexico, 1970-87

Year	Total ('000 m. tons)				Per capita (kg)			
	Maize	Wheat	Rice	Sorghum	Maize	Wheat	Rice	Sorghum
1970	9,639	2,636	284	2,729	188.3	51.5	5.5	53.3
1971	9,530	1,922	244	2,475	180.2	36.3	4.6	46.8
1972	9,001	2,434	255	2,858	164.7	44.5	4.7	52.3
1973	9,723	2,798	319	3,283	172.1	49.5	5.7	58.1
1974	9,129	3,746	392	3,926	156.5	64.2	6.7	67.3
1975	11,104	2,842	473	4,941	184.6	47.3	7.9	82.1
1976	8,927	3,348	306	4,071	144.0	54.0	4.9	65.7
1977	12,122	2,887	371	5,040	189.0	45.2	5.8	79.0
1978	12,347	3,264	206	4,945	188.0	49.7	3.1	75.3
1979	9,203	2,434	362	5,251	136.3	36.1	5.4	77.8
1980	16,561	3,684	389	6,941	238.7	53.1	5.6	100.0
1981	17,504	4,316	519	8,717	245.5	60.5	7.3	122.3
1982	10,381	4,775	418	6,367	142.3	65.4	5.7	87.3
1983	17,692	3,854	275	8,155	237.1	51.6	3.7	109.3
1984	15,376	4,841	490	7,721	201.5	63.5	6.4	101.2
1985	15,672	5,516	699	8,804	201.1	70.8	9.0	113.0
1986	13,386	4,994	360	6,015	168.2	62.8	4.5	75.6
1987	15,850	4,444	381	8,035	195.3	54.7	4.7	99.0

Source: Data provided by the Mexican Government.
Notes: Apparent consumption equals production plus imports, less exports. Rice is milled rice. Figures are gross. No adjustment is made for loss or wastage. Per capita consumption includes grains fed to livestock.

milled rice consumption actually declined slightly. Part of the reason for this stagnation is that maize and rice are traditional staples for lower-income families, especially in rural areas. As incomes rise and the population migrates to urban areas, tastes change and consumption of other grains, such as wheat, increases. Except for cleaning and milling, rice is used without further processing, while maize is mainly eaten in the form of tortillas.[6] Wheat is consumed as bread, pastries, cookies and noodles, while sorghum is eaten indirectly in the form of pork, eggs, and chicken. The consumption of wheat and sorghum grew more rapidly than that of maize and milled rice because middle- and upper-income families prefer these foods.

Income distribution is the main determinant of who consumes maize and milled rice versus processed grain products, such as bread, and meats and the animal products derived indirectly from grain. In 1977, the poorest 50 percent of the Mexican population earned only 17 percent of total income.[7] The bottom tenth spent nearly 62 percent of their income on food, while the top 10 percent spent only 22 percent. Although the poorest 10 percent of the population spent over one-half of their income on food, this accounted for less than 2 percent of national food expenditure. In comparison, the richest 10 percent of the population accounted for 25 percent of national food expenditure.

In 1976-77, the most recent year for which detailed data are available, Mexicans consumed an average of just under 1,300 grams of food daily (Secretaria de Agricultura y Recursos Hidralicos, vol.5). Of this, 450 grams or 35 percent was derived from maize, wheat, rice, dry beans, soybeans, and byproducts. Maize accounted for two-thirds of this subtotal with 326 grams, followed by wheat with 103, and rice with 22. Other important foods based on feedgrains were meat (including chicken) at 74 g/person, and eggs at 26 g/person. With respect to calories, grains provided 60 percent of the total (1,337 calories out of 2,289). Maize furnished 46 percent of total calories, followed by wheat with 13 percent, and rice with 2 percent. Out of 43 grams of protein eaten by the average Mexican, grains contributed 74 percent of the intake (34 g). Maize supplied 60 percent of protein intake, wheat 16 percent, and rice 1 percent.

Consumption of sorghum has grown rapidly since the early 1960s. In 1980-83, sorghum consumption averaged 8 million metric tons, a more than fifteen-fold increase over the early 1960s. An expansion in the number of hogs, dairy cattle, and chickens; the replacement of maize and wheat as feedgrains by sorghum; and the spread of more intensive forms of animal production, including broiler, egg, and pork operations, have all contributed to the rapid growth in sorghum use (Roberts and Mielke, 1986).

Income and Price Elasticities of Demand

Little work has been done on estimating demand elasticities for grains in Mexico. The estimates that exist suffer from the same data-induced problems as supply elasticities. Some recent estimates, summarized in Table 12.3, suggest that, for maize, consumption expands about 1 percent for each 1 percent increase in income, although response to changes in price is only moderate. Even though more maize is already consumed than any other food item, most Mexicans would eat more given a higher income. Because of the relative importance of maize compared to other grains, along with its high nutritional value and low cost, consumption would decline only moderately if prices rose. A recent study (Salcedo, 1985) suggests that the consumption of wheat is less responsive to changes in income than maize but is more responsive to changes in price. Rice has a negative income elasticity. The price elasticities for all grains are relatively low (between 0 and -0.33), reflecting the fact that grains in Mexico are basic commodities with no close substitutes.

Grain Marketing in Mexico

Grain marketing in Mexico is complex and costly (Lustig and del Campo, 1985). There are several factors that contribute to inflated costs, including numerous market intermediaries, antiquated handling facilities, and transportation monopolies. Grain marketing is primarily managed by private enterprise, but the state agricultural marketing agency (CONASUPO – Compania Nacional de Subsistencias Populares S.A. – National Subsistence Popular Company), which also owns and manages processing, storage, and wholesale and retail facilities, plays a key role (Figure 12.2).

CONASUPO: The Government Marketing Corporation

CONASUPO is a decentralized public corporation controlled by the Secretary of Commerce and Industrial Development (SECOFI – Secretaría de Comercio y Fomento Industrial). CONASUPO's main purposes are to protect producer income by preventing price swings (primarily through the operation of a floor price scheme) and to ensure that low-income consumers can purchase sufficient grain (by selling grain at subsidized prices through its retail outlets). CONASUPO includes the parent corporation (CONASUPO proper) and fourteen associated entities, which handle grain processing, storage, and other facilities. The director of CONASUPO is appointed by the President, and the budget is subject to the approval of the legislature, the Secretary of Programming and Budget, and the Secretary of the Treasury.

CONASUPO is an ubiquitous presence in the Mexican economy. Among the measures that the organization employs to control domestic grain markets are purchases from producers at guaranteed minimum prices, imports and exports of grain, the maintenance of subsidized storage facilities, wholesale and retail sales of grains (the latter through a chain of retail stores), milling and baking enterprises, and the sale of flour and baked products at subsidized prices. CONASUPO, along with its subsidiary enterprises, uses these tools to

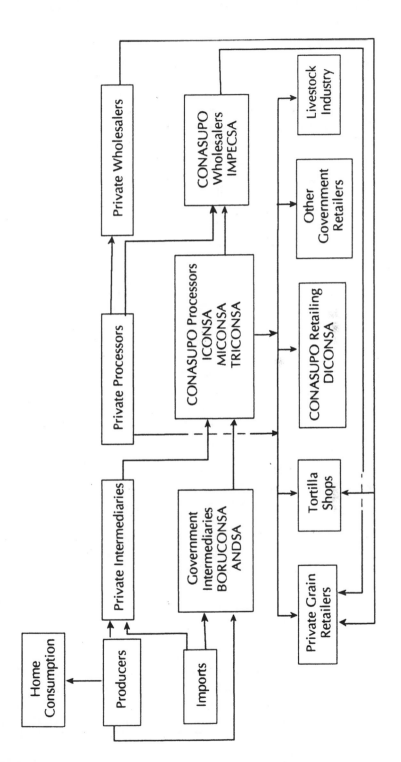

FIGURE 12.2 Grain Marketing Channels in Mexico

regulate the flow of grain to the market and its price, and to maintain reserve stocks. Special attention is given to low-income producers and consumers. When harvests are large and local market prices fall, CONASUPO buys and stores or exports grain to raise prices. When there are shortfalls, it sells stored grain and/or imports, marketing the imported grain at subsidized prices.

CONASUPO buys, stores, and transports grain through its affiliate, Bodegas Rurales (BORUCONSA – Rural Storage of CONASUPO). Other BORUCONSA activities include the supply of production inputs, primarily fertilizer and irrigation, at subsidized prices, the maintenance of local transportation services, and direct sales of grain in rural areas. ANDSA, another CONASUPO corporation, also stores grain belonging to CONASUPO, BANRURAL (Banco Nacional de Credito Rural – National Rural Credit Bank), private mills, and farmers. In 1982, BORUCONSA owned 23 percent of total grain storage in Mexico (estimated at 18 million metric tons), ANDSA accounted for 37 percent, and the private sector 40 percent (Lustig and del Campo, 1985).

Through Maiz Industrializado CONASUPO (MICONSA), Trigo Industrializado CONASUPO (TRICONSA – government-owned baking facilities), and Industrias CONASUPO (ICONSA – government-owned flour mills), CONASUPO processes grains purchased on the domestic and foreign markets into tortillas, bread, pasta, flour, vegetable oil, margarine, and animal feeds. These are sold at subsidized prices, primarily through the organization's retail outlets. In 1981, 78 percent of goods produced by CONASUPO subsidiaries were marketed through CONASUPO's retail and wholesale operations, one percent through other government retail stores, and 21 percent through the private sector (Lustig and del Campo, 1985).

Impulsora del Pequeno Comercio (IMPECSA), a CONASUPO-run wholesaler, supplies small- and medium-size public and private retailers in urban areas with low-priced goods, including grain. Through IMPECSA, CONASUPO provides retail affiliates with its own line of goods, private label products, and low-interest loans. A further branch, Sistema de Distribuidoras CONASUPO (DICONSA – CONASUPO distribution system) maintains 18 thousand retail outlets. DICONSA's stores are located in low-income areas and vary in size from rural corner groceries to large urban stores offering a wide variety of non-CONASUPO food and nonfood items. DICONSA's stated purpose is to help regulate the retail prices of basic food items (such as grains and oils) by selling at prices below those in private stores.

Private Institutions: An Alternative

Because of its limited number of grain receiving points, delays caused by unmechanized and inefficient facilities, slow payment and high quality standards, CONASUPO is not a viable outlet for all producers (Burst, Bredahl, and Warnken, 1984). Small producers often sell to local intermediaries, who generally purchase below the government's minimum prices. CONASUPO's participation in the domestic grain market has declined in recent years, although it continues to maintain considerable control over imports and exports. Current government policy is to encourage private participation in the international market. CONASUPO has established credit programs for grain imports in order to increase the role of the private sector.

There are two basic types of private foodgrain processors; small local businesses (nixtamal[8] mills, tortilla shops and bakeries) and large regional and national processors (wheat, maize and rice mills, and cookie and pasta plants). In 1975, there were almost 45 thousand small concerns (less than 5 thousand pesos in annual sales) producing fresh grain products for local consumption. Medium and large bakeries (over 5 thousand pesos in annual sales), although numbering only 46, were responsible for over 50 percent of total industry value added and revenues. There are few wheat and maize flour mills, and cookie and pasta plants, and those that exist are large in size. There are no large rice millers.

Marketing Feedgrains

Except for processing and distribution, feedgrain marketers are similar to their foodgrain counterparts. Feedgrain processing concerns can be divided into three types: official, independent private firms, and vertically integrated companies. The vertically integrated processors include large producers, producer associations, and agri-business concerns that raise livestock and poultry, as well as mix feed for themselves and others. The private feed industry is dominated by multinationals and large Mexican companies with large production plants. Historically, the trend is toward fewer and larger plants. The official sector is composed of the mixed feed division of CONASUPO and ALBAMEX (Alimentos Balanceados Mexicanos -- Mixed Feed of Mexico), a semi-autonomous government corporation. In 1980, private sector feed millers accounted for 32 percent of production, integrators 62 percent, and the public sector 7 percent (Burst, Bredahl, and Warnken, 1984).

The private feed industry contributes to public decision making affecting feedgrains through CANACINTRA's (Camara Nacional de la Industria de Transformacion) national section on balanced animal feed producers (SENAPABA) and through the National Association of Animal Feed Manufacturers (Asociacion Nacional de Fabricantes de Alimentos Pecuarios A.C.). Among other duties, these two organizations consult with the government on how much feedgrain to import.

Grain Marketing Channels

Grains in Mexico move through many different private and public channels (Figure 12.2). Farmers typically sell harvested grain to private commercial enterprises (intermediaries and processors) or directly to CONASUPO, or the grain is consumed at home as food or feed. Intermediaries may sell stocks to

CONASUPO or to the grain-processing industry. CONASUPO holds grain for future sales, markets to the grain-processing industry, and further processes grains for resale in its own facilities. In 1981-85, CONASUPO purchased, on average, 20 percent of maize production, 38 percent of dry bean output, 45 percent of the wheat crop, 13 percent of rice production, 22 percent of the soybean harvest, and 25 percent of sorghum output.[9] As already stated, government participation in grain marketing has declined from past levels, implying a gradual reduction of consumer subsidies for basic foodstuffs.

The CONASUPO food processing divisions sell processed products to small- and medium-size private processors, private wholesalers, their own wholesale association, IMPECSA (which in turn markets to private retailers), government-run retail stores, private retailers, tortilla shops, and their chain of retail stores (DICONSA). The private food processing industry sells other finished products to the same institutions as the official food processing industry. In 1982, CONASUPO manufactured 40 percent of all maize flour produced in Mexico, 11 percent of cooking oil, 5 percent of soup noodles, 5 percent of wheat flour, and 2 percent of animal feeds (Lustig and del Campo, 1985).

The feedgrain industry uses distributors to sell feed to large nonintegrated producers. Large, vertically integrated operations generally produce feed for their own use, but some, especially producer associations, sell feed on the open market. The mixed feed industry and integrators distribute products through retail feed outlets, pharmacies, and veterinary supply stores. These are generally small-volume local outlets, which retail to smaller livestock producers.

Grain Imports: A Case of Continued Growth

In the 1960s, Mexico was a net grain exporter. The situation has since changed dramatically, and large quantities of maize, wheat, and sorghum are now imported to satisfy domestic demand. Growth in consumer income, a rapidly increasing population, urbanization, government food subsidies, and a shift in consumer preferences toward grain-based animal products all contributed to larger grain imports. Further, the expansion of domestic grain production has been insufficient to keep pace with the rapidly expanding demand.

Maize imports grew continuously from the early 1970s, exceeding 1 million metric tons in most years (Table 12.5). During the 1980s, imports averaged nearly 2.5 million metric tons, surpassing 4.5 million metric tons in 1983. Wheat import trends are similar to those for maize. After 1971, imports increased rapidly, exceeding 100 thousand metric tons in every year except one. In the 1980-86 period, imports averaged over 500 thousand metric tons, and in 1981, were over 1 million metric tons. Until 1974, sorghum imports were minimal, but have grown steadily since, reaching over 3 million metric tons in 1983. Imports have begun to decline because of increased domestic production, and

TABLE 12.5 Mexican Grain Imports, 1970-87

Year	Net imports ('000 m. tons)				Per Capita imports (kg)			
	Maize	Wheat	Rice	Sorghum	Maize	Wheat	Rice	Sorghum
1970	759	-41	16	-18	14.8	-0.8	0.3	-0.3
1971	-256	91	1	-41	-4.8	1.7	<0.1	-0.8
1972	-222	625	-11	246	-4.1	11.4	-0.2	4.5
1973	1,114	707	22	13	19.7	12.5	0.4	0.2
1974	1,281	957	67	426	22.0	16.4	1.2	7.3
1975	2,655	43	a	815	44.1	0.7	0.0	13.5
1976	910	-16	a	44	14.7	-0.3	0.0	0.7
1977	1,984	431	-3	715	31.1	6.8	<0.1	11.2
1978	1,417	479	-60	752	21.7	7.3	-1.0	11.5
1979	745	147	36	1,263	11.0	2.2	0.5	18.7
1980	4,187	899	95	2,252	60.3	13.0	1.4	32.5
1981	2,954	1,123	93	2,631	41.4	15.8	1.3	36.9
1982	256	313	22	1,650	3.4	4.3	0.3	22.6
1983	4,631	394	a	3,308	62.1	5.3	0.0	44.3
1984	2,444	335	171	2,746	32.0	4.4	2.2	36.0
1985	1,715	309	165	2,254	22.0	4.0	2.1	28.9
1986	1,665	224	0	1,182	20.9	2.8	0.0	14.9
1987	1,750	435	0	1,829	21.6	5.4	0.0	22.5

Source: Data provided by the Mexican Government.
aLess than 1.
Note: A negative number denotes net exports.

in 1986, stood at 750 thousand metric tons. Sorghum imports in 1980-86 averaged 2 million metric tons annually. Several predictions for grain production and consumption to the year 2000 in Mexico all suggest that imports will continue to grow, even if government efforts to expand domestic production are successful (Reyes, 1981).

Trade Partners

As might be expected with a large and powerful agricultural neighbor, Mexico purchases the bulk of its imported grains from the United States. The US government encourages this by providing easy credit terms, assisting in securing basic commodities, issuing permits for the use of USDA facilities to hold public tenders, and working jointly to resolve transportation problems (Roberts and Mielke, 1986). The Mexican government has attempted to diversify its import sources for geopolitical reasons, but has met with little success because of the higher costs of obtaining supplies from other countries.

Mexico has typically purchased most of its maize from the United States. In the 1970-86 period, the United States supplied between 50-100 percent of Mexico's annual import needs. Secondary suppliers include Argentina. Until recently, 80-100 percent of Mexico's wheat purchases originated in the United States, although Canada also sold some wheat to Mexico. Since 1983, however, the US supplied only 7 percent of Mexico's import needs because the European Economic Community (EEC), Canada, Australia, and Argentina took over US market share. In 1986-87, prices were so low that imported wheat was used for feed, and the above countries generally provide more price-competitive feed wheat than the United States. The United States has consistently dominated sorghum exports to Mexico, with 38 to 99 percent of annual market share since 1970 (an average of 78 percent). Argentina is the second largest supplier with 577 thousand metric tons in 1981.

Importance of International Trade
in Grains for the Domestic Economy

Since the 1970s, imported grains have become increasingly significant in the Mexican grain economy. The absolute size of imports does not capture their full importance. Comparisons of apparent per capita consumption (defined as production plus imports less exports) and per capita net imports help to clarify the relationship over time (tables 4 and 5).

International trade plays a particularly important role in the domestic maize economy. In the late 1960s, Mexico made net maize exports of 15 kg per capita, with a maximum that surpassed 28 kg in 1967. By 1973, Mexico was a net importer, with per capita imports of over 20 kg or about 10 percent of apparent consumption, rising to 26 percent (61 kg) in 1983. Per capita net imports of maize averaged roughly 33 kg in the 1980-86 period, or slightly less than 16

TABLE 12.6 Major Instruments of Grain Policy in Mexico

	Commodity			
Instrument	Maize	Rice	Sorghum	Wheat
PRODUCTION/CONSUMPTION				
Producer guaranteed price	X	X	X	X
Deficiency payments				
Government purchases	X	X	X	X
Production quota				
Input subsidies				
- credit	X	X	X	X
- fertilizer/pesticides	X	X	X	X
- irrigation	X	X	X	X
- machinery/fuel	X	X	X	X
- seed	X	X	X	X
Crop insurance	X	X	X	X
Controlled consumer price	X	X	X	X
TRADE				
Imports - tariff				
- quota				
- subsidies				
- licensing	X	X	X	X
- state trading	X	X	X	X
Exports - taxes				
- restrictions				
- subsidies				
- licensing	X	X	X	X
- state trading	X	X	X	X
OTHER				
Marketing subsidies				
- storage	X	X	X	X
- transport	X	X	X	X
- processing	X	X	X	X
State marketing	X	X	X	X

percent of per capita apparent consumption. Wheat imports are also important. From 1972 to 1986, per capita net imports exceeded 10 kg in six years, reaching 17 kg or 25 percent of apparent consumption in 1974. In 1980-86, per capita net imports averaged 7 kg, or over 11 percent of apparent consumption per capita. For sorghum, as for maize and wheat, Mexico is reliant on imports to meet domestic needs. In 1983, net imports peaked at 44 kg/capita, about 41 percent of apparent consumption (sorghum is used exclusively as a feed). Since then, per capita net imports were consistently above 20 kg. In 1980-86, net imports per capita averaged over 29 kg, or about 28 percent of apparent consumption.

Mexican concern about growing dependence on food imports is related to perceptions of the risks inherent in international grain trade. Factors beyond the country's control include changes in the agricultural support programs of exporting countries, the lack of long-term supply agreements with exporters, and uncertainty about the availability of foreign exchange.

Government Participation in Grains: Public Policy

Government participation in the production and marketing of grains in Mexico dates from the colonial era (1521 to 1810), although the level of activity greatly increased since the end of World War II. Current government policies involve numerous programs, state corporations, credit institutions, the executive branch, and government secretariats. Major agricultural policy instruments are listed in Table 12.6.

In the mid-1970s, agricultural policy came under attack because of the inability of domestic production to meet growing demands for basic foodstuffs. As a result, in March 1980, the government launched the Mexican Food System (SAM) program. This program concentrated public resources on low-producing, rainfed areas. Agronomists surmised that, because farmers used low levels of inputs, these areas had the greatest potential to respond to improved management. SAM set goals of self-sufficiency in maize and dry beans by 1982 and in other basic commodities by 1985. In attempting to meet these targets, nominal guaranteed prices were raised, credit was channeled to the agricultural sector at preferential interest rates, input subsidies were increased, and crop insurance premiums lowered.

With the deteriorating economic situation in the early 1980s, the costs of maintaining SAM proved too expensive. It was abandoned in 1982 and replaced in 1983 with the National Food Program (PRONAL). Like SAM, production assistance is concentrated on the rainfed areas in PRONAL, although higher-yielding irrigated areas are also included. Whereas SAM's goal was complete self-sufficiency, PRONAL's is "food sovereignty," which implies securing access to food, whether domestically produced or imported. PRONAL personnel emphasize the need to improve efficiency in governmental and private marketing systems.

The government administrations has continued to maintain the broad goals of food sovereignty, while acknowledging that growing imports of foodstuffs will be necessary to meet the nutritional demands of the Mexican population. Under the current program, "Rural Modernization," the government aims to improve support for the agricultural sector through the direct participation of farmers' organizations in policy decision making and, in some cases, in the partial administration of governmental agencies such as CONASUPO, BANRURAL, FERTIMEX (Fertilizantes Mexicanos S.A.—Mexican Fertilizer Company), and PRONASE (state-owned seed agency).

The Formation of Agricultural Policy

The President, as the chief executive officer, heads the agricultural policy-making apparatus, which includes secretariats, decentralized government corporations, and the Agricultural Cabinet. Legally, the president can form cabinets, commissions, and any other groups that he feels are necessary to carry out agricultural policy.

To help the President formulate agricultural policy, an advisory body, the Agricultural Cabinet, exists. The President and the Secretary of Agriculture preside over the cabinet. Other secretaries that pertain to economics and agriculture are members, as well as the heads of the government rural credit bank (BANRURAL), the government fertilizer corporation (FERTIMEX), the government seed company (PRONASE), and CONASUPO. Farmers' organizations also participate in this cabinet.

Supporting the Agriculture Cabinet is a technical body made up of commissions and work groups usually associated with a specific policy or policy instrument. The most important group is the Price Support Committee. Proposals from the committees often reach the Agricultural Cabinet level. Members of these committees are usually sub-secretaries or secretariat department heads. Once the technical committee reaches consensus, the policy is presented to the Agricultural Cabinet for final consideration. When policy decisions are made, the corresponding secretariat or government corporation must formulate a program to enact the policy. For example, for guaranteed prices, this would be CONASUPO, and for credit, BANRURAL.

Government Policies that Affect Grain

Although emphasis has varied over time, the basic instruments of Mexican grain policy have remained the same since the presidency of Lazaro Cardenas (1934-40) and the immediate post-war era. Four major policy areas include (a) price policies and marketing; (b) credit; (c) subsidized inputs; and (d) the construction, maintenance, and operation of irrigation facilities.

In the case of grain prices, the government sets guaranteed (support) prices at the producer level for basic food crops (maize, wheat, rice, and beans), sorghum, and soybeans. In setting guaranteed prices, the Agricultural Cabinet takes supply and demand, relative domestic prices, international prices, and production costs into consideration. From 1960-85, real guaranteed prices (base 1970) of the basic grains followed similar patterns. Prices were highest in 1960 and fell continuously until 1972. In the early 1960s, domestic grain prices were well above world levels, and CONASUPO subsidized exports heavily. To reduce treasury costs, nominal grain prices were frozen from 1962-72. In 1973-74, large nominal price increases were ordered (Figure 12.3). In the case of rice and sorghum, real prices were raised above their 1960 levels. Real prices were then generally allowed to fall until the early 1980s, when they were boosted

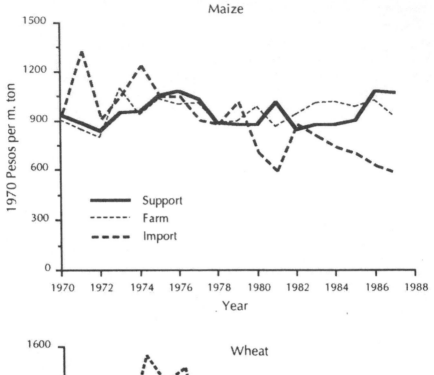

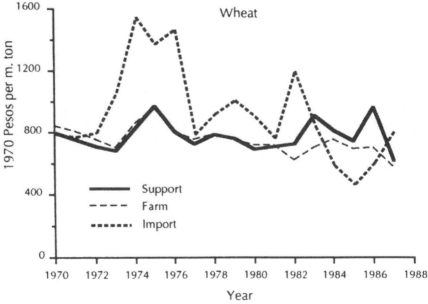

FIGURE 12.3 Real Prices of Maize and Wheat in Mexico. *Source:* Based on data from the Bank of Mexico and the Mexican Ministry of Agriculture.

under the SAM and PRONAL programs. The new prices, however, were still well below their 1960 and 1973/74 levels. In relative terms, the guaranteed price policy failed to make grain production more profitable, although the switch of grain farmers to other crops was checked (Table 12.2).

BANRURAL and another government credit agency, the Agricultural Fund (FIRA -- Fondos Instituidos en Relacion a la Agricultura -- BANAMEX), form twin pillars of agricultural credit. Their basic functions are to introduce modern production technology, encourage the organization of producer groups based on credit needs, transfer income to the agricultural sector, and help farmers achieve greater liquidity. The two institutions gear their loans to small- and medium-size producers and give priority to loan requests involving the basic grains (maize, wheat, and rice) and beans. Producers receive loans, mainly in the form of inputs, from BANRURAL. FIRA guarantees agricultural loans from commercial banks, gives discounted credit to commercial banks that grant agricultural loans, and aids private banks with the formation of their own technological and economic assistance programs. Since the 1970s, credit programs have been oriented to basic foodstuffs and nonirrigated areas, as laid out in the goals of the SAM and PRONAL programs.

In real terms, agricultural credit provided by the government grew steadily from 1945 to 1974, but stagnated from 1974 to 1978. With the introduction of SAM in 1980, expenditures reached a post-war peak of over 12 billion pesos (1960 pesos). The budget cuts caused by the economic crisis, caused a decline to less than 8 billion pesos in 1983 and 1984. Agricultural credit granted by government agencies as a percentage of total governmental credit varied over the years. Before 1960, agriculture accounted for 12 to 16 percent of total governmental credit. Between 1970 and 1980, agriculture's share was generally less than 10 percent. With the establishment of SAM, the share of the credit budget reached 16 percent in 1980 and 12 percent in 1981. As with real expenditure, agricultural credit as a percentage of total governmental credit fell sharply after 1981.

Over the last forty years, the government has tried to maintain stable nominal input prices, chiefly by subsidizing fertilizers, chemical sprays, and improved seeds. In 1976, several government corporations merged to form the national petrochemical corporation, FERTIMEX. FERTIMEX produces 90 percent of the nation's fertilizer output and 30 percent of its chemical sprays. PRONASE, the national seed company, accounts for about one-quarter of seed production, while private seed companies supply the rest. Through pricing, import control and exports, the government regulates the fertilizer, chemical spray, and seed markets. Government-produced inputs are distributed by BANRURAL in the form of loans and by BORUCONSA (the CONASUPO storage company). These programs give priority to the production of staple grains.

The use of agricultural inputs, particularly chemical fertilizers, has grown rapidly. From 1950-80, the use of blend (NPK), nitrogen, and phosphorus fertilizers increased more than one-hundred fold. Meanwhile, the real prices of fertilizers declined steadily over the 1965-83 period. The per metric ton price of anhydrous ammonia fell from 1,800 pesos (1960 pesos) in 1965, to less than 250 pesos in 1983, and the price of urea dropped from over 1,200 pesos, to 350 pesos over the same period. The use of improved seed also increased rapidly. National production for all seeds was less than 7 thousand metric tons in 1960, but in the 1975-77 period, averaged 300 thousand metric tons. The National Seed Corporation (PRONASE) provided about one-third of the total, producing mainly staple grains and bean seeds. Available data show that from 1974-84, there was no noticeable change in the total supply of chemical sprays, although dependency on imports fell. Whereas Mexico produced only about 50 percent of its spray needs in the former period, it produced about two-thirds in the latter.

The government plays a major role in agricultural investment, primarily in the development of irrigation. In the post-war era, the government was virtually the only investor, receiving some external funding from the Interamerican Development Bank and the World Bank. Mexican irrigation policy is characterized by (a) heavy state investment; (b) organization of water use by river basins (although some irrigation systems obtain water from deepwells or springs); (c) dominance of large-scale projects; and (d) increases in irrigated area through the improvement of existing infrastructure. Irrigation projects are supposed to be self-funding through user fees and amortization of construction costs by users. It is estimated, however, that government subsidies, on average, account for between 30 and 55 percent of construction and maintenance costs (SARH, vol.12, 1982).

Except for a decrease in the 1960s and during the current fiscal crisis, real agricultural and irrigation investments increased steadily in the post-war period. Agricultural investment as a percentage of total public investment and irrigation investment as a percentage of public agricultural investment remained relatively stable. Except for 1960 and 1965, when agriculture's share as a percentage of total public investment dropped to 8 percent, it has claimed between 13 and 22 percent of the investment pie. The economic crisis does not appear to have changed this ratio. In some years, irrigation accounted for over 90 percent of government investment in the agricultural sector. Generally, funds spent on irrigation fluctuated between 70 and 98 percent of government agricultural investment.

Policy Objectives

Post-war Mexican agricultural policy can be divided into two periods. In the first (1940-75), agriculture played a subordinate role relative to the industrial sector and was mainly viewed as a source of cheap food for urban areas. Inexpensive food helped keep labor costs low and stable, and agriculture provided low-priced raw materials for industry. Attempts were made to shore up rural incomes with government assistance, including direct transfers, income security through guaranteed prices and subsidized crop insurance, and the construction of rural infrastructure (SARH, vol.12, 1982). Domestic grain producers were protected from international competition through import licensing, import taxes, governmental control of imports and exports, and import and export subsidies.

As grain imports grew in the 1970s, agricultural expenditures were redirected. The prevailing view held that if the agricultural sector was to progress (as it had in the 1940-65 period), the impetus would have to come from dryland farmers with smallholdings. The SAM and PRONAL programs were established to reach this group (Roberts and Mielke, 1986). When these programs were enacted, government expenditures were increased, guaranteed prices for basic foodstuffs were raised, and emphasis was reoriented to the nonirrigated, small-scale sector. In contrast to the former policy period, there was a keen desire to reduce grain imports by improving domestic production and to better the livelihood of small farmers, the most economically disadvantage group.

A Continuing Conflict Between Self-sufficiency and Cheap Food

In the 1940-65 period, Mexican farmers increased production rapidly, and a modern agricultural sector emerged. Cultivated land area expanded, extensive exports generated foreign currency, and export taxes filled government coffers. On the consumer side, nutritional levels rose, and diets became more varied and higher in protein (SARH, vol. 13, 1982). A combination of strong government programs and prevailing market conditions resulted in an agricultural supply, especially in grains, that met demand.

As time passed and the commercial sector matured, gains in productivity slowed. When the government froze grain prices in 1962-73, commercial farmers switched to more lucrative export and industrial crops, over which the government exerted little control. This left maize and beans, the most important foods for the majority of the population, in the hands of peasant producers. As a result, the Mexican agricultural sector bifurcated into a modern, highly productive commercial sector and a traditional subsistence and semi-subsistence sector. Although the survival of small, inefficient production units played a major role in the development of current grain deficiencies, the rapid proliferation of large units also contributed by accelerating migration to urban areas. Thus, demand for basic foodgrains grew more rapidly than domestic supply.

In spite of emphasis on low-income producers, the government's programs only reached certain zones (mainly larger farms in the more progressive irrigated areas in the north, northwest, and central pacific), certain crops (chiefly export and industrial crops), and commercial producers (Roberts and Mielke, 1986). The campesino sector was unable to take advantage of government programs because of its inability to repay loans. This was due to low productivity and crop losses; traditional agreements to sell crops to intermediaries in return for loans; difficulties in exploiting the package of modern inputs; a lack of extension services and educational and technical support; problems in using machinery efficiently on small farms; and the minimal response to government price incentives for grain produced mainly for home consumption.

Many view the public program focus on the northwest, north, and central pacific irrigated zones as detrimental to the country's rainfed areas, which contain the majority of the farm population and area. In these areas, mainly in the central and southern plateau, yields are lower, growth in reclaiming new area has been slow or negative, and diversification away from maize and bean production has not occurred. In some years, high support prices encouraged staple crop production on irrigated land at the expense of potentially high-value fruit and vegetable export crops. Agricultural policies have resulted in a less efficient allocation of economic resources and have failed to set the foundation for sustained growth in the agricultural sector.

The agricultural sector was also affected by macroeconomic policies. For many years, high import tariffs, which benefitted the manufacturing sector, encouraged industrial rather than agricultural investment. In the 1970s, an overvalued currency kept agricultural imports cheap and acted as a further disincentive to investment. Although the government followed a "cheap consumer food" policy through subsidization, there was no targeting of subsidy programs, and it is debatable whether much of this assistance reached low-income consumers. Devaluations after 1976 helped to redress investment disincentives. Decreasing subsidies, coupled with devaluations, resulted in strong increases in real food prices. This, along with government budget deficits, has been a major source of inflation and falling real wages.

At the heart of the grain deficit is Mexico's cheap food policy and the need to maintain a large positive balance of payments to service foreign debt. To keep food cheap for urban consumers, producer prices must be kept low or heavily subsidized. Because of budgetary constraints on producer subsidies, grain self-sufficiency is unlikely, and grain imports will be necessary to meet consumer demand. In addition, high-value export crops are needed to earn hard currency. If grain production is to expand, growth will have to come in part at the expense of export crops or by increasing production in rainfed areas as proposed under the SAM program. Given the special circumstances of Mexican agriculture, including differences among irrigated and nonirrigated lands and peasant and commercial producers, there are structural rigidities that will hamper substantial increases in agricultural production.

The basic thrust of Mexican agricultural policy has changed over the last few years toward the reduction of consumer subsidies and government intervention in the grain market. However, the government will continue to face strong pressures to return to a cheap food policy, with attendant pressure to provide inexpensive food through increased imports. In pursuing its production goals, the government will have to find a balance between food sovereignty and imports; that is, the encouragement of methods that support the production of basic foodstuffs, but also allow export production to increase.

Notes

1. The authors would like to acknowledge the assistance of Elida Treviño, Jose Moreno and Oscar Ramirez in the processing of data for this chapter.

2. The geographical regions of Mexico and their component states are (i) Northwest: Baja California Norte, Baja California Sur, Sonora, Sinaloa, Nayarit; (ii) North: Chihuahua, Durango, Coahuila, Nuevo Leon, Tamaulipas, Zacatecas, San Luis Potosi, Aguascalientes; (iii) Central Pacific: Jalisco, Colima, Michoacan, Guanajuato, Queretaro; (iv) Central: Mexico, Puebla, Morelos, Tlaxcala, Hidalgo, Veracruz, Mexico City; (v) South: Guerrero, Oaxaca, Tabasco, Chiapas, Campeche, Yucatan, Quintana Roo.

3. Settlements with 2,500 or more inhabitants are considered urban, those with fewer people, rural. Roughly 50 percent of the population is concentrated in the Central and Central-Pacific regions.

4. In 1981, there were roughly 25 pesos to the US dollar. In 1988, the corresponding figure was roughly 2,300.

5. This and subsequent harvested areas include double-cropped area.

6. A tortilla is an unleavened bread "pancake" made from corn meal.

7. Information is taken from the 1977 National Survey of Family Expenditure and Income. This is the most recent survey of its type (SARH: Vol. 13, 1982).

8. Nixtamal is boiled maize in kernel form used to make tortillas.

9. Calculated from data provided by CONASUPO.

References

Aguirre, V. *Efecto del la Politica Gubernamental de Compra y Venta sobre los Saldos de Comercio Exterior de Maiz y Sorgo.* M.S. thesis, Colegio de Postgraduados, Chapingo, Mexico, 1984.

Avila, J. *Reaccion de los Agricultores de México a la Politica de Precios Agricolas.* M.S. thesis, Colegio de Postgraduados, Chapingo, Mexico, 1982.

Barkin, D. and B. Suarez. *El Fin de la Autosuficiencia Alimentaria.* Mexico City: CECODES - Nueva Imagen, 1982.

Batalla, A. *Geografia Economica de Mexico.* Mexico City: Trillas, 1980.

Burst, A. C., M. E. Bredahl, and P. F. Warnken. *Growth and Structure of the Mexican Feedstuffs Industry.* College of Agriculture International Series 7, Special Report 315. Columbia, MO: University of Missouri, 1984.

Fernandez, D. *La Intervencion del Estado en la Regulacion del Mercado del Arroz en México.* M.S. thesis, Chapingo, Mexico, Colegio de Postgraduados, 1986.

Fernandez, P. *La Intervención del Estado en la Regulación del Mercado del Arroz en México.* M.S. thesis, Colegio de Postgraduados, Chapingo, Mexico, 1986.

Hernandez, J. *La Intervencion del Estado en la Regulacion del Mercado: La Politica de Precios del Maiz en México.* M.S. thesis, Colegio de Postgraduados, Chapingo, Mexico, 1983.

Lustig, N. and A. Del Campo. "Descripcion del Funcionamiento del sistema CONASUPO." *Investigacion Economia* July - September (1985): 215-43.

Mielke, M. J. *Government Intervention in the Mexican Crop Sector.* Washington, D.C.: USDA, ERS, Agriculture and Trade Analysis Division, 1989.

Reyes O. S. "Producción y Consumo de Alimentos en México: Una Imagen al Año 2000, Energéticos, Alimentos y Proyecto Nacional." Mexico City: 1981 (mimeograph).

Roberts, D. H. and M. J. Mielke. *Mexico: An Export Market Profile.* Foreign Agricultural Economic Report Number 220. Washington D.C.: USDA, 1986.

Salcedo, S. *Economic Analysis of Cheap Food Policies: The Wheat Sector in Mexico.* M.S. thesis, University of Wisconsin, Madison, 1985.

Secretaría de Agricultura y Recursos Hidrálicos, (SARH). *El Desarrollo Agropecuario de México: Pasado y Perspectivas.* Vol. 12 *Política Agrícola* and Vol. 13 *Problemática Alimentaria.* Mexico City: Secretaría de Agricultura y Recursos Hidráulicos, 1982.

13

North-South Grain Markets and Policies: A Synthesis

David Blandford[1]

The case studies provide a substantial amount of varied and important information on grain market structures and policies of relevance to other countries. This information is the focus for this final synthesis chapter. Policy objectives and instruments are discussed in the first part, followed by lessons learned from the successes and failures of the study countries. The second part of the chapter deals with more general issues, particularly the ways in which government intervention affects domestic and international grain markets.

Part I: A Cross Country Comparison of Grain Policies

Objectives

Many countries include a general statement of agricultural policy objectives in enabling legislation, which often directly pertains to grains. The objectives of US policy, for example, as set out in the Food Security Act of 1985, are "to extend and revise agricultural price support and related programs, to provide for agricultural export, resource conservation, farm credit, and agricultural research and related programs, to continue food assistance to low income persons, to ensure consumers an abundance of food and fiber at reasonable prices, and for other purposes" (U.S. Congress, 1985:1). Sometimes stated objectives may be difficult to reconcile with one another. The European Community's agricultural policy, for example, as spelled out in Article 39 of the Treaty of Rome, states that the aim is to increase producer earnings and to guarantee "reasonable prices" for consumers (Harris, Swinbank, and Wilkinson, 1983:35). These objectives appear to be in conflict, because the methods used to increase producer earnings necessitate raising consumer prices substantially above world market prices (see Chapter 4).

Stated objectives are also not always mirrored by country actions. For various reasons, governments may be unable to pursue consistently a specific set of objectives. The political system under which policies are formed may make such consistency impossible (see for example the chapter on the United States). Furthermore, governments' ability to influence grain markets may be limited by the way in which these markets operate. In several of the study countries, attempts to impose border taxes and control cross-border or internal shipments of grain have been thwarted by the ability of private traders to avoid taxes and controls. Regardless of the measures taken by governments, national grain markets are seldom isolated completely from the international market. If government intervention creates profit opportunities, some individuals will be prepared to respond to these, even under the threat of legal sanctions.

Despite these difficulties, inferences can be made about the "revealed" objectives of policy from the types of measures governments use, and their willingness to accept the outcomes of these measures. If, for example, a country uses a pricing policy that consistently favors consumers rather than producers, it is reasonable to infer that consumer welfare is a more important policy concern than producer welfare, regardless of any public pronouncements to the contrary. Policy preferences are classified into four broad areas: (a) production; (b) welfare; (c) stability; and (d) international trade (Table 13.1). In many cases, these overlap. Trade objectives may simultaneously influence welfare and stability objectives, for example. All classifications are arbitrary, and this table is no exception. Nevertheless, it provides a framework within which to evaluate what governments may attempt to achieve in managing the grain market.

Food security is an objective shared by every country listed in Table 13.1, although how this concept is interpreted and the degree of priority attached to it differs widely among countries. Because grains are such a basic commodity in the food supply, all countries are concerned about maintaining consumer access to an adequate supply of grains. Some countries in the South associate food security with self-sufficiency, although their policies have not always been effective in achieving this objective. Self-sufficiency is unimportant to the Northern countries included in this study, excepting the European Community. There are other Northern countries, however, such as Japan and the Scandinavian countries, that place a high priority on self-sufficiency in food staples (OECD, 1983).

Many Southern countries, where domestic production does not meet demand, are vitally concerned with improving grain sector productivity. Because the agricultural sector plays a critical role in economic growth and development, increased grain production is viewed as an essential element in the food security equation. Few Southern countries are willing or able to rely primarily on international markets for grain supplies. Expanding populations mean that grain sector productivity must rise if the ratio of imports to consumption is to remain static or decline. In the North, where grain supplies more often outstrip

TABLE 13.1 Summary of Major "Revealed" Grain Policy Objectives[a]

	Production		Welfare		Stability		Food security	Trade		
	Increase	Diversify	Producer	Consumer	Prices	Producer income		Foreign exchange	Market share	Self-sufficiency
NORTH										
Australia			X				X		X	
Canada			X			X	X		X	
EC			X		X		X		X[b]	X
US			X			X	X		X[b]	
SOUTH										
Cameroon	X			X	X		X			X[c]
China	X		X	X	X		X	X		X
Colombia	X		X		X		X			X[c]
Dom. Rep.	X			X	X		X			X
India	X		X		X		X			X
Indonesia	X			X	X		X			
Mexico	X	X		X	X		X			
Tanzania		X		X	X		X			

[a]Objectives inferred from policy instruments chosen and their effects.
[b]With increasing trade conflict between the European Community and the United States.
[c]Basic foodgrains only.

demand, leading to surplus stocks, managing or controlling surplus production receives priority rather than boosting output.

Northern countries tend to be biased toward producer rather than consumer welfare. These countries support producer prices, make direct income support payments to producers, and use a variety of other measures to favor producers and rural inhabitants over consumers and urban inhabitants. In Chapter 1, it was shown how the real price of grains on international markets has tended to decline over time. Much of this drop can be attributed to the pace of technological change. In fact, the rate of increase in supply has consistently outstripped demand. This has occurred in most Northern countries as grain production systems have become increasingly sophisticated technologically. Downward pressure on prices subsequently exerts downward pressure on farm incomes. Although wealthy Northern countries can afford to either support prices or make direct payments to producers to tackle the "farm income problem," most frequently, governments choose to support prices, even though this is an inefficient way of transferring income to producers. Price supports aid large, rich farmers more than small, poor ones because the benefits reaped from higher prices depend on production volume. When producers raise output to profit from price supports, surpluses frequently result. As production expands, variable costs rise, and only a diminishing portion of the higher price is translated into additional net income. In the long run, price supports can lead to inflated prices for inputs, particularly land. The supply of farmland is largely fixed. As the price of farm output rises, farmers compete to obtain more land to increase total earnings and profits. Subsequently, land prices rise (see, for example, meteoric rises in land prices in the Midwestern United States grain belt in the 1970s, which was at least partially fueled by high international grain prices (Blandford and Schwartz, 1986), further reducing the long-term benefits farmers derive from price supports.[2]

In the South, in contrast, governments tend to implement policies that improve consumer and urban welfare over producer and rural welfare. Many Southern countries attempt to keep the consumer price of grains low. The economic argument for this is that food is also an important "wage good" in poorer societies. Maintaining cheap food prices may help to sustain low industrial wages and promote urban employment and industrial growth. Whether this actually occurs is debatable. A far more important reason why governments keep the consumer prices of grains low is to satisfy their vocal and politically potent urban populations. A few Southern countries maintain a producer welfare bias in their policies. India is probably the most dramatic example; price and other policies have been largely oriented towards producer and rural welfare (see Chapter 10). In recent years, the Chinese government has used its complex grain pricing system to try to maintain a balance between policies that favor the rural population and those that favor the urban population (see Chapter 8).

Price stability is a major concern in the South but is not awarded as much importance in the North. This is because of the potentially serious political consequences of sharp swings in staple food prices in low-income societies. Thus, consumer price stability is generally an important part of the food security equation in most Southern countries. Producer price stability has been viewed as complementary to rural development objectives in a number of countries, most notably India. Few countries articulate farm income stability as an explicit objective, but it is important in some Northern countries, particularly Canada.

Country attitudes to food security and self-sufficiency frequently have important implications for international trade. Two additional trade concerns that apply relate to foreign exchange earnings or expenditures and market share. Increasing foreign exchange earnings through the sale of domestic grains is a policy objective in some Southern countries, most notably China. For many net-importing Southern countries, however, the need to conserve foreign exchange places fiscal constraints on grain trade policies, linking self-sufficiency with food security. Historically, policies that help to maintain export market share have been pursued by the Canadian Wheat Board. However, all Northern exporters are increasingly concerned with market share as growing grain surpluses have led to the expanded use of export subsidies. This is particularly true for the European Community and the United States since the early 1980s.

Instruments

Only a small number of objectives appear to underlie the grain policies of the study countries, but most countries use a wide array of instruments to influence domestic production, prices, marketing, and trade (Table 13.2). The choice of instruments is influenced by a number of factors, including country size and income. Most countries maintain state agencies that participate in domestic grain marketing and in international trade. Only in Cameroon, the European Community, and the United States is state intervention in domestic grain marketing limited, and there is no state control over grain exports or imports (although these are influenced by other policy instruments).

Producer Measures

Input measures, such as fertilizer or water subsidies, are used in many of the study countries but are not a major grain policy instrument. From a political standpoint, input subsidies have the disadvantage of being less visible to producers than price supports, even though they can be more effective in increasing grain output. Their relative effectiveness depends on the subsidized input's contribution to grain production and its specificity. If, for example, chemical fertilizer use on grain crops is low, subsidizing fertilizer could lead to

TABLE 13.2 Summary of Principal Grain Policy Instruments

| | Production/consumption | | | | | | | Trade | | | | Domestic market | | |
| | Inputs | | Price guarantees or control | | Direct payments | | Con-sumer sub-sidiesᵇ | Quotaᶜ | Import tariff | Export sub-sidiesᵈ | State trade | Subsidyᵉ | Margin controlᶠ | Gov. market-ing |
Country	Sub-sidies	Crop insur-ance	Prod.ᵃ	Cons.ᵃ	Defic.	Income stabil-ization								
NORTH														
Australia	*		X	X	*			*		*	X	*		X
Canada		*	*	*		X		*		*	X	X		*
EC	*		X					X		X				*
US	*	*	X		X					*				X
SOUTH														
Cameroon	Xᵍ		Xᵍ	*				*ʰ	X			Xᵍ	*	*ʰ
China	*		X	X			X				X			X
Colombia	*ⁱ		X				*	X	X	*	X	*	*	X
Dom. Rep.	*	*	X	X			*		*		X	*	*	X
India	*		X	X			X	*			X			X
Indonesia	X		X				X					*		X
Mexico	*ʲ	*	X	*			*				X	X	*	X
Tanzania			X	*							X	*		X

X = major policy instrument.

* = measure is used, but is less important.

ᵃTypically accompanied by government purchases/sales.

ᵇ"Targeted measures," including rationing, fair price shops, food-for-work, food coupons, and food entitlements.

ᶜIncludes import/export restrictions, prohibitions, restrictive licensing, and variable levies.

ᵈIncludes concessional credit and food aid.

ᵉIncludes storage, transport, and processing subsidies.

ᶠIncludes procurement, storage, distribution, and processing.

ᵍRice only.

ʰParallel domestic purchase requirement for rice and wheat flour.

ⁱConcessional credit and water.

ʲFertilizer and water.

a significant growth in total output. However, fertilizer can easily be applied to other crops, and farmers may choose to use at least some of the subsidized fertilizer on these, particularly if they yield more income than grain. This "leakage" may be one reason why policy-makers prefer other measures, such as output subsidies, to encourage higher grain output. A further factor in many Northern countries is growing public concern about the overuse of subsidized inputs, such as water for irrigation, and the environmental consequences of encouraging high applications of chemical fertilizers and pesticides.

In virtually every country, producer price guarantees or controls are major instruments of grain policy. The use of consumer price controls is less common. Producer price guarantees are highly visible and therefore have political advantages. However, if the objective is to promote output, increasing output prices can be less effective than using input subsidies in developing countries. This is particularly true when a significant proportion of domestic grain output is used for home consumption. The focus on subsidizing the use of fertilizer on rice in Indonesia and water and other inputs for maize in Mexico are examples where this is evident. Further, controls on output prices often create unwanted distortions elsewhere in the economy. High grain prices have a direct negative effect on the livestock sector by increasing feed costs, resulting in major problems in the European Community and in Southern countries, such as the Dominican Republic and Mexico, where the use of grain for livestock feed is growing rapidly. Output price controls typically require complementary interventions in trade, and these can create conflict with other countries. A notable example is the increasingly acrimonious dispute between the European Community and the United States over grain import barriers and export subsidies.

Price supports are the most popular mechanism for supporting producer incomes. Few governments tackle the problem through direct payments because of the heavy burden these place on budgets. Many Southern countries could not afford such payments, even if they wished to use them. In Northern countries, where direct payments could be employed, the high visibility of their costs is a limiting factor politically. Indirect producer transfers are less apparent to the general public when they are hidden in higher consumer prices. Nonfarm voters generally react more negatively to government subsidies than to hidden taxes in the price of bread or rice. For example, there seems to be greater domestic political pressure to reduce expenditures on direct payments in the United States, than to cut back the substantial tax on European Community consumers created by the Common Agricultural Policy. In any case, producers prefer to receive income transfers indirectly through higher prices, rather than directly through government payments because of "welfare stigma." Despite the political limitations, direct payments hold advantages over price supports from both equity and efficiency perspectives. Price supports benefit larger, more efficient farmers most, whereas direct payments can be targeted to smaller farmers. Although policy-makers in Northern countries frequently voice concern about the welfare of small or medium-sized farms, larger farms are the principal beneficiaries of payments linked to production (Blandford, 1987). Price supports often encourage excess production; direct payments need not if they are not linked to the requirement to produce. For these and other reasons, some economists argue that current price support programs in Northern countries should be replaced by direct payments (Blandford, de Gorter, and Harvey, 1989).

Consumer Measures

Grains are used in a variety of ways. In the North, most grains are consumed in highly processed forms or indirectly as livestock products. In the South, more grain is consumed in lightly processed forms. In the North, consumer prices of grain and grain-based products tend to be set above comparable world prices. In the European Community, for example, producer price supports lead to higher consumer prices because imports are taxed. In the South, the opposite is generally the case, reflecting the emphasis on consumer welfare over producer welfare. Imports are often subsidized, usually by maintaining an overvalued exchange rate. In richer countries, consumer price controls are sometimes employed as a means of raising small tax revenues (e.g., in Australia and Canada) or to transfer income to producers without government expenditures (e.g., the European Community). In the South, price controls are often adopted to keep staple foods cheap to further industrial development objectives or to help control urban unrest, often driven by availability and price of food. Price controls are a less efficient means than direct consumer subsidies for improving consumer welfare, often distort consumption decisions, and tend to benefit rich and poor alike. In the North, where consumer price controls lead to higher prices, grain consumption declines (directly as food and indirectly through livestock feeding), and imports are reduced. If the country is grain surplus, exports increase (see, for example, the European Community). In the South, where controls generally result in lower consumer prices, grain consumption rises, with subsequent pressure for higher imports to fill escalating demand. This has occurred in China and the Dominican Republic.

In addition to general consumer price controls, some Southern countries use selective consumer subsidies. These are generally targeted to particular groups. In India, for example, the "fair price" shops distribute grain to poorer consumers at low prices. Richer consumers buy higher-priced grain on the open market. Subsidies are frequently directed toward urban consumers, even though rural consumers may be poorer. The degree or accuracy of targeting in special distribution schemes varies substantially. India, for example, targets distribution of subsidized grains to the poor to a greater extent than in Indonesia, but none of the study countries has a systematic method for allocating subsidized grain to the poorest or most vulnerable groups. The equity and efficiency arguments for using targeted consumer subsidies rather than a general price subsidy are similar to those for direct income payments to producers.

Trade Measures

Nontariff barriers (NTBs) in the form of quotas, licensing, or state trading are the most common trade measures. Most countries regulate trade directly to maintain stable domestic prices, rather than employing tariffs. The costs of tariffs are more visible to consumers than NTBs, making them less politically attractive. NTBs are popular even though the government forgoes tariff revenue. When domestic prices exceed world prices, "rents" or excess profits are associated with trade barriers such as quotas. An individual licensed to import under a quota scheme purchases the commodity at a lower international price and sells at a higher domestic price, making quota rents monetarily and politically important. When licenses are allocated, select and powerful groups are often favored. Consequently, strong pressures may exist not to change the status quo in the grain import system.

The handling of trade by a state agency provides an alternative to NTBs and allows the government more direct control over trade. State trading allows the government to capture excess profits associated with imports, as it would through a tariff. In some cases (e.g., Cameroon), governments use tariff or quota revenues to subsidize domestic grain production, but substantial monies can also be consumed by state trading. For example, China loses on wheat imports but makes a profit on rice exports. As wheat imports have grown, net costs have risen (see Chapter 8). Where frontiers are porous and the difference between domestic and international prices is large, it may be difficult to prevent the entrance of low-priced imports. This has been the case in Cameroon and Colombia. As the Dominican Republic's experience illustrates, it may also be difficult to prevent low-priced domestic grain from being exported to higher-priced areas.

Export subsidies are widely used by the Northern exporters. The European Community employs subsidies because its domestic prices exceed world prices, and, as a result, surplus grain cannot be marketed internationally without assistance. In other exporting countries, such as the United States, subsidies are used to protect or increase market share. There are, however, problems with this approach. Subsidies increase export volume most effectively if targeted to markets where price response is greatest. However, importers in less responsive markets may also demand parallel preferential treatment. This happened with the US export enhancement program when Colombia and Mexico sought subsidies as a condition for continuing to purchase grain from the United States. Furthermore, the effectiveness of subsidies is reduced if they are matched by competitors. The principal beneficiaries of export subsides are exporting firms, whose export volume and profits increase, and importers, who pay lower prices.

Export subsidies can result in short-run gains to Southern importers, as they have for Colombia and Mexico. However, depressed world prices can create long-run problems for the domestic grain industries of importers. Some countries restrict low-priced imports to soften or prevent their domestic price-depressing effects (e.g., India) or impose tariffs to offset the subsidies of Northern exporters (e.g., Cameroon). Low world prices created by rich country subsidies can increase costs for countries seeking self-sufficiency in grains. The disposal of periodic surpluses, at depressed world market prices, can mean substantial financial losses for governments attempting to achieve self sufficiency (e.g., India and Indonesia).

Domestic Marketing Measures

Domestic marketing subsidies for storage, transportation, and processing are not major measures in most countries, despite the fact that grain marketing infrastructure and services are often poorly developed in Southern countries. Expenditures to promote or improve infrastructure could be justified on "public good" grounds, because they provide general benefits to the community. However, marketing infrastructure tends to be awarded a low priority in the South, and budgetary allowances are often reduced or eliminated during times of economic adversity. Unfortunately, marketing subsidies are politically unattractive because their benefits are not highly visible.

Government involvement in domestic procurement and distribution of grains is widespread in both the North and South, but for different reasons. In the North, government participation is driven by the surpluses generated by producer price supports. In the South, involvement is dictated by the need to guarantee consumer access to cheap food, or to countervail natural monopolies in the marketing system. Government marketing can also be used to raise revenue. Direct state control over the physical allocation of grain is politically important in many countries, particularly in the South. Frequently, government marketing activities are complementary to other measures, such as price controls and consumer subsidies. Governments may also be forced to adopt subsidies and price controls because of secondary effects generated by other policy measures. For example, if a government chooses to implement a flat support price that does not vary through space or time, some grain will probably have to be purchased directly. There may be insufficient private incentives to encourage purchases in localities distant from primary markets or to encourage storage, since the costs of storage will not be reflected in an increase in prices during the off-season.

Successes and Failures

The case studies in this book highlight the successes and failures of various management strategies for grain markets adopted by the study countries. A summary of these indicates the breadth and width of choices taken and provides a useful backdrop for future policy decision making in these and other countries (Table 13.3).

Some countries have had considerable success in improving grain output by promoting the adoption of new technology, such as improved varieties or methods of production. China, India, and Indonesia stand out among countries of the South in this regard. As indicated above, this success is based on the existence of a technology package suitable to their particular conditions, developed under the aegis of government support. The countries in the North have also undergone rapid technological change. High and stable producer prices in the European Community stimulated adoption of new techniques that increased yield. In the United States, substantial public investment in research, particularly through the university system, has had an important impact on the generation of new technologies.

The experience of the countries of the North demonstrates the importance of grain marketing systems that function efficiently. These countries maintain well-developed infrastructure in terms of transportation and storage. Public expenditures on rural roads, bridges, and rail lines help to ensure that grain can move rapidly from producers to consumers, as well as provide needed production inputs to farmers. Northern countries have grades and standards for grain, with inspection services to police these. This facilitates trading of large volumes of grain "sight unseen" and lowers marketing costs. Some Southern countries have successfully improved their grain market infrastructure (e.g., Colombia, India, and Mexico). However, in many countries of the South, marketing is neglected relative to other competing concerns.

A number of countries have achieved substantial domestic price stability by insulating the domestic market from international markets (e.g., the European Community, India, and Indonesia). This has been achieved at the cost of surplus production. Only Canada has directly addressed the problem of farm income stability through the use of farmer/government income stabilization funds.

Some Southern countries have begun to reform their grain policies to provide greater incentives to producers (e.g, China). Frequently overlooked in the past, there are still examples of continued failure to provide appropriate incentives (e.g., the Dominican Republic). In many countries, extensive regulation of the grain marketing system and resulting price distortions cause depressed output and government wrought changes in production and consumption. These problems are often compounded by inappropriate macroeconomic policies, particularly overvalued exchange rates, which cause a deterioration in the terms of trade of domestic grain producers.

TABLE 13.3 Positive and Negative Achievements of Grain Policies in the Study Countries

Country	Positive	Negative
Australia	Domestic price stabilization for producers and consumers	Pooling masks price signals Excessive regulation of domestic market
Cameroon	Growth in production with limited government intervention in market	Failure to implement a sustainable policy for domestic rice
Canada	Direct approach to farm income stabilization through stabilization payments	Pooling and transportation subsidies mask price signals
China	Use of economic incentives to increase production and raise rural incomes Large investment in new technologies leading to rapid technological change	High budgetary costs of urban cheap food policy Distortions in consumption created by ration system
Colombia	Market stabilization through limited intervention Focus on improving grain marketing structure and institutions	Macroeconomic policies discriminate against agriculture
Dominican Republic	Rice production through technological change fostered by government policy	Extensive regulation of the grain marketing system with price distortions reinforced by unfavorable macroeconomic policies Increasing budgetary and foreign exchange costs of policies
European Community	Price stability promotes structural and technological change	Failure to adjust real prices for changes in productivity Depression and distortion of demand (particularly feed) from high prices. Use of export subsidies with consequent depression of world prices

India	Systematic economic approach to price setting, generating substantial domestic economic stability	Long-run trend to structural surplus in some grains
	Facilitation of technological change	Costs incurred through buffer stocks to stabilize domestic prices
	Improvement in marketing structures and institutions	
Indonesia	Rice production increased through technological change	Use of costly input subsidies to intensify production and full self-sufficiency policies
Mexico	Trend toward liberalization of domestic grain markets	Failure to improve productivity of peasant dryland maize
	Improvement of domestic market structures	Market instability generated by changes in policy
United States	Large public stocks to buffer world market fluctuations	Use of production and export subsidies with depressing effects on world markets
	Some attempts to address income support directly through deficiency payments	

Despite the widespread use of price supports and controls, few countries have successfully implemented a systematic approach to setting internal prices. A notable exception is India, where economic factors are carefully taken into account. The activities of the Price Commission, described in the Indian chapter, demonstrate the amount of information and sophistication of human resources required to administer grain prices publicly without creating substantial distortions. The European Community and the United States (to some degree) have been relatively unsuccessful in this regard because of the strong influence of politics on support prices. Even where politics intrude less (e.g., Australia and Canada), policy choices have tended to mask price signals to producers.

Many countries now find their grain policies increasingly costly. Primary culprits are policies designed to achieve or maintain self-sufficiency (e.g., India and Indonesia), policies that keep consumer prices low (e.g., China and the Dominican Republic), and policies that protect producer income (the European Community and the United States). In a number of countries, policy instability itself has a negative effect on domestic grain markets (e.g., the Dominican Republic). Large and erratic changes in policy can create as much instability for producers and consumers as uncontrollable, weather-induced instability does in others (e.g., Australia and India).

Part II: Issues Raised by the Case Studies

A number of important issues are raised by the case studies in this volume. The first is the practical realities and constraints that governments must take into account when implementing grain policies. If policies are to achieve their objectives, it is essential that these constraints be articulated. Second, there is the issue of how much government intervention is actually needed in the grain sector. Although this harbors heavy ideological overtones, an attempt is made to provide an objective evaluation. Third is the question of grain market stability and food security. These related issues have emerged repeatedly throughout this volume and also have ideological underpinnings. Again, an attempt is made to provide an objective view of some of the major considerations. Finally, it is increasingly clear that we live in an interdependent world. The overlapping effects of domestic grain policies among countries have been referred to in a number of the studies in this volume. The final part of this chapter briefly addresses the implications of this interdependency.

Structural Realities and Constraints on Grain Policies

In many of the countries in this study, governments are currently re-evaluating their role in domestic grain markets. In some cases, the role of the private sector has been increased and that of the public sector reduced. The fundamental importance of grains in all economies, both for agriculture and for economic and political stability as a whole, means that the public sector will continue to play an important part in grain markets. Exactly how large this role should be is open to debate and is influenced by many economic and political factors. The extent of government involvement in grain marketing will inevitably differ among countries. Thus, the focus in this section is on the structural realities and constraints that influence the effectiveness of public policies, whatever the role a government chooses to adopt. These constraints are vividly illustrated by the case studies.

Demand

The structure of demand for grains and changes in this structure must be taken into account in attempting to achieve policy objectives. The pattern of demand influences the effectiveness with which government can meet consumption, price, and production goals. Further, changes in food preferences have a major impact on the grain industry globally. In many Southern countries, wheat and rice are increasingly the grains of choice for direct human consumption, displacing other grains or nongrain foods as staples. Consumers often exhibit high price and income elasticities for wheat and rice because of their palatability and convenience, particularly in urban areas. Livestock sector developments also affect the grain sectors of most countries. The growing demand for meat, eggs, and dairy products is encouraging the production of feedgrains (particularly for poultry and hogs). Demand for feed also pressures supply in terms of quantity and quality. Finally, in some countries, the industrial use of grains has grown rapidly, for example in the United States for sweetener and fuel alcohol (both are policy related), and in China for alcoholic beverages.

Many Southern countries work to promote consumption of domestically produced grains in preference to imports. Sometimes this leads to the production of grains unsuited to the local climate, soils, or farming systems. If a country is best suited to producing maize, rather than, say, rice or wheat, maize production and consumption should be promoted, rather than wheat and rice production. If the promotion of other grains is not a viable option, other food crops or industrial crops may be an alternative. A country will be forced to import grains it cannot produce efficiently, but this is preferable to wasting domestic resources.

Consumers in the South are more sensitive to changes in grain prices than in the North because of their income levels and the relative importance of grain in their diets. This sensitivity is felt not only through the use of grains for food but also through demand for livestock products. The price elasticity of demand is higher for lower-income groups than higher-income groups within Southern countries. This has important distributional implications for pricing policies and for the pressures placed on domestic supply or imports. If governments keep the consumer prices of foodgrains low, poorer people will benefit, but demand for grains will also increase. If domestic production does not respond to the rise in demand, either because of technical constraints or low producer prices, excess demand will have to be met with imports. If governments keep the consumer price of grains used for livestock feed low, the principal beneficiaries will be wealthier individuals who can afford to buy livestock products. The increased use of livestock feed may also stimulate imports. When grains are used both as food and feed, pricing policy becomes even more problematic. In this case, keeping prices low will stimulate both food and feed demand, benefit both rich and poor, and require increased domestic production or imports.

Production

An important, but often neglected, fact is that grain producers readily respond to changes in prices. This is vividly demonstrated in the case studies for China and India. Pricing policy has been a key element in stimulating grain production in both of these countries. Changes in output prices and the relative prices of inputs and output can cause producers to switch between grains, even if the total supply elasticity of grain is low. This is the reason why the Indian Prices Commission pays close attention to price relatives among grains and between grains and competing crops in forming its recommendations for producer prices. The responsiveness of output with respect to input prices can be important. This is demonstrated in Indonesia, where subsidized fertilizer played a critical role in expanding rice production. Price elasticities of supply across regions or types of farms will vary depending on production alternatives. For example, the responsiveness of wheat output in Mexico is much greater than that for maize. Wheat is an irrigated commercial crop, and producers can easily grow alternative crops on the land. Maize is an unirrigated subsistence crop, and there are limited production alternatives in the areas where it is grown.

Paradoxically, the response of grain supply to price tends to be higher in the South, particularly in the short run. The fixity of capital in the capital-intensive production processes of the North limits response. In the United States, government programs introduce additional rigidities in the reallocation of land among grains and between grains and competing crops. In the South, grain supply is a marketable surplus, production is less specialized, and farmers own fewer fixed assets. Even though subsistence needs will limit the extent to which the area devoted to grains will vary, partial dependence on commercial sales can make the amount of grain marketed highly responsive to changes in prices. This must be taken into account when setting prices.

Important intra-country differences exist among grain farms in both the North and South. In the North, there can be substantial differences in farm size, patterns of ownership, and capital intensity of production. In the South, such differences also exist, but in addition, economic dualism -- the parallel existence of "modern" and "traditional" farms – is particularly prevalent. Dualism leads to a marked variation in production response to incentives, creating different production/marketing requirements (e.g., technological, storage, transportation, and credit needs). These differences must be taken into account in attempting to implement production objectives. Dualism can also have important distributional implications. In Mexico, for example, the greatest productivity increases have occurred on modern wheat farms. Efforts to increase productivity on traditional maize farms have been expensive and achieved only limited success. In allocating credit and other resources, Mexican policy-makers have had to try to balance the need to increase grain output to feed a rapidly expanding population, with the need to improve the welfare of small peasant farmers.

In the case studies, there are notable examples of countries that succeeded in improving productivity. Their success is attributable to the availability of an appropriate technology package (improved seeds, fertilizers, and irrigation) teamed with producer access to markets (for both inputs and outputs) and appropriate price incentives. Many of the new grain varieties require additional inputs of fertilizer and improved water management to realize their full potential. In India, for example, irrigation and the availability of fertilizer were crucial elements in increasing productivity. Yield increases are sometimes more pronounced on large or commercial farms (e.g., Colombia), but can also be achieved on small semi-subsistence farms (e.g., the Dominican Republic). The development of a specialized seed industry and the use of certified seed has been important, particularly in Northern countries, and hybrids have played a role in some cases (e.g., sorghum in Colombia, rice in China). The use of hybrids has only been feasible in these countries because an infrastructure to produce and distribute hybrid seed was developed.

Maize yields have not increased to nearly the extent of wheat and rice yields in the South. This is because technological advances in open-pollinated maize have not kept pace with those for wheat and rice. As indicated in Chapter 1, relatively little effort has been devoted to improving yields of open-pollinated maize. The early development of maize hybrids meant that plant breeders in Northern countries neglected open-pollinated maize. The international research system has devoted most of its efforts (and had most of its success) in developing rice and wheat varieties. The need for a sophisticated seed production and distribution system to handle hybrid maize seed, which along with fertilizer and chemical inputs is largely responsible for the dramatic yield increases in the North, is also a bottleneck. The lack of a technology package for grains suitable for African conditions is a particular problem for the development of grain production systems in Africa. Maize production in Cameroon, for example, has managed to keep pace with demand primarily through an expansion of cultivated area. With continued rapid population growth, the land frontier will eventually be exhausted. If consumer prices of maize or imports are not to increase dramatically, the country will be faced with the challenge of improving maize productivity on the existing area through the use of fertilizer on open-pollinated varieties, or the introduction of high-yielding hybrids.

To stimulate output successfully, technology packages must be readily available to producers. Packages must be appropriate to country conditions, given farm and domestic input market structure. The development of simple indigenous technologies and delivery systems are often neglected. Technological change can have important consequences for income distribution because it is rarely neutral with respect to farm size. Several of the countries in this book (e.g., India and Mexico) have had to face trade-offs between efficiency and equity in increasing grain production. There are no simple solutions to this problem, but it is important for policy-makers to recognize its existence. It is also important to realize that ensuring sustained increases in productivity through technological change requires a substantial long-term financial commitment by governments and continuous government support for the diffusion of new techniques.

Markets and Marketing

The importance of infrastructure to an efficient grain market cannot be overstated, yet it is often ignored because infrastructural investment yields indirect, longer-term results, which are less popular with politicians. Improvements in transportation, storage, and processing should be high priority items in most Southern countries. It is virtually impossible to meet grain policy objectives without adequate infrastructure, which benefits both producers and consumers. As in the case of production technology, infrastructure should be "appropriate" in terms of scale, location, and foreign exchange requirements.

Grading and readily available, consistent market information, are important factors in improving the efficiency of grain markets. Market efficiency is important in meeting domestic grain objectives at minimum cost. Institutions for the development of efficient markets are often poorly developed in the South; for example, mechanisms for the self-regulation of trading on local markets or competitive exchanges where large quantities of grain can be traded on a spot or forward basis. In some cases (e.g., India and Colombia) the government has taken an active role in developing such institutions. This is an important part of improving the efficiency of grain markets in Southern countries, particularly when the private sector has a major role.

Appropriate incentives for all involved parties are required at the marketing and production levels. For example, unless prices change throughout the year to reflect seasonal harvests, allowing marketers to make a profit from storing grain, private storage will not be offered. If the government cannot allow prices to fluctuate, then it must undertake responsibility for storage. Quality differentials must be recognized through grading, with higher prices paid for more desirable grades. Governments have an important role in ensuring that the level and structure of marketing incentives are not distorted.

Other Economic Factors

The macro environment (rate of nonagricultural growth, exchange rates, interest rates, and inflationary trends) can have important and sometimes unforeseen effects on domestic grain markets. Macroeconomic policies (crafted by those who may be unfamiliar with the grain economy), particularly exchange rate policies, can have very marked effects on grain markets. Sometimes these effects outweigh sector-specific policies. In Colombia and the Dominican Republic, for example, overvaluation of the exchange rate has had a major impact on the competitive position of domestic grain producers, often outweighing the effects of price supports in the Colombian case, and strengthening the effect of low domestic prices in the Dominican Republic's case. In the South, where agriculture is often still a major determinant of economic health, developments in the grain economy can have important macroeconomic implications. The economies of China and India, for example, are affected significantly by changes in grain prices and in the incomes of grain producers. Grain policies must remain flexible, particularly in the case of administered prices, when domestic macro variables are unstable. The experience of the United States during the early 1980s demonstrates the importance of this lesson. Rigid and high support prices for grains essentially priced US exports out of world markets when the value of the US dollar appreciated significantly against other major currencies.

Unlike macroeconomic policies, other exogenous factors, such as debt crises or rapid increases in oil prices, may be largely outside the control of policymakers. But these can have an important impact on grain markets by altering the terms of trade, costs of inputs, availability of foreign exchange, rate of growth of domestic income, government investment priorities, and the cost and availability of capital. Commodity market problems, to which Southern countries are especially susceptible, particularly trade policies in other countries, can be an important constraint for both exporters and importers because market access is limited and price trends are changed. Despite the distortions created by domestic policies, international prices determine the opportunity cost of domestic grain policies and are the only external indicator available to guide domestic pricing policy.

Additional domestic constraints on grain policies, important in both the South and North, are created by the ways existing institutions operate and the rents created by some policy instruments. There are often substantial vested interests in maintaining the status quo. Further international constraints can be created by foreign policy considerations. Many Southern countries, because of their relatively small size, cannot influence the international market very much, although China and India are notable Southern exceptions. Countries may

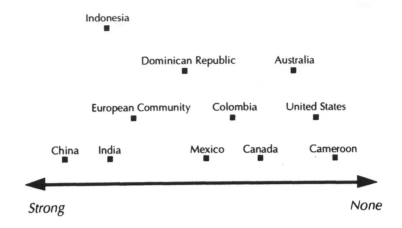

FIGURE 13.1 Government Intervention in the Grain Market

have difficulty in dealing with these constraints, but the limitations they place on policy objectives and the choice of policy instruments need to be acknowledged.

How Much Government Intervention Is Necessary?

The case studies illustrate many of the different methods governments employ to manage production, consumption, trade, and prices in domestic grain markets. In this section, some of these methods are discussed, as well as what is implied by different levels of government intervention.

The Existing Level of Intervention

Evaluating the level of intervention in a country's grain sector is difficult because objective measurements cannot easily be made. Quantitative measures, for example, of the level of protection, can give an idea of how much the government intervenes directly or indirectly in price formation, but these measures are difficult to compute on a comparable basis across countries. In this study, a subjective assessment of the level of intervention is made based on factors such as the number and type of instruments employed and their impact on production, consumption and marketing, the role of the private sector in the grain marketing system, and the degree of linkage between domestic and international markets.

Based on these factors, a schematic representation indicates the relative range of intervention among the study countries (Figure 13. 1). At the extreme no-intervention end of the spectrum, which includes none of the study countries, no domestic market or trade measures are employed, and there is no direct or indirect intervention in grain pricing. Countries that do not intervene in their grain sectors might undertake public good activities, such as establishing grades, standards, or inspection services and probably also supply infrastructure, particularly roads. At the other end of the spectrum, strong control implies that prices are set and regulated by the government, trade is managed by the state, and all marketing functions are provided by state agencies. The study countries are placed on the spectrum on the basis of a subjective assessment of their relative level of intervention. The scale is ordinal rather than cardinal, and distances between countries on the chart are of no significance other than to indicate a country rank with respect to other countries. Thus, US grain policies are judged to be less interventionist than those in Australia, which are less interventionist that those in the European Community, but the distance between countries on the chart does not imply, for example, that EC policies are five times more interventionist than Australian policies.

None of the study countries are located at either extreme, although Cameroon is relatively close to the no-intervention end. It exercises limited control over domestic markets, primarily by influencing import prices. Among Northern countries, the United States is furthest along the right-hand side of the spectrum. Unlike countries such as Australia or Canada, it lacks a marketing board, and unlike the EC it maintains less direct control over domestic grain prices and international trade.

Closest to the other end of the spectrum is China, where international and domestic grain trade are tightly regulated. China has historically followed the socialist model of state pricing and single channel marketing, although the situation is changing. India and Indonesia are further to the right than China. There, the state plays an important role in managing the grain sector, as well as in pricing and regulating international trade. However, the private sector maintains a significant, legally sanctioned role in domestic marketing. In terms of the level of intervention, these countries are broadly similar to the European Community. All three have an explicit self-sufficiency objective and can muster the domestic resources to achieve it. In the middle left of the spectrum is the Dominican Republic, which exerts less control over its domestic grain economy than India, for example, but more than Mexico and Colombia.

Northern and Southern countries occupy all positions on the spectrum. None conforms to the extreme of complete free trade (not even Australia, although New Zealand, which grows little grain, would probably lie closest to the far right if it were included in this study). Likewise, none adheres completely to the command economy model, although, before the recent reforms, China came close.

A country's success in maintaining a strong grain industry and in achieving basic policy objectives, such as food security, does not depend on its location on the spectrum. Cameroon has been relatively successful in meeting its food security objectives with minimal intervention. On the other hand, the United States has had limited success in stabilizing grain prices and farm income, particularly when program costs are taken into account. Yet both countries lie on the right-hand side of the spectrum of intervention. Indonesia and India have been relatively successful in achieving their food security objectives while exercising more control than the Dominican Republic or Mexico, for example. One of the lessons to be drawn from the case studies is that the degree of success in meeting grain policy objectives does not depend on how much, or how little, the government intervenes in the domestic grain economy. But success is dependent on what types of measures governments use, and how appropriate these are to achieving specific policy objectives.

The European Community and the United States, two large and internationally significant grain traders, are located at opposite ends of the intervention spectrum. Perhaps this explains the escalating conflict between the two on grain policy issues. As indicated in the EC chapter, Community policies have led to surpluses, which have been exported using export subsidies. US policymakers responded to this new competitor by offering cheap grain and by introducing a subsidy program to counter EC subsidies. For the Community, the international market is simply a means of disposing of surpluses that are a byproduct of keeping domestic prices high to support producer incomes. For the EC, international markets play a residual role, while for the United States, they are an integral part of the domestic market, and prices on both are inextricably linked. Because both countries view the international market differently, it is difficult to reconcile the respective grain policies.

Choosing the Level of Intervention

The case studies demonstrate that domestic grain markets can be managed in many ways to achieve a variety of objectives. No single set of domestic instruments is superior in terms of achieving specific objectives at minimum costs. Policies that employ less direct market intervention can be just as expensive as those involving more. The contrast between US and EC grain policies provides a nice example.

The choice of domestic grain policies is a domestic matter, although international repercussions may result. A particular set of policies cannot be advocated as superior or inferior to another. However, to help maximize the chances of success, it is possible to identify the implications for domestic policy choice of occupying a particular position on the spectrum. There are several criteria for judging success that cannot easily be collapsed into a single indicator. However, some of the most important include maintaining an adequate supply of food to consumers at reasonable prices, providing sufficient incentives and access to input and output markets to maintain rural incomes, achieving a level of domestic output consistent with comparative advantage, and controlling public expenditures and taxes. The basic policy components that meet the

TABLE 13.4 Necessary Components of Successful Grain Policies at Three Different Levels of Intervention

Low	Medium	High
1. Investment in research and development to promote technological change in grain production.	1. Investment in research and development to promote technological change in grain production.	1. Investment in research and development to promote technological change in grain production.
2. Investment in basic infrastructure to reduce marketing costs.	2. Investment in basic infrastructure to reduce marketing costs.	2. Investment in basic infrastructure to reduce marketing costs.
3. Implementation of grades and grading, rules and regulations, and dispute settlement procedures for private transactions.	3. Implementation of grades and grading, rules and regulations, and dispute settlement procedures for private transactions.	3. Implementation of grades and grading, rules and regulations, and dispute settlement procedures for private transactions.
4. Use of targeted producer/consumer subsidies to achieve distributional objectives.	4. Use of targeted producer/consumer subsidies to achieve distributional objectives.	4. Use of targeted producer/consumer subsidies to achieve distributional objectives.
	5. Limited purchases and sales by a state agency to stabilize domestic prices.	5. Substantial purchases and sales by a state agency to control domestic prices set at a level to generate sufficient producer incentives.
		6. State operation of most of the grain storage, and at least some of the distribution system.
		7. Regulation of international trade, i.e., through state trading agency.

above criteria for three levels of intervention, namely low, medium, and high, are indicated in Table 13.4. These correspond to the range for the study countries, from Cameroon (low) through Colombia and Mexico (medium), to China (high) discussed earlier.

If a country wishes to pursue the low intervention route, it must invest in productivity enhancing research and development to increase grain output, keep consumer prices low, and increase rural incomes. Controlling consumer costs involves reducing marketing costs by improving basic infrastructure, such as roads. To enhance the efficiency of price formation and promote competition, governments must establish rules and regulations that govern private transactions. Facilitation of the establishment of a system of grades and grading and a method for resolving disputes is necessary. Many of these activities could be labelled "public goods."

In the low intervention scenario, the private sector primarily purchases, transports, and stores grain. It handles grain processing and trade. Government participation is limited to establishing and enforcing rules and procedures that allow the private sector to operate competitively, thus controlling costs. However, because distributional objectives underlie most country grain policies, the minimally interventionist option also typically involves the use of targeted subsidies, for example to promote the modernization of smaller farms in Southern countries, for income support in richer countries, and to meet basic needs or nutritional objectives in the general population.

In the medium intervention alternative, similar needs lead to the formation of government policies not unlike those found in the low intervention route. Analogous policies are necessary to provide public goods and promote competition. However, most medium intervention country governments take a more active role in managing the internal market. For example, a state marketing agency might be involved in buying and selling grain domestically to stabilize domestic prices. In this scenario, the private sector is still the major vehicle for grain marketing and handles most grain storage and transportation.

Under the high intervention option, government policy must fulfill public good needs as in the other two alternatives. Sufficient infrastructure, technological change, and efficient markets are still vital to providing an assured supply of grain at reasonable prices. However, at this level of intervention, the public sector takes a more active role in domestic price formation. State agencies are responsible for many grain purchases and sales to meet domestic pricing objectives. The trade interface is managed through the same domestic marketing agency or a related organization. The state is also actively involved in ensuring that sufficient domestic storage is available. It probably handles at least some facets of grain distribution, for example, imports. Pricing policies are established to maintain domestic production incentives, and well-developed accounting procedures and cost control systems are enforced to replace the economic pressures exerted in a competitive situation.

The higher the level of public control over domestic grain markets, the more that variables must be balanced against each another. Greater skill and resource commitments are required to manage markets publicly; otherwise, distortions result. If the state exercises a substantial influence over domestic prices, it will inevitably become more involved in marketing and international trade, and thus will be forced to provide adequate domestic management of these functions. The state must also ensure that basic infrastructural and institutional needs, such as those identified in Table 13.4, are met, because these ultimately affect the economic health of the grain industry and the consumer cost of grain.

The general summary of measures in Table 13.4 is not accompanied by a list of the particular types of instruments needed under each level of intervention, for example, what kinds of pricing policies should be employed, or whether a system of variable taxes/subsidies is preferable to import licensing. There are economic arguments in favor of particular alternatives, some of which were outlined in the first part of this chapter. The case studies provide the best, real life examples of successful strategies, including advantages and disadvantages. Economists would argue that economic planners should carefully weigh all policy choices, but ultimately domestic political and institutional circumstances are likely to have a large hand in determining policy choices.

Grain Market Stability and Food Security

Most of the countries represented in this study and elsewhere, whether exporters or importers, view instability in grain prices as a major problem. Most take some measures to promote domestic price stability for producers and consumers.[3] In some cases, price instability is created by uncontrollable domestic weather, which can severely depress production (e.g., India); in others, fluctuations in world market prices play a large role (e.g., Cameroon). Countries choose many different ways to tackle instability; contrast the Australian, Canadian, EC, and US approaches discussed in this volume for example. In most Northern countries, the primary emphasis is on stabilizing producer prices, although this does not necessarily stabilize revenue or income. When countries are large, prices may move inversely with production, thus helping to reduce variability in revenues. Even without this, when variability in domestic production is the chief source of instability, price stabilization may have little effect. Prices are the most visible and easily regulated variable in the stability equation, affecting both producers and consumers. Consequently, price stability tends to be the primary focus of most stabilization policies. Only Canada manages income instability directly through a farm income stabilization scheme. The Western Grains Stabilization program, and other provincial programs, are designed to stabilize producer income rather than prices (see Chapter 3).

Most countries, particularly those in the South, are more concerned with food security than price stability.[4] Although precise definitions vary, put simply, food security is provision of an adequate supply of, and access to, food on a continuing basis for the general population. In reality, this is often difficult to achieve because price and income together determine access. Food security is as much or more a poverty problem in most Southern countries, as it is a production or marketing problem (e.g., India). In general, most of the countries in this study approach the food security issue using one or a combination of the following: (i) increase production through technological change, reinforced by output price supports or input subsidies; (ii) boost consumption by controlling consumer prices; and (iii) use a combination of production, consumption, and trade measures to seek some measure of self-sufficiency. Few countries use targeted interventions (such as selective food subsidies) as part of their food security package, despite the potential cost effectiveness of such measures. It is more popular politically to spread the benefits of food security policies among society at large than to direct these to the most vulnerable groups.

As noted by McCalla (1989), a nation's long-run food security is a national responsibility. He points out that food security does not require that a country be self-sufficient in its food security policies. According to McCalla, there are six food strategies available to a country as summarized in Table 13.5. They range from total import dependence through various degrees of self-sufficiency to commercial exporting.

The information contained in Table 13.5 needs to be supplemented with the following points made by McCalla. First, a country's overall food strategy is likely to consist of targets with respect to particular commodities. For example, a country might target to be a commercial exporter of one particular commodity in order to finance, say, a strategy of partial import dependence with respect to another. Second, attitudes about what constitutes domestic costs prohibitive enough to warrant 'total import dependency' status are likely to vary, both across countries with respect to a particular commodity and across different commodities within any one country. It is clear, for example, that India and Mexico have been prepared to incur high costs to avoid import dependency with respect to rice than it has in the case of maize. Decisions about acceptable cost levels will obviously depend on financial resources and, no doubt, on the political strength of various interest groups. Third, each of strategies (2), (3), and (4) involve strategic choice in the light of the degree of production variability. For example, in choosing 'best-year' self-sufficiency status, a country experiencing a range of production of ±10 percent of average production would target for 90 percent self-sufficiency. Fourth, there is a discontinuity on the spectrum of alternatives when a country switches from an exporting to an importing position because transportation and transaction costs move from one side of the market to the other. If it costs $10 per tonne to import from, or export to, the nearest market, there would be a $20 range in which neither exports nor imports would be profitable.

Table 13.5 Food Security Strategies

Strategy	Rationale/characteristics
1. Total import dependency	A country may be unable to produce a product or the costs of doing so are prohibitive (e.g., wheat in Indonesia).
2. Partial import dependency	Regular importation of a portion of food requirements, with various possible criteria about how much to import (e.g., import quantities that can be financed from agricultural exports—called "sectoral self-relaince"—which seems to have been the strategy followed by China during the 1960s and 1970s in relation to rice exports and wheat imports; allocate some proportion of foreign exchange earnings to food imports).
3. Best-year self-sufficiency	Targeting of production so there are imports in most years but complete self-sufficiency in best years. Imports could replace stockholding and be used to stabilize prices in years of shortage.
4. Average self-sufficiency	Targeting of production to fully meet food needs on the average, with exports (or stock accumulation) in good years and imports (or use of carryover stocks from good years) in years of shortage. Characteristic of India's food trade during the 1970s and 1980s.
5. Absolute self-sufficiency	Target production to fully meet food needs even with the historically worst crop so that there are exports in most years.
6. Commercial exporter	Comparative advantage or subsidization of exports makes the country a regular exporter (e.g., wheat in the case of Australia, Canada, and the United States).

Source: A summary of food security strategies described by McCalla (1989).

McCalla (1989) also discusses the complexity of choosing an appropriate strategy. Among the important points he makes are the following:

a. The economic issue involved in choosing a strategy is the relative long-run costs of domestic production versus the costs of imports. The choice will be conditioned by comparative advantage and domestic considerations such as demand patterns, available technology and infrastructure.

b. Potential choices are not static. They vary with world prices and differential changes in technology which alter patterns of comparative advantage. A country has to form expectations about world price levels in setting targets. Because the marginal social costs of domestic production can be expected to increase as production (and the self-sufficiency level) increases, higher self-sufficiency targets on the part of small developing countries acting as price-takers would be rational if there were expectations of higher world prices. Correct anticipation of future prices is itself a difficult task.

c. Figuring the relative costs of expanding domestic production through research and technical change on the one hand and increasing imports on the other is not a simple matter. For example, one has to discount the payoff from research investment because of time lags involved and the payoffs themselves are uncertain. How does one measure 'psychic' costs of greater import dependency, those costs deriving from exposure to fluctuating international prices and availability of imports, among other things? The tripling of wheat prices in 1972/73 and its negative imports on countries that chose a permanent import strategy (e.g., Colombia for wheat) is notable in this regard. The activities of international agencies also affect the choice. Financing facilities, such as the IMF's Cereal Facility help countries meet increased food import costs. On the other hand, the Consultative Group on International Agricultural Research (CGIAR), the Food and Agriculture Organization (FAO) and most bilateral aid agencies assist countries to meet self-sufficiency goals.

The difficulties involved in choosing a national food strategy, McCalla argues, is not sufficient justification for choosing self-sufficiency (the simplest choice). Choices should be made on the basis of economic considerations and it is imperative that economic policy analysts monitor the costs accompanying particular choices. It is most unlikely on economic grounds that the optimal strategy for a country will involve complete isolation from world markets. On the other hand, it is inefficient for a country to adjust long-term research investments on the basis of annual or monthly fluctuations in world prices, although this is what Colombia and Mexico seem to have done, in contrast to countries such as India and Indonesia.

Two final points should be made to help demonstrate that maintaining a successful balance in the food security equation is extremely complex and is not simply a matter of becoming self-sufficient in grain production. First, grain is cheap relative to other foods and achieving self-sufficiency in grain production may not do much to increase rural employment levels or rural incomes. Second, focusing on self-sufficiency might not result in *food security* since the latter is partly a poverty problem. Few of the case study countries use targeted interventions, such as selective subsidies, to ensure access to basic foods by the poorest groups in society.

North-South Grain Issues and Problems

Grain policies everywhere are driven by domestic objectives and considerations. Decision makers rarely take international effects into consideration, except in so far as the interface with international markets constrains domestic policy choice. This is true even for Northern exporters such as the United States, although countries that export a significant portion of domestic production, such as Australia and Canada, are more restricted by international market conditions than others when choosing domestic policy. In most Southern countries and the European Community, the international market serves as a convenient place to "export" the effects of fluctuations in domestic production. Only a few countries (the United States, Canada, India, and possibly Indonesia) hold sufficient stocks to assimilate a significant amount of their own production instability. Only in the United States (and to some extent Canada) can these stocks be accessed by other countries, thus helping to stabilize international market prices. For most countries, international markets are residual markets. Because they are treated as residuals, international markets absorb the effects of the distortions created by the domestic policies in both North and South. This is particularly noticeable in the case of countries with a large annual output, such as the United States, European Community, India, and Indonesia. The effect of domestic policies in these countries on their production, consumption, and trade can have significant implications for international market prices. However, when their combined impact on the volume of trade is considered, small countries, such as Cameroon, the Dominican Republic, and Colombia, can also have an important effect on international markets.[4]

Intervention in grain trade by both Northern and Southern countries is a worldwide phenomenon and contributes to instability in international markets. In Northern countries, intervention is often necessitated by domestic farm income support policies, while in Southern countries, food security policies may require it. With so much interference, world grain prices are inevitably more distorted and unstable than they would be under free trade conditions. Many Northern countries would like to export more grain, particularly to Southern markets. But, their own policies create artificially depressed prices that threaten rural sector stability in the South. Further, many Southern countries are unwilling to rely on world markets for grain supplies, enter world markets only as a last resort, and try to minimize purchases as much as possible. Yet, world market prices are the only available external indicator that can be used as a basis for domestic pricing policies in the South. Importing countries' national food security is placed at risk by the uncertainty and instability inherent in international grain markets, although their own national food security policies are also major contributors.

As indicated in Chapter 1, the axis of grain trade is increasingly shifting to a North/South orientation. With this change, the interrelationships between North/South grain policies are becoming even more significant. All countries need to "get their policies right." To some extent, this can be achieved through the adjustment of individual national country policies. Many Southern countries, for example, have taken steps to raise incentives and improve efficiency in domestic grain markets because of the unsustainable level of domestic expenditures generated by existing policies, or because of external pressures emanating from a large foreign debt. But unilateral changes in national policies

David Blandford

alone cannot be expected to solve problems of North/South interdependency. Recently, efforts have been taken by the world community to tackle international problems created by domestic agricultural policies through multilateral negotiations under the General Agreement on Tariffs and Trade (GATT). However, emphasis has been almost totally on the policies and concerns of Northern countries, particularly the European Community and the United States.

Southern interests are largely neglected in international negotiations on agricultural policy, despite the fact that Southern countries are a major force in international agricultural trade. Resolving conflicts between the United States and the European Community over income support policies and, in particular, reducing the trade distortions that these policies create, is important, not the least for the future of international grain markets. But the income support issue is only one of many problems. The reform of Southern country policies is at least as important for the future of international grain markets as is reform in the North.

When Southern countries reform domestic grain policies to reduce international market distortions, their concerns, particularly over food security, must be directly addressed. Issues such as the level of, and access to, international grain stocks, limitations on the unilateral restriction of exports, availability of emergency food aid, technology transfer, and accessibility to financial resources to develop domestic market infrastructure are important problems that must be acted on if Southern countries are to be expected to adopt domestic policies that open their grain markets to world market forces. The degree of openness must be linked to the level of development. Parallel concessions on access to Northern markets for Southern products is also needed.

The studies in this volume demonstrate important lessons to be drawn from the experiences of different countries in managing their domestic grain markets. Because of their vital role in the food supply, grains will continue to be a major focus of policy in most countries. Many of the policy objectives are similar, and a wide array of instruments are used. Policy-makers can learn much from the successes and failures of their peers in other countries. In the future, achieving important policy objectives, such as food security, in an increasingly interdependent world will require not only that grain markets and policies in each country become more efficient, but also more closely linked to markets and policies in other countries.

Notes

1. Discussions with the authors of the case studies were a major ingredient in this chapter. Another was a paper presented by Alex McCalla to the Recife meeting of the KIFP/FS and subsequently published in *1988 World Food Conference Proceedings* (see McCalla, 1989). I am particularly indebted to Colin Carter for his help in organizing

and interpreting the material.

2. Land sellers will reap benefits from higher land prices. Consequently, persons who retire from farming or leave the industry for other reasons will tend to gain. For new entrants to farming or those who wish to expand the size of their farms, higher land prices simply increase the costs of production.

3. Note that we are defining "stabilization" as the reduction of short-term price fluctuations, those within a single season or between two seasons, not the practice of maintaining consistently high prices to transfer income from producers, or consistently low prices to transfer income from producers to consumers.

4. This might require investment in infrastructure to handle the distribution of imported grain.

5. The relationship between domestic policies and international trade and prices is extremely complex. An analysis of some of the principal types of policies is contained in Zwart and Blandford, 1989.

References

Blandford, D. "Distributional Impact of Farm Programs and the Dilemma of Adjustment." *American Journal of Agricultural Economics* 69(1987): 980-87.

Blandford, D., H. de Gorter and D. Harvey. "Farm Income Support with Minimal Trade Distortions." *Food Policy* 14(1989): 268-273.

Blandford, D. and N.E. Schwartz. "Effects of Changes in the Domestic and International Environment on US Agriculture." In US Department of Agriculture, Economic Research Service. *Embargoes, Surplus Disposal and US Agriculture.* Washington, DC, 1986.

Harris, S., A. Swinbank, and G. Wilkinson. *The Food and Farm Policies of the European Community.* New York: Wiley, 1983.

Hayami, Y. and M. Kikuchi. "Investment Inducements to Public Infrastructure: Irrigation in the Philippines." *Review of Economics and Statistics* 6(1978): 70-77.

McCalla, Alex F. "Developing Country Productivity and Trade: Complementary or Competitive." In John W. Helmuth and Stanley R. Johnson (eds.), 1988 World Food Conference, Proceedings Vol. II: Issues papers. Ames: Iowa State University Press, 1989.

Organization for Economic Cooperation and Development. *Review of Agricultural Policies in OECD Member Countries, 1980-82.* Paris, 1983.

U.S. Congress. *The Food Security Act of 1985.* Public Law 99-198. Washington, D.C.: U.S. Government Printing Office, 1985.

Zwart, A.C. and D. Blandford. "Market Intervention and International Price Stability." *American Journal of Agricultural Economics* 71(1989):380-88.

About the Contributors

Rodolfo Alvarado, Ministry of Agriculture, Colombia

Ralph Bierlen, Centro de Economia, Colegio de Postgraduados, Mexico

*David Blandford, Head of the Trade Analysis Division, Directorate for Food Agriculture and Fisheries, OECD, Paris (formerly professor at Cornell University)

*Colin A. Carter, Professor, Department of Agricultural Economics, University of California, Davis, California, U.S.A.

Harry de Gorter, Assistant Professor, Cornell University, Ithaca, N.Y., U.S.A.

Brian Fisher, Executive Director, Australian Bureau of Agricultural and Resource Economics, Canberra, Australia

Madeleine Gauthier, Economist, Agricuture Canada, Ottawa, Canada

Sarah Lynch, Economic Consultant, Ithaca, New York, U.S.A.

*Jaime Matus-Gardea, Professer, Centro de Economia, Colegio de Postgraduados, Mexico

Kevin Parton, Senior Lecturer, Department of Agricutural Economic and Business Management, University of New England, Australia

*Roley Piggott, Senior Lecturer, Department of Agricutural Economic and Business Management, University of New England, Australia

*Norberto A. Quezada, Sigma One Corp., Research Triangle Park, No. Carolina, U.S.A.

*Alvaro Silva Carreño, General Manager, IDEMA, Ministry of Agriculture, Colombia

*Davendra Tyagi, The late Chairman of Commission for Agricultural Costs and Prices, India

*Simei Wen, Professor, Department of Agricultural Economics, South China Agricultural University, China

*KIFP/FS Fellow

About the Book and Editors

This valuable collection of case studies examines the domestic policies of selected countries that relate to the most basic of agricultural commodities—grains. Contributors discuss the many approaches governments take in managing their domestic grain markets, evaluate the choices of alternative policies governments face, offer insight into the effective implementation of policy choices, and draw clear and practical observations about the conditions necessary to ensure a healthy and dynamic domestic grain market.

Although acknowledging that domestic policies need to be viewed in a global context, contributors to this volume deliberately emphasize domestic objectives—the primary concern of policymakers—rather than meeting international commitments. They adopt a market-oriented philosophy, arguing that governments are more likely to achieve their aims when they work with, rather than against, market forces. The book ultimately provides a balanced assessment of the pros and cons of government intervention when developing a domestic grain trade policy and will prove valuable to trade policymakers, agricultural economists, and other scholars and policymakers concerned with a nation's food supply.

David Blandford is head of the Trade Analysis Division, Directorate for Food Agricultural and Fisheries at the Organization for Economic Cooperation and Development, Paris. **Colin A. Carter** is professor of agricultural economics at the University of California at Davis. **Roley Piggott** is senior lecturer in agricultural economics and business management at the University of New England, Australia.